Make: Wearable Electronics

Kate Hartman

Make: Wearable Electronics

by Kate Hartman

Printed in Canada.

Published by Maker Media, Inc., 1005 Gravenstein Highway North, Sebastopol, CA 95472.

Maker Media books may be purchased for educational, business, or sales promotional use. Online editions are also available for most titles (*http://safaribooksonline.com*). For more information, contact O'Reilly Media's corporate/institutional sales department: 800-998-9938 or *corporate@oreilly.com*.

Editors: Brian Jepson and Emma Dvorak
Production Editor: Kara Ebrahim
Copyeditor: Jasmine Kwityn
Proofreader: Amanda Kersey
Indexer: Ellen Troutman
Cover Fashion Designer: Angella Mackey
Interior Designer: David Futato
Illustrator: Rebecca Demarest
Cover Photographer: Henrik Bengtsson
Cover Model: Jenny Andresson
Technical Editors: Rob Faludi, Erin Lewis, Pearl Chen, and Lynne Bruning
Custom Illustrator: Jen Liu
Research Assistant: Hillary Predko

August 2014: First Edition

Revision History for the First Edition:

2014-08-05: First release

See *http://oreilly.com/catalog/errata.csp?isbn=9781449336516* for release details.

ISBN: 978-1-449-33651-6

[TI]

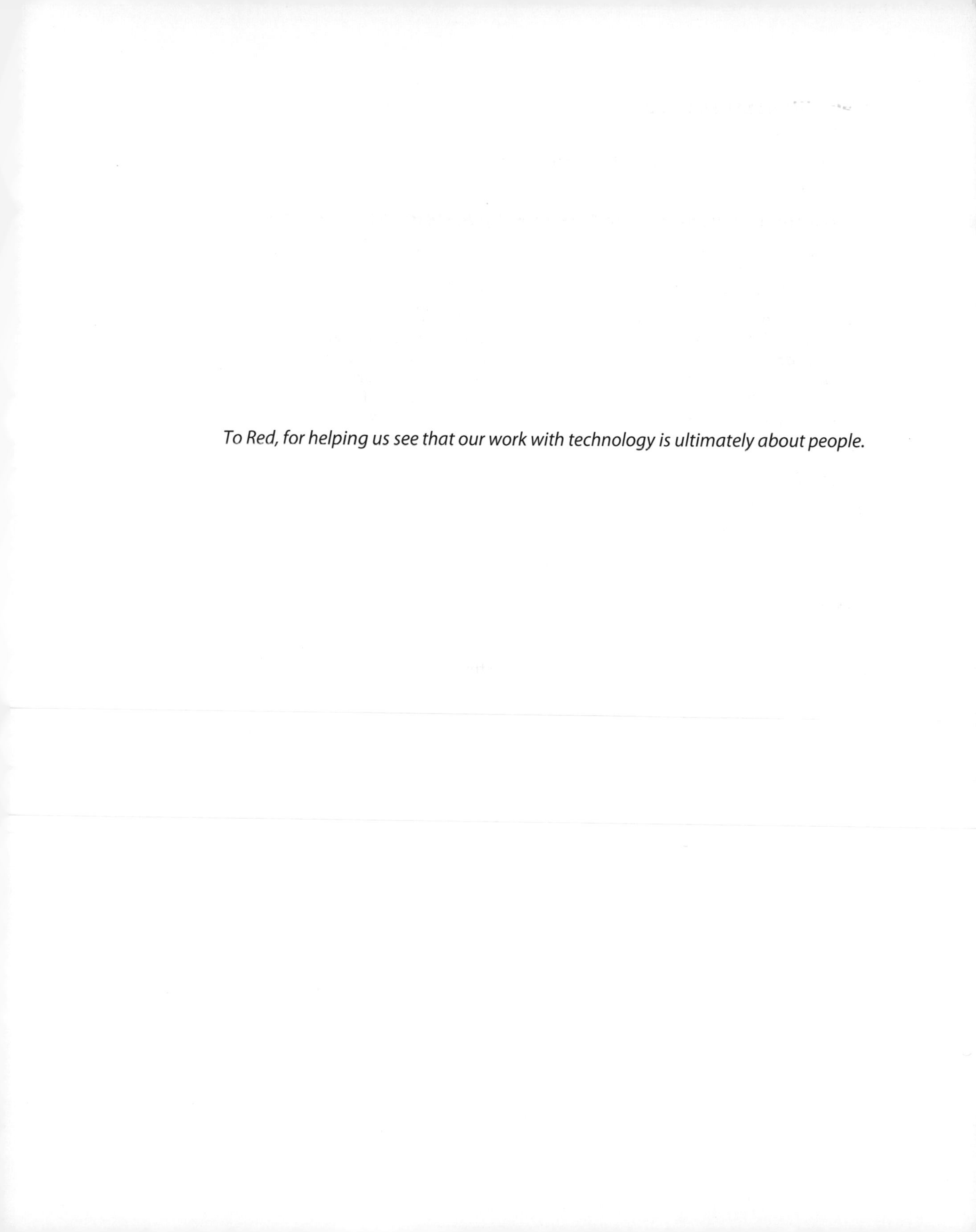

To Red, for helping us see that our work with technology is ultimately about people.

Table of Contents

Preface

About The Book

Figure P-1. *"Monarch," a muscle-triggered kinetic textile created by the Social Body Lab*

Our bodies are our primary interface for the world. Interactive systems that live on the body can be intimate, upfront, and sometimes quite literally in your face. They sit close to your skin, inhabit your clothing, and sometimes even start to feel like part of you. This makes wearable electronics an exciting, challenging, and inspiring area to work in.

On one level this book is about how to make wearable electronics. It will introduce you to the tools, materials, and techniques necessary to create interactive electronic circuits and embed them in clothing and other things that can be worn.

On another level, this book is asking you: *"What's next?"*

We're living in a moment where wearable technologies are just starting to become a part of our everyday lives. They live on our wrists and in our glasses. They track our activities and transport us into virtual worlds. But this is just the beginning. There is still a lot that has yet to be revealed.

This book is inviting you to join the conversation about the future of wearable and body-centric technologies. What do we need? What do we want? And what should be avoided?

In the last 10–15 years, the technology that lives in our pockets has dramatically transformed. In the next decade, we can expect to see great strides in the development of the technology that lives on our bodies and in our clothes. It's a good time to ask questions and express opinions. This book will hopefully help you get started with that.

Who This Book Is For

This book is for people who want to roll up their sleeves and make some wearable electronics. This includes students, researchers, hackers, makers, fashion designers, engineers, industrial designers, developers, costume enthusiasts, artists, and textile mavens.

There are two perspectives from which you might be approaching this book.

The first: you know some stuff. There's a broad range to this. Maybe you're someone who has used an Arduino to blink an LED at a workshop once upon a time. Or maybe you run a design firm that produces massively robust interactive installations in museums and now you've got a client who wants you to generate a prototype that's wearable. Either way, you know enough to have a sense of what universe you're in. This book will help you build upon what you already know and might even lead you into some areas you didn't expect!

The second: when it comes to electronics and programming, you're a bit of a n00b. Maybe you're a fashion designer that realizes that interactivity in clothing is something you should wrap your head around. Or perhaps you're a sociologist who is developing a data-collection system that includes sensors that live on the body. Or maybe you're an artist with a newfound interest in self tracking. In any case, there are likely many things in this book that you may not have heard of before. If you're in this category, take this advice: be brave. It's OK if things are new to you or if you don't understand it on the first go. This book might be your gateway to a whole new list of things you didn't realize you wanted to learn. Stick with it—it's interesting stuff!

What You Need to Know

This book covers most of the basics, but it does assume that you understand soldering and basic hand sewing. If either of these things are new to you, check out Appendix C for resources where you can learn more. It is possible to complete most of the examples in the book with one or the other, but I do encourage you to learn both.

How This Book Is Organized

This book will take you on a journey that starts with circuit basics and ends with how to make interactive wireless wearables. In between, you'll learn about materials, microcontrollers, sensors, and actuators, and how these things fit into the world of wearable electronics. Here is what lies ahead:

Chapter 1, Circuits
: This chapter introduces you to circuit basics and then will show you six different ways to build the same circuit using different conductive materials.

Chapter 2, Conductive Materials
: Here you will take a deep dive into the range of conductive materials that we can use to construct circuits.

Chapter 3, Switches
: On, off, and beyond! This chapter provides an overview of switch basics and explains how to create your own.

Chapter 4, E-Textile Toolkits
: This chapter reviews the different electronic textile toolkits that are available for use in your wearable electronics projects.

Chapter 5, Making Electronics Wearable
: Making a circuit is one thing, but wearing it is another. This chapter goes through factors to consider when designing wearable electronics.

Chapter 6, Microcontrollers
: This is where the brains come in. This chapter provides an introduction to both the hardware and software aspects of getting up and running with microcontrollers.

Chapter 7, Sensors
: Sensors are what microcontrollers use to listen to the physical world. This chapter provides an introduction to the basics of working with sensor data and presents a variety of sensors that are useful in the wearable context.

Chapter 8, Actuators
: Actuators make things happen! From light to sound to motion, this chapter introduces you to actuators that can be employed in your wearable electronic designs.

Chapter 9, Wireless
: Time to bust out! This chapter introduces three approaches for wireless communication, meaning your project can send and receive data without being tied down.

Appendix A, Tools
: This provides an overview of the electronics and sewing tools that you might need for your studio, workshop, or lab.

Appendix B, Batteries
: Power it up! Here you'll find details of battery options for your wearable electronics projects.

Appendix C, Resources
: Want to learn more? Here's a list of resources that will take you above and beyond what's covered in this book.

Appendix D, Other Neat Things
: This is a selection of materials and processes that might help you make your wearables happen.

Appendix E, Microcontroller Options
: Here you'll find a more comprehensive list of microcontroller options to use in your wearable electronics projects.

About the Title

Make: Wearable Electronics does indeed cover how to make electronics that are wearable. More broadly, it provides a non-traditional approach to constructing electronic projects. The tools and techniques that are covered can also be applied to textiles, tapestries, toys, and more!

About Experiments and Projects

Throughout the book, we'll walk through *experiments* that will get you going and take a look at real-world *projects* that will serve as inspiration. A deliberate gap has been left between the two.

Some wearable electronics and e-textile books show you exactly how to build a particular project. This is not one of them. Instead, this book provides the building blocks that will help bring your own ideas to life.

About the Examples

Here are some technical notes about the examples presented in this book:

Connections

Most of the example circuits presented in this book can be created using alligator clips. Alligator clips can always be replaced by conductive thread, soldered wires, or other conductive materials as desired.

Power

All of the analog circuits can be powered using CR2032 batteries. Except where noted, the microcontroller circuits can be powered either by 1,000 mAh rechargeable lithium polymer batteries or via the microcontroller's USB connection. For alternative power options, see Appendix B.

Code

All code can be found here: *https://github.com/katehartman/Make-Wearable-Electronics*

About Part Numbers

Throughout the book, you will see part numbers that are preceded by a supplier code. These are the codes that will be used:

- AF: Adafruit Industries
- DK: Digi-Key
- IV: Inventables
- LE: Less EMF
- MS: Maker Shed
- RO: RobotShop
- RS: RadioShack
- SF: SparkFun Electronics

You can learn about these suppliers and more in Appendix C.

What Was Left Out

This book does not attempt to replicate existing resources. Take note of the references and project examples that are woven into each chapter, as well as the materials provided in Appendixes C and D. These breadcrumbs will lead you to a world of smart and talented thinkers, makers, and visionaries working in this and intersecting fields.

Experiment: Imagined Wearable

An experiment in the preface? That's right! The best time for you to start prototyping wearable electronics is right now. Sometimes it's easier to work through ideas before you even know what technologies you might use to create them.

Imagine something intended to be worn on your body (a garment or accessory) that would help you better relate to the world around you. It could be something practical, possible, or desirable. Or it could be something ridiculous, outlandish, annoying, or invasive. The technology that your garment utilizes does not have to actually exist and can be one of your own invention.

Once you've imagined your wearable, create a physical, wearable prototype or mock up that demonstrates what it might look like and how it would work. To make it, you can modify something that already exists (t-shirt, sneakers, top hat, etc.) or create something new from raw materials. It doesn't have to be fancy. Sometimes paper, duct tape, and Sharpies will do just fine.

This is a conceptual prototype—you do not need to implement any technology. Instead, focus on the design of the piece as well as the story behind it. Feel free to be creative, playful, and inventive. Try creating supporting materials such as instructions for use or user scenarios to help develop the story of your wearable. If you need some inspiration, check out the sidebar on this page.

Conventions Used in This Book

The following typographical conventions are used in this book:

Italic
: Indicates new terms, URLs, email addresses, filenames, and file extensions.

`Constant width`
: Used for program listings, as well as within paragraphs to refer to program elements such as variable or function names, databases, data types, environment variables, statements, and keywords.

`Constant width bold`
: Shows commands or other text that should be typed literally by the user.

`Constant width italic`
: Shows text that should be replaced with user-supplied values or by values determined by context.

This icon signifies a tip, suggestion, or general note.

This icon indicates a warning or caution.

Shape-Shifting Glasses

While I hate that I need to wear glasses, I love the glasses that I own. I interchange two pairs of prescription glasses, and my choice of glasses before I leave the house is influenced by what I happen to be wearing, where I am going, who I will be seeing, and how I am feeling. These factors actually create a wide variety of potential glasses, but sadly, I only have two pairs.

The Shape-Shifting Glasses would be the product of an advanced form of nanotechnology (I think?) and would be moldable to any shape, while maintaining accurate prescription. A button in one of the arms (think of the size of a inset reset button on a small electronic, the kind you have to press with a pin) would activate the moldability/solidification of the glasses. When activated, you would simply squish and stretch the lenses to the preferred size and shape, set the button again, and they would then remain in the desired fashion.

-Elijah Montgomery

Figure P-2. *Elijah Montgomery's Shape-Shifting Glasses, an imagined wearable*

Using Code Examples

This book is here to help you get your job done. In general, you may use the code in this book in your programs and documentation. You do not need to contact us for permission unless you're reproducing a significant portion of the code. For example, writing a program that uses several chunks of code from this book does not require permission. Selling or distributing a CD-ROM of examples from Make: books does require permission. Answering a question by citing this book and quoting example code does not require permission. Incorporating a significant amount of example code from this book into your product's documentation does require permission.

We appreciate, but do not require, attribution. An attribution usually includes the title, author, publisher, and ISBN. For example: "*Make: Wearable Electronics* by Kate Hartman (Make). Copyright 2014 Kate Hartman, 978-1-4493-3651-6."

If you feel your use of code examples falls outside fair use or the permission given here, feel free to contact us at *bookpermissions@makermedia.com*.

Safari® Books Online

Safari Books Online is an on-demand digital library that delivers expert content in both book and video form from the world's leading authors in technology and business.

Technology professionals, software developers, web designers, and business and creative professionals use Safari Books Online as their primary resource for research, problem solving, learning, and certification training.

Safari Books Online offers a range of product mixes and pricing programs for organizations, government agencies, and individuals. Subscribers have access to thousands of books, training videos, and prepublication manuscripts in one fully searchable database from publishers like Maker Media, O'Reilly Media, Prentice Hall Professional, Addison-Wesley Professional, Microsoft Press, Sams, Que, Peachpit Press, Focal Press, Cisco Press, John Wiley & Sons, Syngress, Morgan Kaufmann, IBM Redbooks, Packt, Adobe Press, FT Press, Apress, Manning, New Riders, McGraw-Hill, Jones & Bartlett, Course Technology, and dozens more. For more information about Safari Books Online, please visit us online.

How to Contact Us

Please address comments and questions concerning this book to the publisher:

Make:
1005 Gravenstein Highway North
Sebastopol, CA 95472
800-998-9938 (in the United States or Canada)
707-829-0515 (international or local)
707-829-0104 (fax)

Make: unites, inspires, informs, and entertains a growing community of resourceful people who undertake amazing projects in their backyards, basements, and garages. Make: celebrates your right to tweak, hack, and bend any technology to your will. The Make: audience continues to be a growing culture and community that believes in bettering ourselves, our environment, our educational system—our entire world. This is much more than an audience, it's a worldwide movement that Make is leading —we call it the Maker Movement.

For more information about Make:, visit us online:

> Make: magazine: *http://makezine.com/magazine/*
> Maker Faire: *http://makerfaire.com*
> Makezine.com: *http://makezine.com*
> Maker Shed: *http://makershed.com/*

We have a web page for this book, where we list errata, examples, and any additional information. You can access this page at:

> *http://bit.ly/wearable-electronics*

To comment or ask technical questions about this book, send email to:

> *bookquestions@oreilly.com*

Acknowledgments

In 2004, I attended an information session about the Interactive Telecommunications graduate program at New York University. Red Burns (who was the long-standing chair of the program at the time) candidly told us, "If you think you know what you're going to do here, you're wrong." She was, as always, right.

I never expected to become a wearable technologist, nor did I anticipate I would write a book about such things. But here we are. My opportunity to participate in all of this would never have happened without the support, hard work, and enthusiasm of some truly fabulous individuals.

To the following people I would like to offer my most heartfelt thanks:

To Brian Jepson, who first put this ball in my court and has been integral to this process from beginning to end.

To my editors, Meghan Blanchette, Shawn Wallace, and Emma Dvorak, who provided the guidance needed to sculpt these pages into something printworthy.

To my technical editors, Rob Faludi, Erin Lewis, Pearl Chen, and Lynne Bruning, who lent their attentiveness and depth of knowledge.

To Rob Faludi, also, for telling me I should write this thing.

To my research assistant, Hillary Predko, for her skill, zest, and speed.

To Jen Liu, whose delightful illustrations ignite the imagination.

To Angella Mackey, whose work is featured on the cover, for showing us that electronics can possess mystery, allure, and sass.

To the constellation of brilliant artists, designers, and makers whose projects are featured throughout this book for challenging our expectations and showing us what's possible.

To Leah Buechley for boldly challenging the way we think about designing electronics.

To the folks at SparkFun Electronics and Adafruit Industries for leading the way in making e-textile gear available to the masses.

To Becky Stern, Hannah Perner-Wilson, and Mika Satomi for setting the bar for ninja documentation skills.

To Syuzi Pakhchyan for nuturing our community.

To Despina Papadapoulous for the inspiration and the on-ramp.

To Dan O'Sullivan for telling me to just make the hats.

To Tom Igoe for helping me to become a teacher.

To OCAD University for taking a leap of faith and making me an assistant professor.

To the Digital Futures crew, especially to Suzanne Stein, Emma Westecott, Barbara Rauch, Paula Gardner, Caroline Langill, Tom Barker, Adam Tindale, Nick Puckett, Simone Jones, and Jeff Watson, for the collegiality and the camaraderie.

To my students for being brave and making beautiful things.

To Sara Diamond, Monica Contreras, and Helmut Reichenbächer for developing and sustaining a foundation upon which I could build.

To the rockstar research assistants who helped transform the Social Body Lab from an empty room to a vibrant ecosystem: Boris Kourtoukov, Borzu Talaie, Calliope Gazetas, Erin Lewis, Gabe Sawhney, Hazel Meyer, Hillary Predko, Izzie Colpitts-Campbell, Jackson McConnell, Julian Higuerey-Nuñez, Ken Leung, Oldouz Moslemian, Rachael Kess, Rickee Charbonneau, Rob King, Ryan Maksymic, and Stewart Shum. The pages of this book are painted with your talent and effort.

To the nice folks at the White Squirrel Coffee Shop for keeping me caffeinated and giving me a place to think.

To Gabe Sawhney for listening to my rants.

To Kati London for the candor.

To Tony Wong, Ted Redelmeier, and John Rose for holding the rope when I need to muster my courage.

To Carrie Schulz for being a stellar fellow cartographer.

To Jason Bellenger for the patience and for the adventures.

And finally…

To my parents, who though they also never anticipated that I would become a wearable technologist, somehow still knew how to support me every step of the way.

Circuits

1

Welcome to the world of wearable electronics! Before diving into designing complex, body-based, interactive projects, it is important that you have an understanding of basic circuits. In this chapter, you will learn about both how circuits work as well as how to construct them using a variety of tools and materials.

Figure 1-1. *"Connection and Motion" by Izzie Colpitts-Campbell; this wearable circuit uses stainless steel-coated brass chain to connect LEDs to a battery pack*

Circuit Basics

There are some essential concepts that everyone should know when constructing circuits. These concepts will help guide circuit design and choice of components.

A *circuit* is a closed loop of electricity that contains a power source and a load. Conductors provide pathways for the electricity to travel between components in the loop.

A *power source* provides electricity. You will use a battery or battery pack as the power source for all of the circuits you create in this book. Batteries are a sensible power source for wearable electronics because they are relatively portable and compact.

A *load* is something that makes use of the electricity in the circuit. For the examples in this chapter, you will use light-emitting diodes (LEDs) as the load in your circuits.

A *conductor* is a material that permits the flow of electricity. In this chapter, you will use a variety of conductors to create electrical connections in your circuits.

A *circuit diagram* is a clear, concise representation of the components and connections in a circuit. It helps you understand the electrical connections being made within a circuit. It is not an image of an actual circuit.

In a circuit diagram, each component is represented as a symbol. Figure 1-2 shows an example of a few.

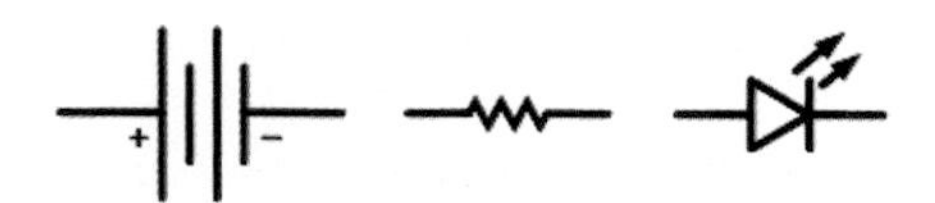

Figure 1-2. *Circuit symbols (left to right) for a battery, resistor, and LED*

Using these symbols, you can draw a diagram of a simple circuit, as shown in Figure 1-3.

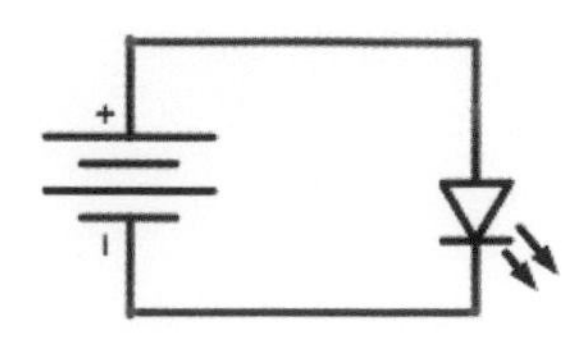

Figure 1-3. *Basic circuit with battery and LED*

Electricity is thought of as traveling from the point of highest electrical potential (*power* or "+") to the lowest (*ground* or "–"). So in this circuit, it flows from the positive terminal of the battery (marked with "+") to the positive terminal of the LED, through the LED, out the negative terminal of the LED to the negative terminal of the battery (marked with "–"), thus completing the loop (see Figure 1-4). Along the way, it will travel through the LED and (assuming it provides the correct power requirements) cause the LED to light.

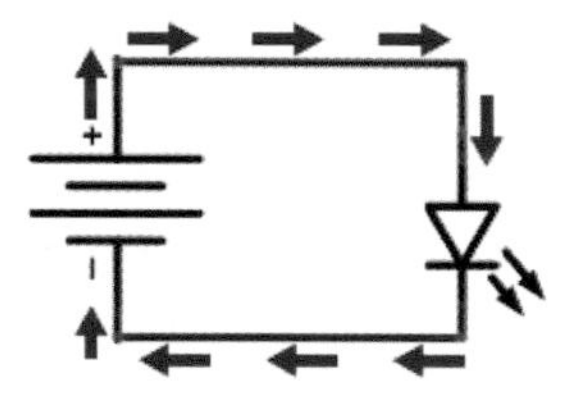

Figure 1-4. *The current (indicated by the red arrows) travels from power (+) to ground (-)*

Electricity likes to follow the path of least resistance. You can think of it as being a bit lazy. If electricity has the option of working to light up an LED or to take a path through a nonresistive material back to the battery, it's going to take the easy road. You can see this in Figure 1-5.

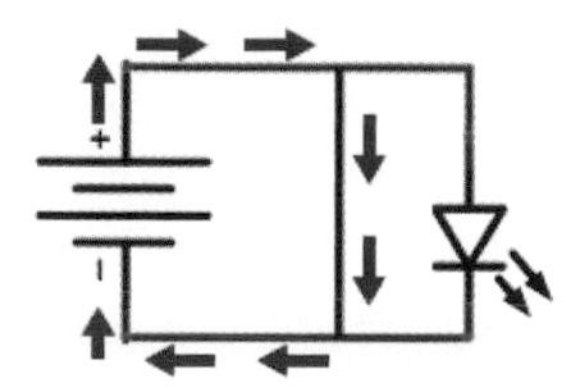

Figure 1-5. *When presented with the opportunity, electricity will always follow the path of least resistance; in this circuit, the electricity does not reach the LED, so the LED will not light up*

The problem with this alternative path is that it creates a *short circuit*. A short circuit is a closed loop of electricity that has a power source but no load.

If electricity is fed from the positive end of the battery directly into the negative, depending on the duration of the short, it will likely drain the battery. In some situations, the results can be more severe, including smoke, melted wires, and damaged components. At minimum, your project won't function properly. No matter what the circumstance, shorts are not good, so it's best to make sure they don't happen in your circuit.

Insulators are materials that do not conduct electricity. They can be used to prevent short circuits.

To see how this circuit might look in real life, you can use components like a 3V battery (CR2032) and a 5mm through-hole LED (see Figure 1-6).

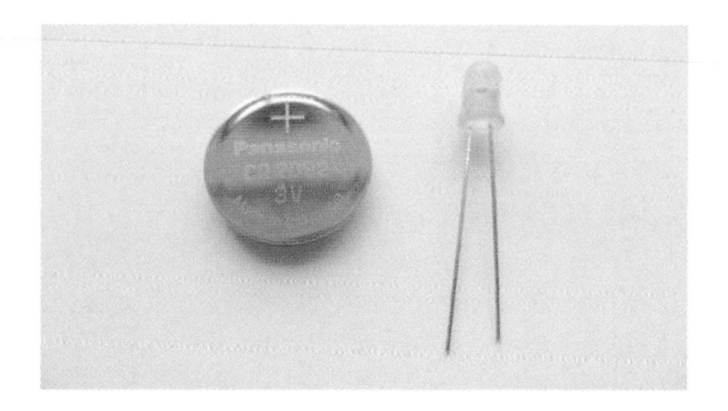

Figure 1-6. *CR2032 3V battery and 5mm LED*

In order to implement the circuit depicted in the circuit diagram, all you need to do is press the positive end of the LED to the positive end of the battery and the same with the negative end of each component, as shown in Figure 1-7.

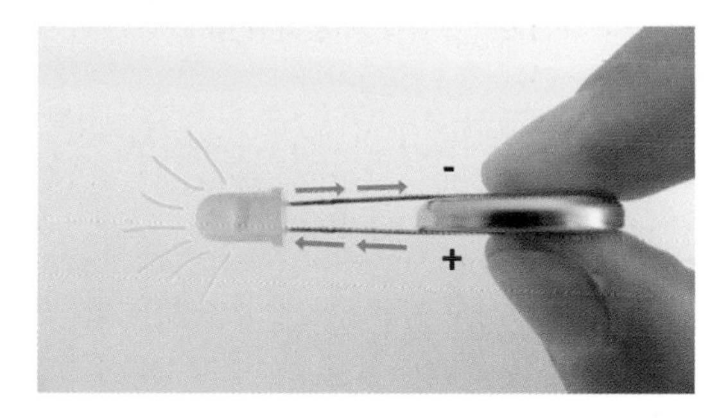

Figure 1-7. *A simple circuit*

And bam! You have a circuit. This configuration allows electricity to flow from the battery, through the LED, and back to the battery giving the LED the power it needs to light up. This technique was used by James Powderly and Evan Roth (Graffiti Research Lab) to create magnetic modules (see Figure 1-8) that could act as light graffiti on buildings and other urban structures.

Figure 1-8. *LED "throwies" (image courtesy of James Powderly and Evan Roth)*

Ohm's Law

The circuit you just created is a quick-and-dirty way to light up an LED, but there are a few more things to learn before you can construct a technically correct and long-lasting circuit.

There are three key pieces of information to pay attention to when designing a circuit:

Voltage
: The difference in electrical energy between two points. It is measured in volts (V).

Current
: The quantity or amount of electrical energy passing a particular point. It is measured in amps (A) or milliamps (mA).

Resistance
: The measure of a material's ability to prevent the flow of electricity. Resistance is measured in *ohms*, which is represented by the ohm symbol (Ω).

Ohm's law states that voltage (V) is equal to current (I) times resistance (R). As with any equation, if you know two of the three variables, you are able to determine the third. All three variations of this equation can be helpful to you as you're learning to construct circuits:

- $V = I \times R$
- $I = V \div R$
- $R = V \div I$

As it turns out, in the circuit you just created with the LED and battery, the LED is actually receiving a bit more than the desirable amount of current. Excessive current can shorten the LED's life or even burn it out. Because this battery supplies a relatively low amount of current, and an LED is not a particularly sensitive or expensive component, you can get away with more of a hacky approach. But ideally you should create a circuit that respects the needs of its components. To do this, you can use Ohm's law to determine how much resistance is needed.

To determine the voltage that needs to be used up, you will need to find the difference between the source voltage (Vs) and the forward voltage (Vf), which is the voltage used up by the LED. So the equation is actually as follows:

- $R = (Vs - Vf) \div I$

The source voltage (Vs) is that which is supplied by the battery. In this case, the CR2032 battery supplies 3V.

You can find the forward voltage (Vf) and the current required by the LED on the LED's datasheet (Figure 1-9). A *datasheet* is a document supplied by a component's manufacturer. This document provides information about the component, including the component's electrical needs and tolerances, mechanical diagrams of its physical packaging, diagrams of any pins or connections, and details on intended use and expected performance.

Absolute Maximum Ratings: (Ta=25℃).

ITEMS	Symbol	Absolute Maximum Rating	Unit
Forward Current	I_F	20	mA
Peak Forward Current	I_{FP}	30	mA
Suggestion Using Current	I_{SU}	16-18	mA
Reverse Voltage (VR=5V)	I_R	10	uA
Power Dissipation	P_D	105	mW
Operation Temperature	T_{OPR}	-40 ~ 85	℃
Storage Temperature	T_{STG}	-40 ~ 100	℃
Lead Soldering Temperature	T_{SOL}	Max. 260℃ for 3 Sec. Max. (3mm from the base of the expoxy bulb)	

Absolute Maximum Ratings: (Ta=25℃)

ITEMS	Symbol	Test condition	Min.	Typ.	Max.	Unit
Forward Voltage	V_F	I_F=20mA	1.8	---	2.2	V
Wavelenength (nm) or TC(k)	$\Delta\lambda$	I_F=20mA	587	---	591	nm
*Luminous Intensity	I_V	I_F=20mA	150	---	200	mcd
50% Viewing Angle	2θ1/2	I_F=20mA	40	---	60	deg

Figure 1-9. *A detail from the LED's datasheet*

According to the datasheet, the forward voltage of this LED is rated as 1.8 to 2.2V. So let's say 2V. The LED requires 16–18mA of current, with a maximum of 20mA. Let's use 17mA or 0.017A for the calculations.

Now that you have all of the necessary information, the equation will play out as follows:

- $R = (Vs - Vf) \div I$
- $R = (3V-2V) \div 0.017A$
- $R = 58.82\Omega$

These calculations tell you that you should ideally add 58.82 ohms of resistance to the circuit.

Finding Datasheets

If you order a component online, there will usually be a link to the datasheet on the web catalog page that you ordered the component from. In rare cases, datasheets will be included in the packaging when you purchase a component. If not, the easiest place to start is the Internet. Open your favorite search engine and type in the part number and the word "datasheet" to find it. You will likely find a PDF of the datasheet on the manufacturer's website, the distributor's website, or in a datasheet database, of which there are many on the Web. If you don't know the part number, you can even try a description such as "5mm yellow LED."

The added bonus is that these days many distributors provide an abundance of additional resources on their parts pages. Not only do they include links to datasheets, but also to tutorials, circuit diagrams, circuit board design files, sample code, and sometimes even example projects.

Understanding Resistors

Now you know how much resistance you need. But how do you add resistance to the circuit?

A *resistor* is a component that resists the flow of electricity. It can be implemented in a circuit to use up extra electricity that is not needed by the load.

You know from the Ohm's law equations that you need approximately 58.82Ω resistance for the circuit. However, resistors come in set values, so there may not always be the exact resistor that meets your needs.

If you don't have the exact resistor you're looking for on hand (or if it doesn't exist), you do one of two things:

- Use the next largest value. In your circuit, this will be a 62Ω resistor.
- Combine two resistors in a row that add up to the correct value (e.g., 56Ω + 3Ω = 59Ω).

For your purposes, go with a 62Ω resistor to reduce the amount of wiring. After adding the correct resistor to your circuit, the circuit diagram will look like Figure 1-10.

If you don't have a 62Ω resistor handy, you can use the more common 68Ω value, or even 100Ω.

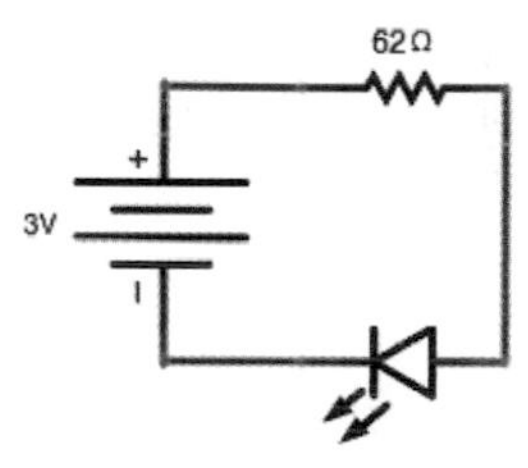

Figure 1-10. *Circuit diagram with 3V battery, LED, and 62Ω resistor*

When looking at the actual component, the value of a through-hole resistor can be determined by the color bands displayed on it. Each color indicates a value. A 62Ω resistor is marked with the colors blue(6), red(2), black(1), and gold(±5%), as shown in Figure 1-11.

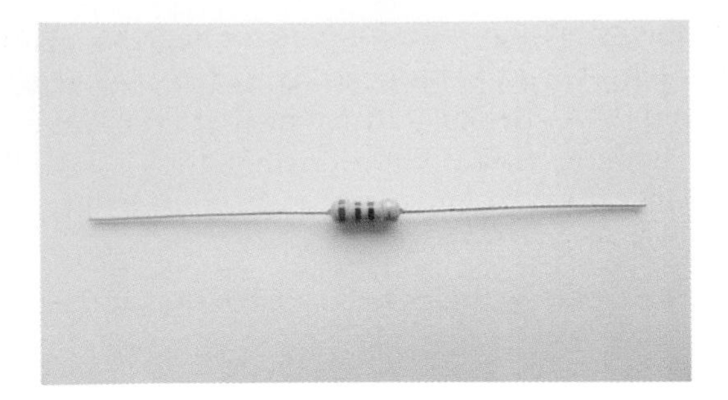

Figure 1-11. *A 62Ω resistor*

You can decode these bands by consulting a resistor color chart, going to a resistor calculator website, or downloading a resistor application for your smartphone. Table 1-1 shows a table you can use to decode a resistor's color codes. Figure 1-12 shows the ResistorCode iPhone app.

When reading the color bands on the resistor, orient it so that the silver or gold band is on the right. Then read the colors from left to right. The first two bands will indicate the first two digits of the number. The third band indicates the multiplier for that digit. The fourth band indicates the tolerance.

Table 1-1. Resistor chart

Color	Value	Multiplier	Tolerance
Black	0	1	-
Brown	1	10	±1%
Red	2	100	±2%
Orange	3	1K	-
Yellow	4	10K	-
Green	5	100K	±0.5%
Blue	6	1M	±0.25%
Purple	7	10M	±0.1%
Gray	8	100M	±0.05%
White	9	1000M	-
Gold	-	1/10	±5%
Silver	-	1/100	±10%
None	-	-	±20%

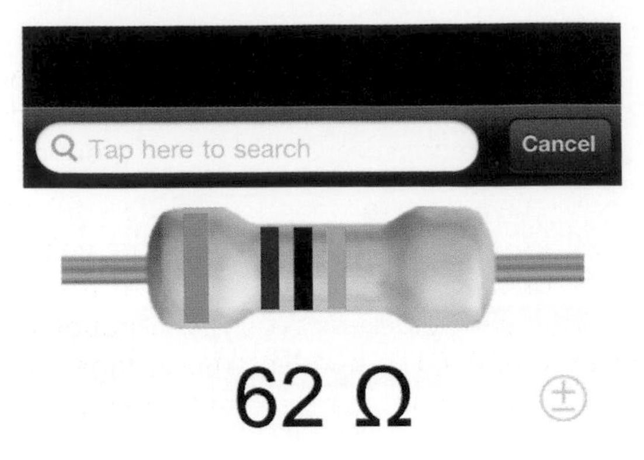

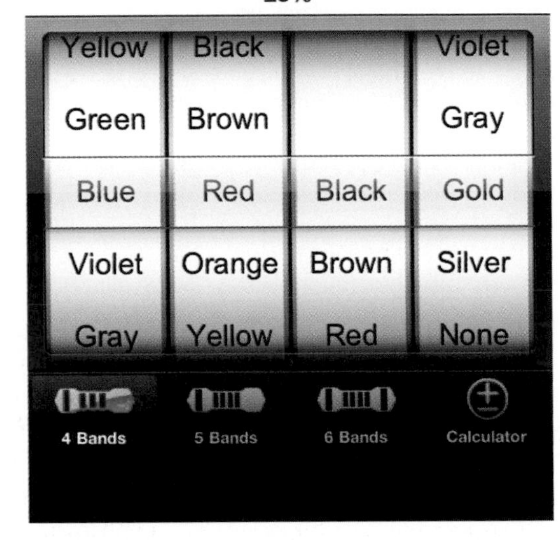

Figure 1-12. *Using a resistor app to decode the colors of a resistor*

Series and Parallel

OK, so you know how to create a circuit with one LED, but how about three? When adding additional components to the circuit, you need to understand the difference between *series* and *parallel*.

Series
: In a series, components, like LEDs, are connected in a row. Electricity flows through one, into the next, and then into the next.

Parallel
: In a parallel configuration, components are connected side by side, each with an independent connection to power and ground.

LEDs can be connected in series (Figure 1-13). But the power source must supply adequate voltage. The factor that needs to be considered is called *voltage drop*. When electricity passes through and gets used up by a component, the voltage drops before it moves on to the next component.

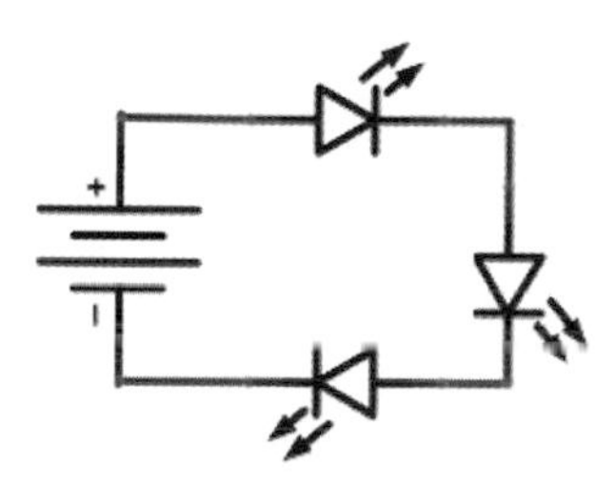

Figure 1-13. *Three LEDs in series*

The LEDs you have been working with in this chapter are rated for a forward voltage drop of 1.8–2.2 volts. This means that if you used a 3V battery as your power source that the first LED would get 2V, the second 1V, and the last no voltage. This obviously won't work. The way to fix this is to use a power source whose voltage could accommodate this voltage drop. Because each of the three LEDs has a voltage drop of around 2V, you would want a battery pack that provided at least 6V.

But there is also another way to connect multiple LEDs to a power source. Figure 1-14 shows three LEDs in parallel.

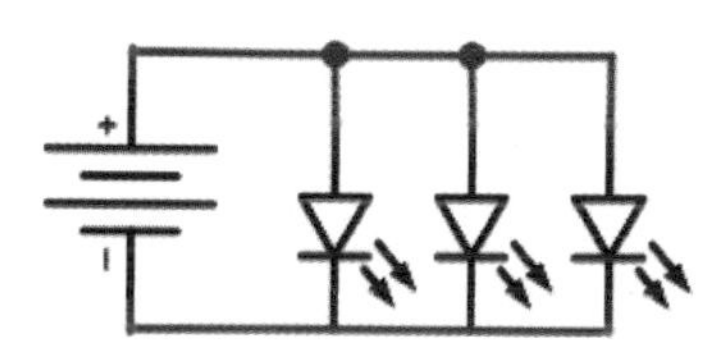

Figure 1-14. *Three LEDs in parallel*

In this situation, each LED receives the same amount of voltage, but the current is divided between them. The only thing missing from this circuit is the resistors. With resistors, the circuit will look like Figure 1-15.

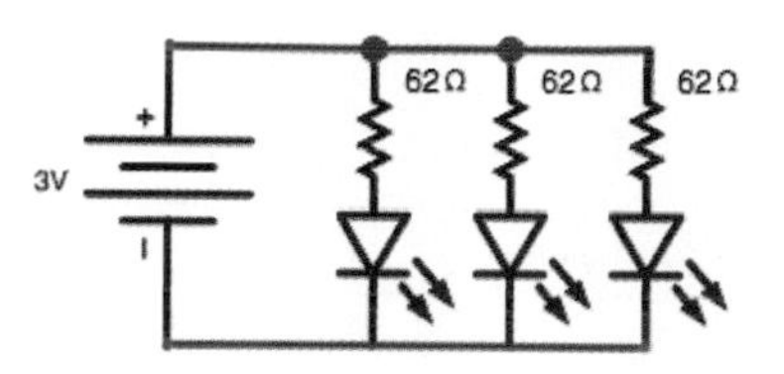

Figure 1-15. *Three LEDs in parallel with resistors*

If all the LEDs are the same, then they will each use the same resistors. However, if you add an LED that requires a different amount of voltage or current, you can use Ohm's law (see "Ohm's Law" on page 3) to calculate which resistor it needs.

Batteries can also be placed in series or in parallel. When batteries are connected in series, the voltage of the two batteries is added together. When they are connected in parallel, the voltage stays the same but their available current is added together.

For instance, AAA batteries usually supply around 1.5V. If two AAA batteries were placed in series (Figure 1-16), their voltage would be added together and the resulting battery pack would provide 3V and the same amount of current as a single AAA battery. If two AAA batteries were placed in parallel (Figure 1-17), the battery pack would supply 1.5V but twice as much current.

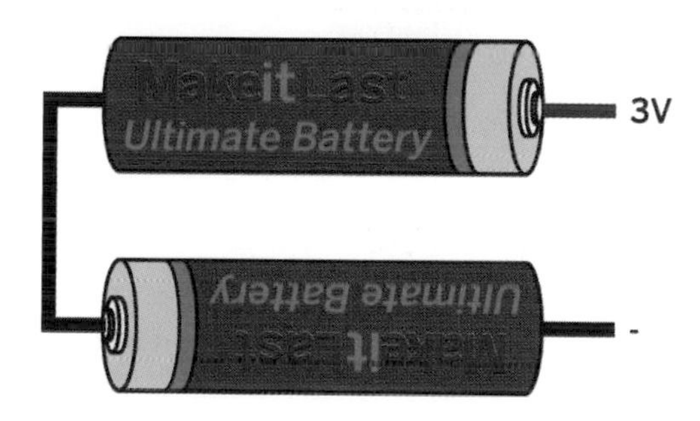

Figure 1-16. *AAA batteries in series*

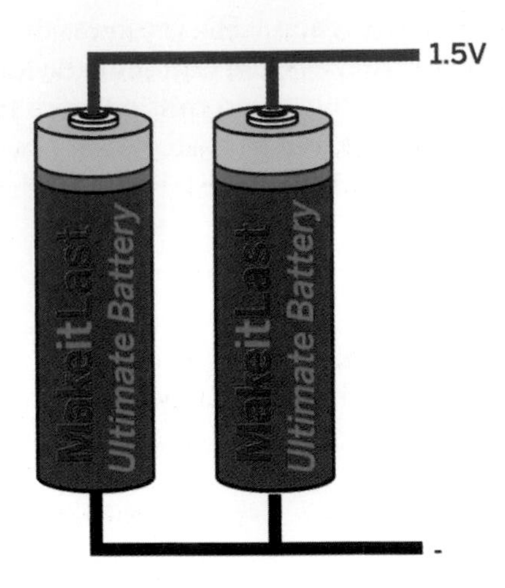

Figure 1-17. *AAA batteries in parallel*

Determining Polarity

Certain electronic components have a predetermined *polarity*. This means that it matters which way the component is connected in the context of a circuit. An LED is a good example.

LED stands for *light-emitting diode*. A *diode* is a component that only allows electricity to pass through it in one direction. If you connect an LED backward, electricity will not be able to pass through it, which means it will not light up.

Depending on the LED, there are four possible ways in which you might determine its polarity (Figure 1-18):

1. With through-hole LEDs, the leg of the anode (positive) side of the LED is usually longer than the cathode (negative).
2. Some manufacturers put a flat spot on the base of the lens of the LED by the cathode leg.
3. If you take a look inside the lens of the LED, you can see that there are two pieces that extend up from the legs. The piece that attaches to the cathode leg is the larger one that extends up and over that of the anode.
4. If all else fails, you can always give it a try on a breadboard or with alligator clips. If it is otherwise properly connected to a power source and doesn't light up, it probably means you have it in the wrong way. Flip it and give it another try.

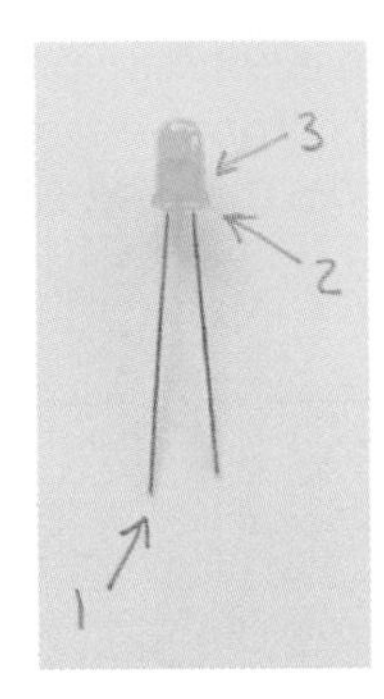

Figure 1-18. *Three places to look when determining the polarity of an LED*

Determining polarity of other components tends to vary, but be on the lookout for a + or – sign, which will indicate the positive or negative side of the component. Red (positive) and black (negative) wires can also be a clue.

Using a Multimeter

Because you cannot see, smell, or hear electricity, you'll need a special tool to detect it. A multimeter, shown in Figure 1-19, can be used to check *continuity* (whether current flows unimpeded through two points) as well as to measure voltage, resistance, and current.

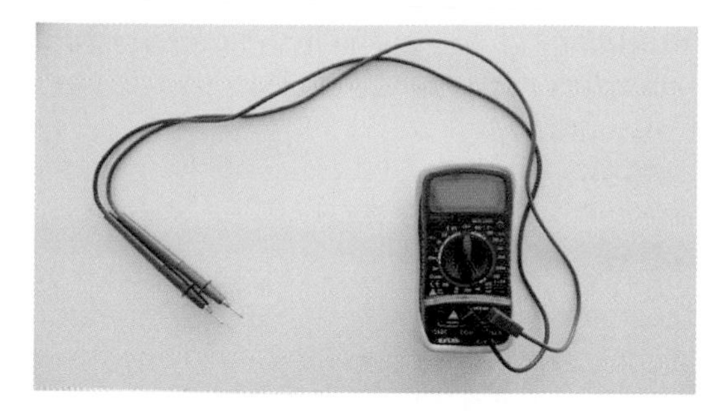

Figure 1-19. *A multimeter*

They usually have a dial (or buttons), shown in Figure 1-20, that are used to select a particular function, and probes (see Figure 1-21) that are used to make connections with whatever it is you are measuring.

Figure 1-20. *The dial of a multimeter is used to select the function*

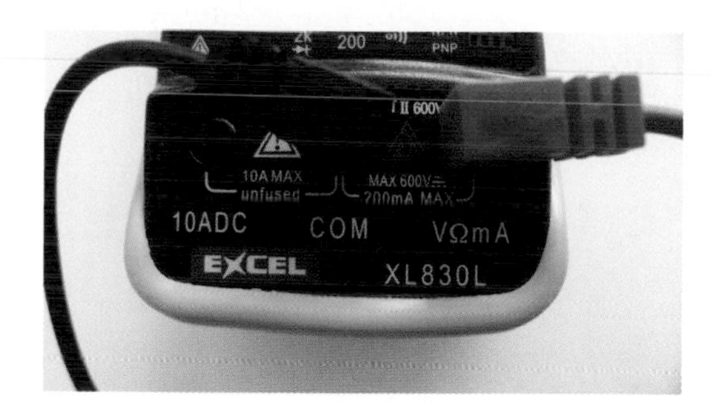

Figure 1-21. *The connection for the probes; on this meter, you will need to move the red probe to the left if you are measuring current higher than 200mA*

There are a variety of multimeter tutorials available in basic electronics books as well as on the Internet. Check out Appendix C for some references. Here we will just cover the basic concept of what a multimeter does and when you might use it when creating circuits.

Continuity

The simplest but perhaps most useful function of a multimeter is the continuity or conductivity test. This is most often marked on the dial with a speaker or audio symbol (see Figure 1-22) because meters will beep when the test is positive. Once the dial is in position, simply place the probes at two locations across which you'd like to test the continuity or conductivity.

Figure 1-22. *The knob set for a continuity test*

This can be used to check the continuity of a questionable connection. Place the probes on either side of the questionable connection and if the meter beeps, you're good to go.

It can also be used to check for short circuits by placing the probes in two places that are *not* supposed to be connected. If the meter beeps, then you know you've got a short circuit somewhere.

Finally, it can be used to check the conductivity of a material. Place the probes at two points on a material and see if you're able to establish a connection across it (see Figure 1-23). This can be especially useful if you're shopping for conductive materials in unusual places like a fabric store. I recommend investing in a pocket multimeter for just these occasions.

If you are testing a material that you *think* is conductive but the meter doesn't beep, the next step is to measure resistance in ohms, just in case it conducts "well enough" for your needs.

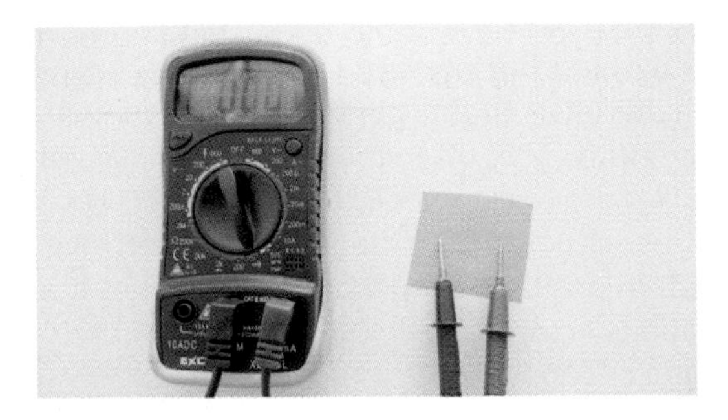

Figure 1-23. *Using a multimeter to test the conductivity of conductive fabric*

Resistance

To measure resistance, turn the knob to the portion of the dial marked with the ohm sign (Ω) and place the probes on either side of the component or material you would like to measure. You can use alligator clips to securely hold the component as shown in Figure 1-24.

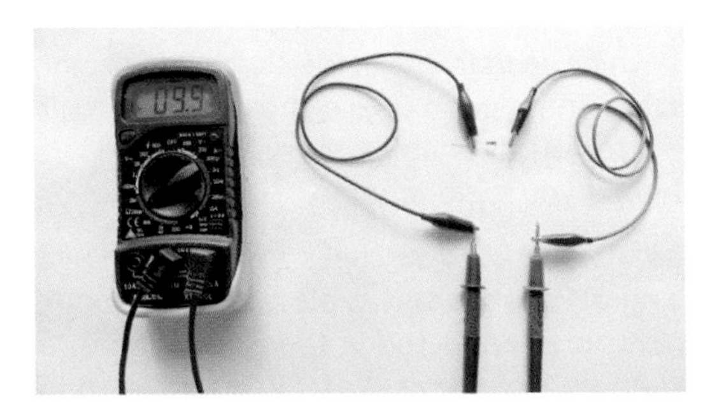

Figure 1-24. *Measuring the resistance of a fixed resistor*

With this setting, you can check the value of a fixed resistor, monitor the changing resistance of a variable resistor, or determine the resistance of a material like conductive thread. If your multimeter is not autoranging, you will have to select a resistance setting that's in the range of what you expect the component to be.

Voltage

Multimeters can also measure voltage. This is helpful for checking the state of a battery or determining if components of a circuit are receiving the voltage that they need. Turn the dial on the multimeter to the "V-" setting, set it to the range of voltage you expect to read, and place the probes on either end of whatever you want to measure (Figure 1-25).

Figure 1-25. *The knob set to measure voltage*

Figure 1-26 shows a battery that is at full strength, and Figure 1-27 shows one that's fading in power.

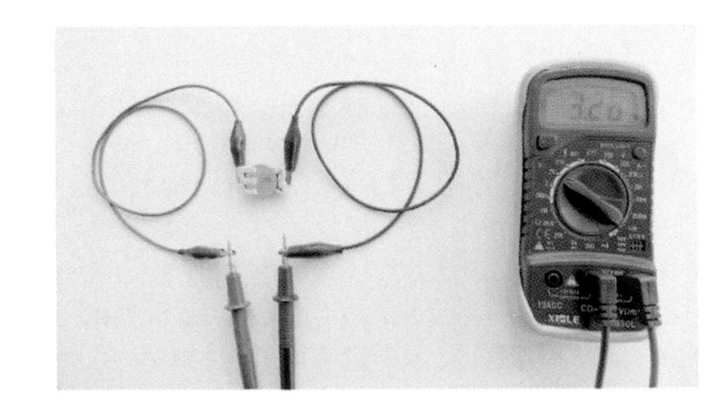

Figure 1-26. *Reading voltage of a fresh CR2032 battery*

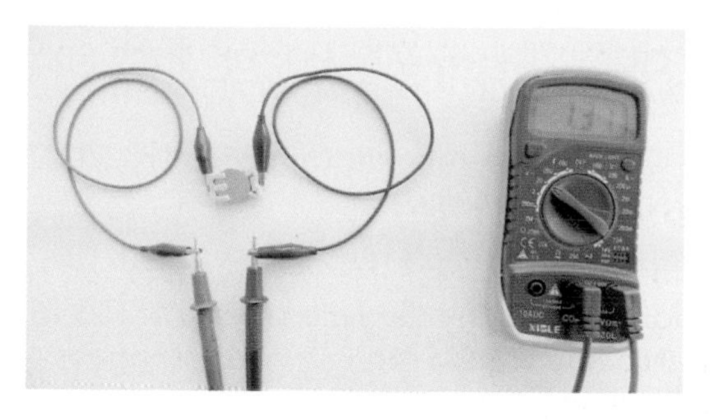

Figure 1-27. *Reading voltage of a CR2032 battery that is fading*

Current

The process for measuring current with a multimeter is a bit different. The meter actually needs to be *in series* with the circuit in order to determine how much current is being pulled. Turn the dial to the "A" or "mA" section and select the appropriate range. With some meters, you may need to move the probe to another terminal at the bottom of the meter. Check your multimeter manual for details. Once you're set up, find a location where you can insert the meter into the circuit and take a reading. Knowing how much current a circuit draws at its peak usage and over time can be extremely helpful in terms of determining which battery to select for your project. More on this in Appendix B.

More About Circuits

The world of circuits is wide and wonderful. This chapter only covers the details that you needed to know in order to build the examples that follow in this book. To learn more, be sure to check out books like *Make: Electronics* by Charles Platt (O'Reilly), *Practical Electronics for Inventors* by Paul Scherz and Simon Monk (McGraw-Hill/TAB Electronics), and *Getting Started in Electronics* by Forrest Mims (Master). Details about these resources and others can be found in Appendix C.

Constructing Circuits

As you learned earlier, conductors or conductive materials are materials through which electricity can pass. When constructing a circuit, conductive materials provide the pathway for electricity to flow from one component to another.

Now that you have some understanding of how circuits work and how to measure different aspects of them, you can start to think about how to physically construct them. This section will illustrate a variety of ways to bring this basic LED circuit to life.

As you move through different iterations of this circuit, you'll see that through the use of different conductive materials it can take on many shapes and sizes. The core electronic components that you work with will be the same in each circuit. What will differ is the materials and tools you use to create the connections between the components.

Figure 1-28 shows the circuit you will create.

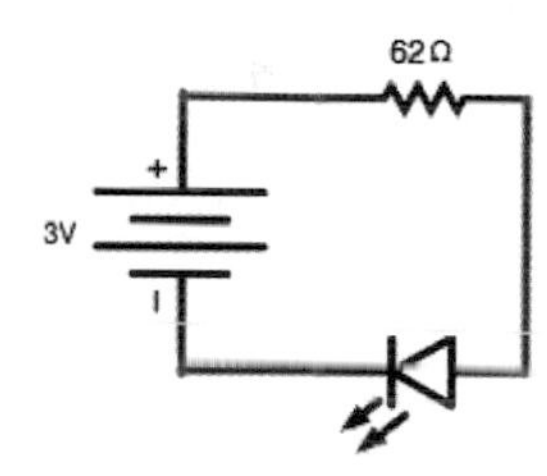

Figure 1-28. *A circuit with a 3V battery, LED, and resistor*

Keep in mind that a circuit diagram shows only the electrical connections. It does not reflect the physical layout or the materials or tools used to create the electrical connections.

The core parts you'll be using are as follows (see Figure 1-29):

- CR2032 battery (AF[1] 654, DK[2] P189-ND, SF[3] PRT-00338)
- CR2032 battery holder (AF 653, DK BA2032SM-ND, SF DEV-08822)
- 62Ω through-hole resistor (DK 62QBK-ND)
- 5mm through-hole yellow LED (DK 160-1851-ND, SF COM-09594)

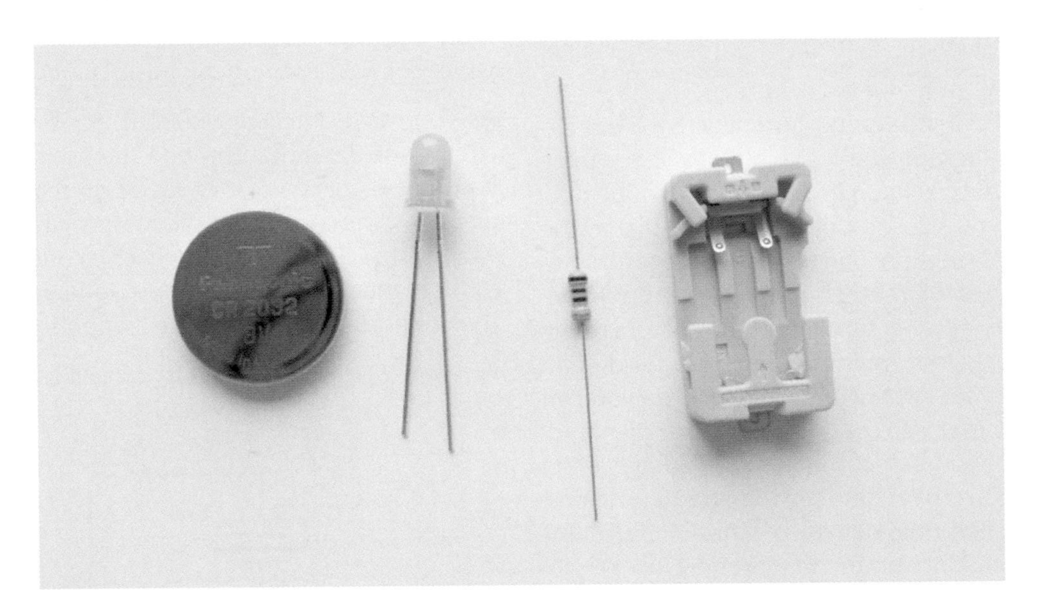

Figure 1-29. *CR2032 battery, LED, resistor, battery holder*

These are all basic and inexpensive components. The LED is a standard one you'd find in any basic electronics kit. The CR2032 battery meets the needs of the circuit both in terms of its voltage and current ratings. It also features a slim profile, which prevents your circuit from getting too bulky. The battery holders you are working with are actually intended for surface-mount electronics, but you will be modifying them for through-hole and soft-circuit applications. Note that the minus sign on the base of the holder shows you which terminal of the battery holder is ground.

Using these components, you will create six different versions of this circuit, constructed with alligator clips, wires, a breadboard, protoboard, conductive fabric, and conductive thread. Let's get started!

1. Adafruit (for a complete list of supplier abbreviations, see "About Part Numbers" on page xiv).
2. Digi-Key (for a complete list of supplier abbreviations, see "About Part Numbers" on page xiv).
3. SparkFun (for a complete list of supplier abbreviations, see "About Part Numbers" on page xiv).

Alligator Clip Circuit

Alligator clips provide a quick way to prototype simple, temporary circuits. This method is used with many e-textile toolkits.

Parts and materials, shown in Figure 1-30:

- (1) CR2032 battery (AF 654, DK P189-ND, SF PRT-00338)
- (1) CR2032 battery holder (AF 653, DK BA2032SM-ND, SF DEV-08822)
- (1) 62Ω through-hole resistor (DK 62QBK-ND)
- (1) 5mm through-hole yellow LED (DK 160-1851-ND, SF COM-09594)
- (3) alligator clip test leads: red, black, and yellow (AF 1008, RS[4] 278-1156, SF PRT-11037)

Figure 1-30. *Parts for alligator clip circuit*

First, clip a red cable to the positive terminal of the battery holder and a black cable to the negative, as shown in Figure 1-31. The use of these standardized colors helps you remember what's what.

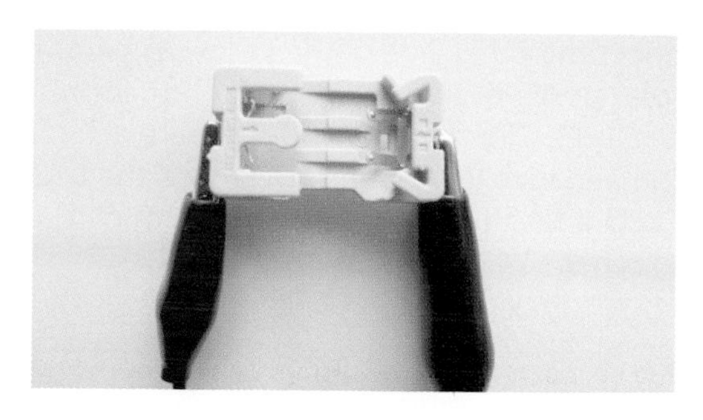

Figure 1-31. *Battery holder with red alligator clip attached to + and black to −*

Next, clip the other end of the red to the resistor. Clip the yellow alligator clip to the other side of the resistor. See Figure 1-32.

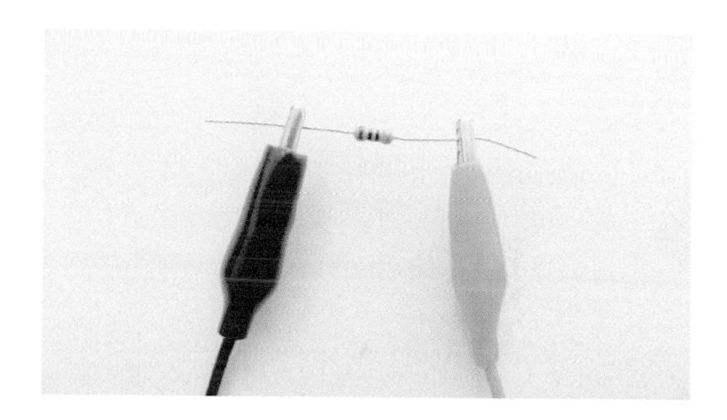

Figure 1-32. *Resistor with alligator clip connections*

4. RadioShack (for a complete list of supplier abbreviations, see "About Part Numbers" on page xiv).

Connect the other side of the yellow alligator clip to the positive side of the LED, as shown in Figure 1-33. Clip the other side of the black cable to the negative leg of the LED. Spread the legs of the LED a bit to ensure that the legs don't touch each other.

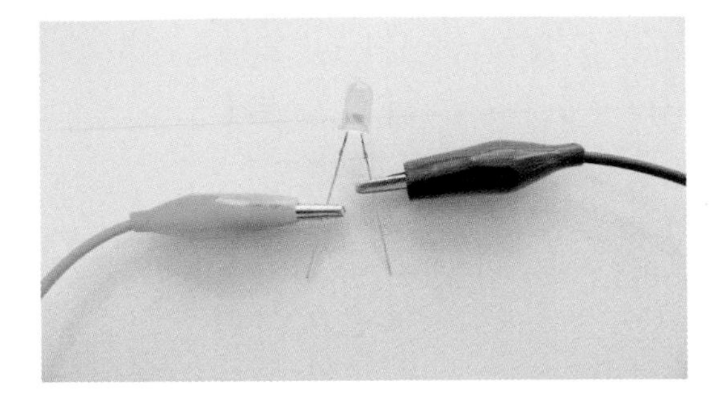

Figure 1-33. *LED with alligator clip connections*

Alligator clips grab on to components nicely but do have a tendency to slide around a bit, so keep an eye out for shorts.

The alligator clip circuit is now complete! Add a battery to light up the LED.

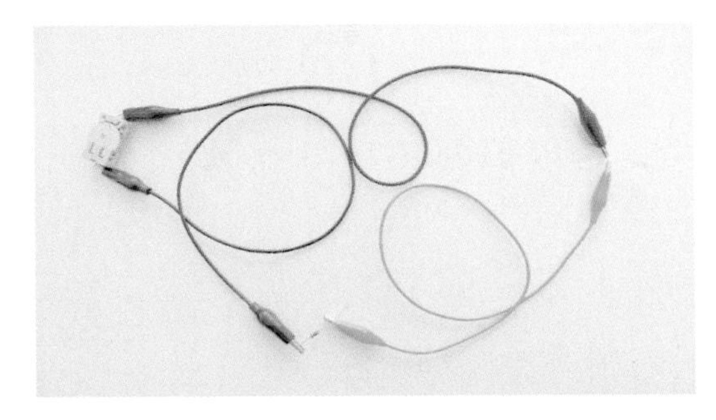

Figure 1-34. *Complete circuit*

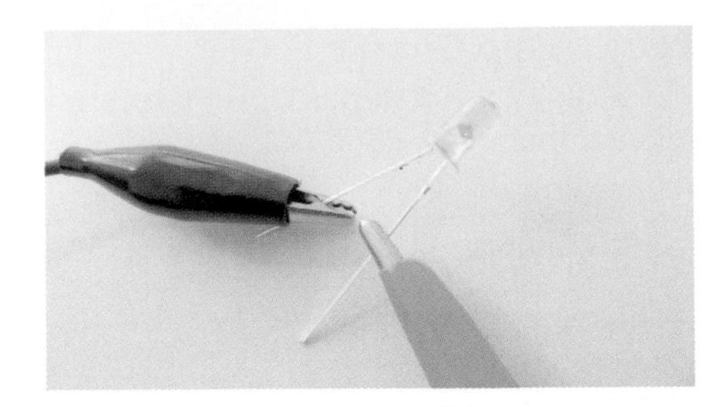

Figure 1-35. *LED lit up*

Wire Circuit

Wires can also be used to create connections between components in a circuit. Twisting, bending, or crimping establish a base physical connection, and then soldering those points establishes a secure electrical connection. In the following example, you will use some 22-gauge hookup wire to to connect components and heat shrink tubing to insulate connections.

Parts and materials, shown in Figure 1-36:

- (1) CR2032 battery (AF 654, DK P189-ND, SF PRT-00338)
- (1) CR2032 battery holder (AF 653, DK BA2032SM-ND, SF DEV-08822)
- (1) 62Ω through-hole resistor (DK 62QBK-ND)
- (1) 5mm through-hole yellow LED (DK 160-1851-ND, SF COM-09594)
- 22 AWG solid-core hook up wire, in red and black (AF 1311, SF PRT-11367)
- Heat shrink tubing (AF 344, RS 278-1610, SF PRT-09353)

Tools:

- Wire stripper
- Soldering iron and solder
- Heat gun

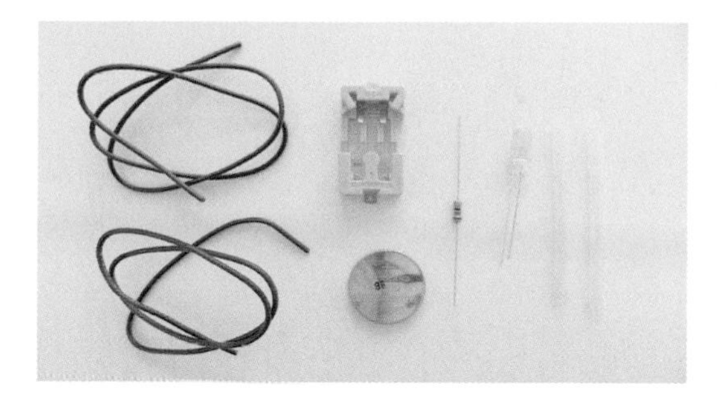

Figure 1-36. *Parts for wire circuit*

Soldering

In this and many of the examples that follow, you will be soldering. If you are new to soldering, take a look at a soldering tutorial to get yourself up to speed. My favorite is the *Soldering Is Easy* (*http://bit.ly/1pEZhUE*) comic book by Mitch Altman, Andy Nordgren, and Jeff Keyzer. Look at Appendix C for a list of more resources.

Start by tightly wrapping the leg of a resistor around the positive leg of an LED (Figure 1-37). Solder it in place (Figure 1-38) and then trim any excess length on either leg (Figure 1-39).

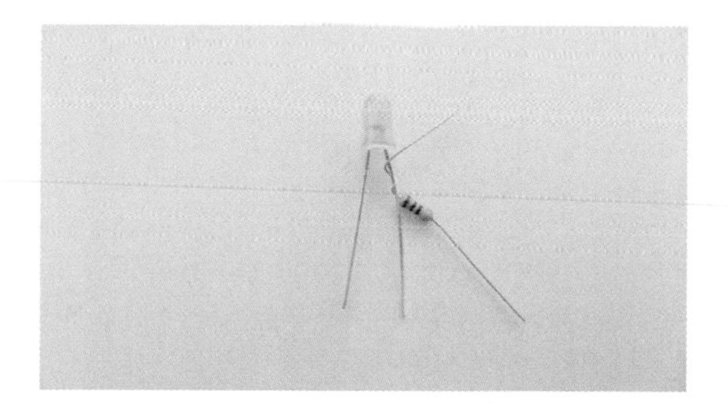

Figure 1-37. *Wrapping resistor leg around positive leg of LED*

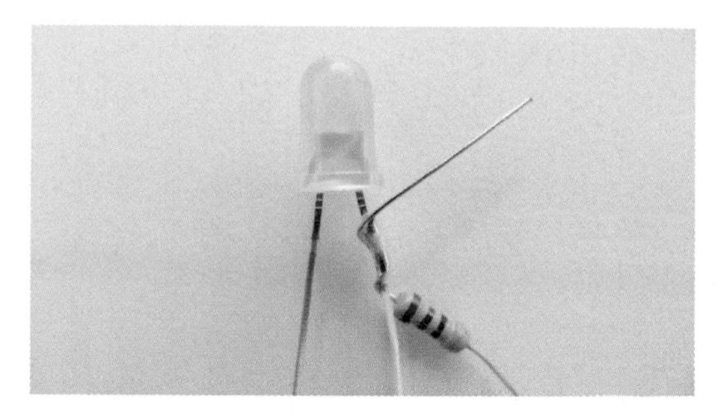

Figure 1-38. *Soldered connection*

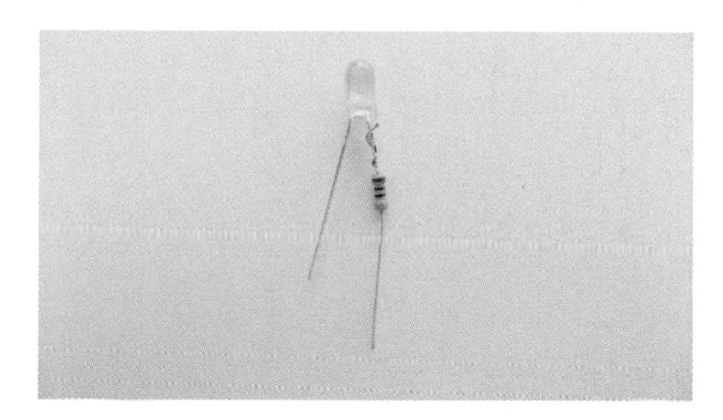

Figure 1-39. *Trimmed connection*

Next, you will need to prepare your wires using wire strippers (Figure 1-40).

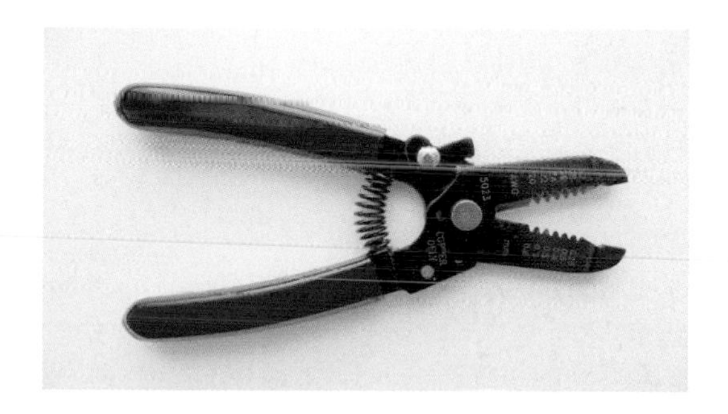

Figure 1-40. *Wire strippers*

Cut a short length of the red wire. Place the wire in the slot of wire strippers marked "22 AWG" or "0.6 mm," about a centimeter from one end of the wire, as shown in Figure 1-41.

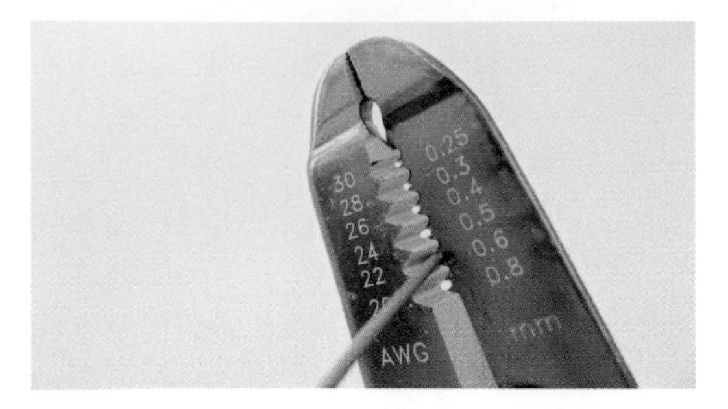

Figure 1-41. *Wire ready to be stripped*

Next, close the wire strippers fully, hold the long end of the wire with your other hand, and pull the strippers gently toward the short end. This will cut and remove the sleeve and expose the wire inside, as shown in Figure 1-42.

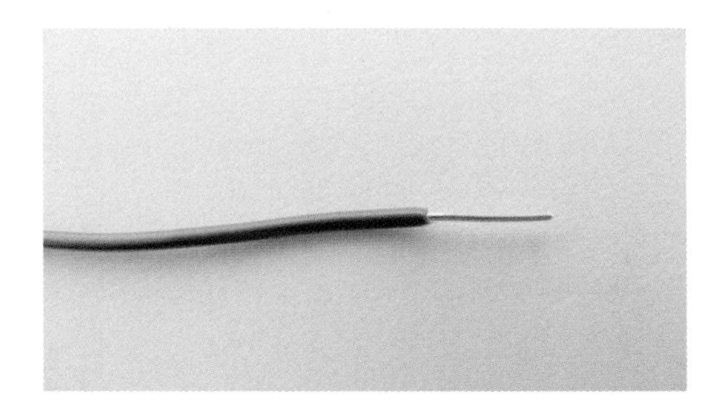

Figure 1-42. *Stripped 22-gauge wire*

Repeat for the other end of the red wire and then do the same to both ends of your black wire. Your wires are now good to go.

Next, wrap the end of the black wire around the negative leg of the LED (Figure 1-43). Wrap the red wire around the resistor leg. Solder both in place and trim (see Figure 1-44).

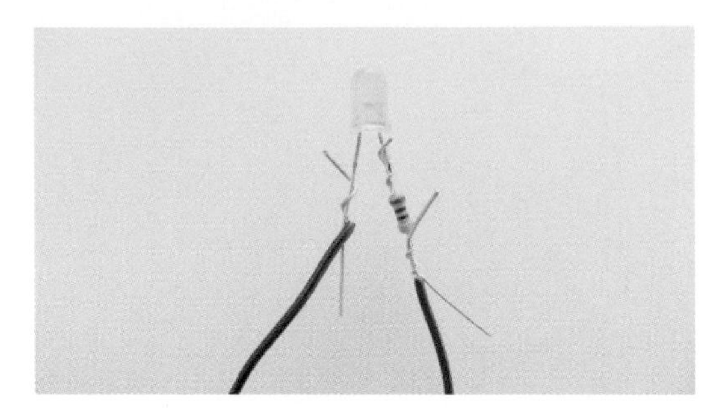

Figure 1-43. *Wrapping black wire around negative LED leg and red wire around resistor leg*

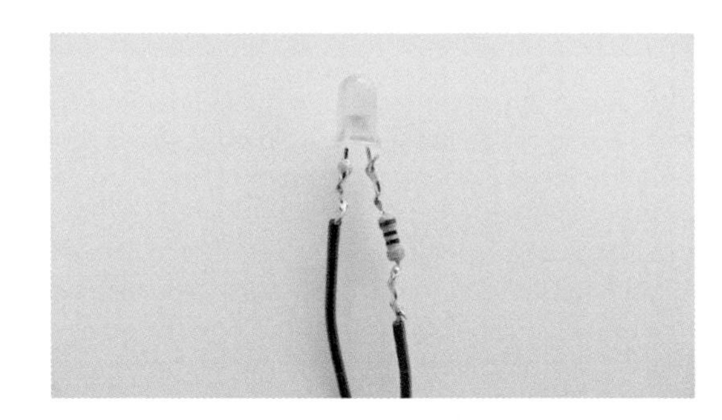

Figure 1-44. *Connections soldered and trimmed*

The position of the exposed connections create a high potential for a short circuit. These connections can be insulated by using some *heat shrink*, non-conductive tubing that shrinks to protect components when exposed to heat.

Place some heat shrink tubing over the exposed connections. Hold everything in place using helping hands, as shown in Figure 1-45. Then use a heat gun to shrink the heat shrink so that it is snug against the connections. This area of the circuit now has proper insulation (see Figure 1-46).

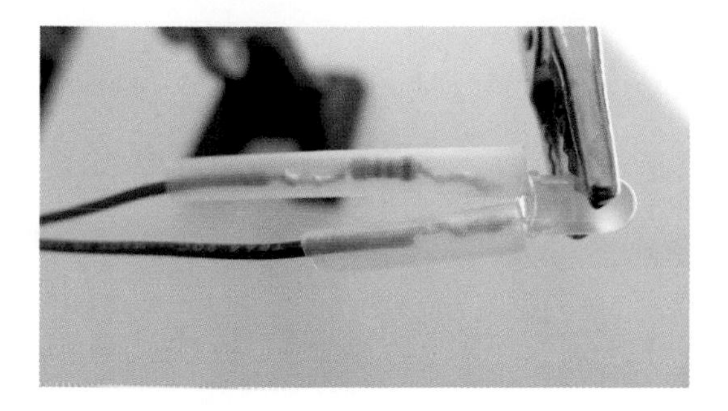

Figure 1-45. *Heat shrink tubing in position*

Figure 1-46. *Heat shrink tubing shrunk*

Next, connect the other end of the red wire to the positive terminal of the battery holder and the other end of the black wire to the negative terminal. Solder in place to secure the connection. Figure 1-47 shows the red wire in place, and Figure 1-48 shows it soldered. Similarly, Figure 1-49 shows the black wire, and Figure 1-50 shows it soldered into place.

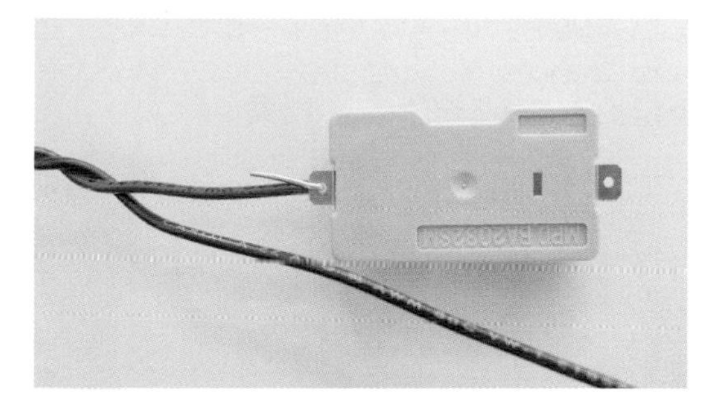

Figure 1-47. *Red wire hooked around positive terminal of battery holder*

Figure 1-48. *Positive terminal soldered*

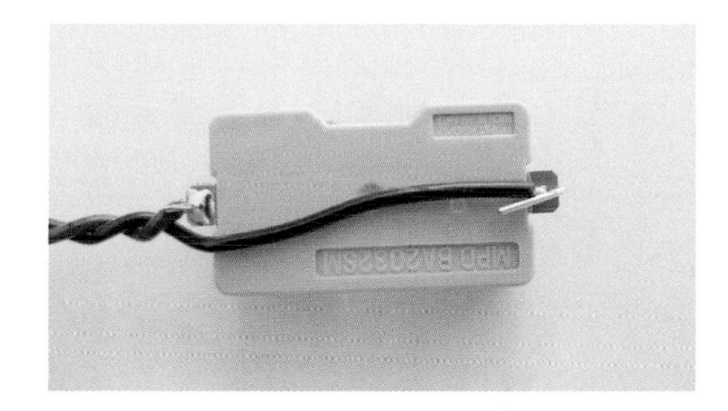

Figure 1-49. *Black wire positioned*

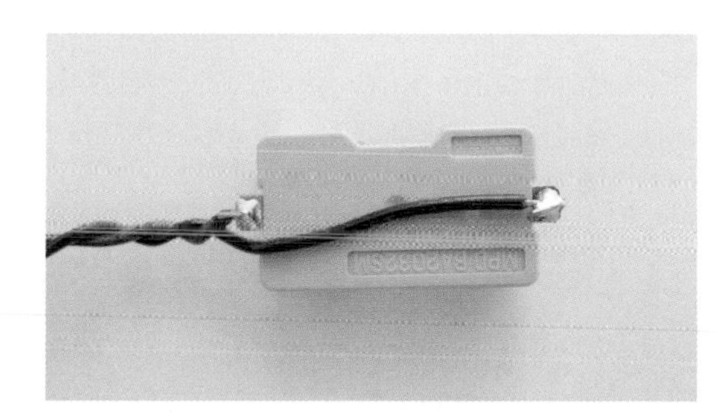

Figure 1-50. *Black wire soldered*

Your freeform circuit is now complete (see Figure 1-51)! You can create your own variations on this technique through the use of different wire types and by varying the length of wire.

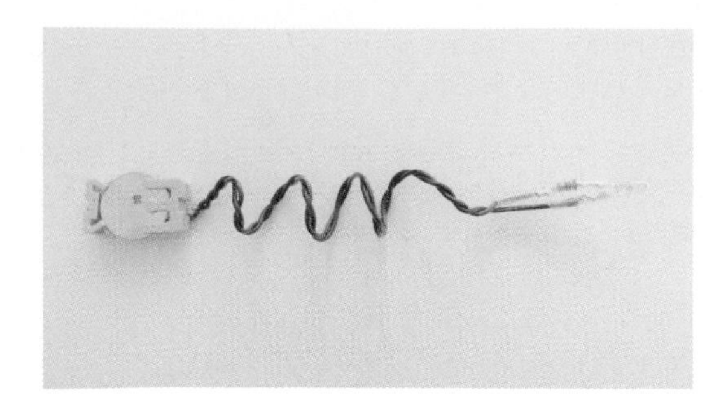

Figure 1-51. *Completed circuit*

Breadboard Circuit

For slightly more complex circuits, breadboards are an excellent solution. They allow you to connect and disconnect through-hole components with ease. Underneath the surface of the breadboard are steel clips that connect wires to conductive traces that run beneath the holes.

Parts and materials, as shown in Figure 1-52:

- (1) CR2032 battery (AF 654, DK P189-ND, SF PRT-00338)
- (1) CR2032 battery holder (AF 653, DK BA2032SM-ND, SF DEV-08822)
- (1) 62Ω through-hole resistor (DK 62QBK-ND)
- (1) 5mm through-hole yellow LED (DK 160-1851-ND, SF COM-09594)
- 22 AWG solid-core hook up wire, in red and black (AF 1311, SF PRT-11367)
- (1) breadboard (AF 64, DK 438-1109-ND, SF PRT-09567)
- (2) break-away 0.1" (2.54mm) straight male header pins (AF 392, JC[5] 103369, SF PRT-00116)

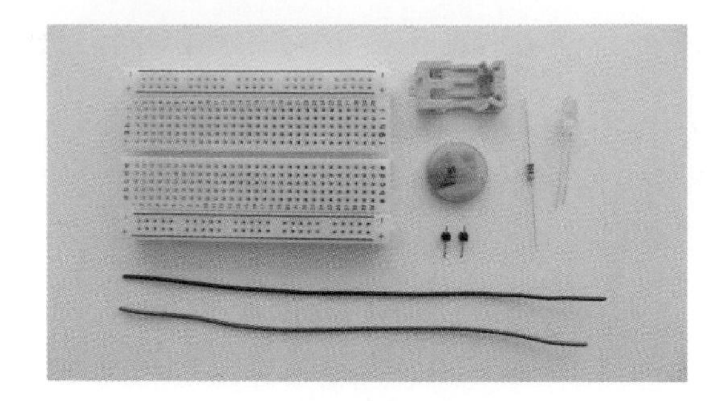

Figure 1-52. *Parts for breadboard circuit*

Tools:

- Wire strippers
- Soldering iron and solder

The circuit you are creating will look like Figure 1-53.

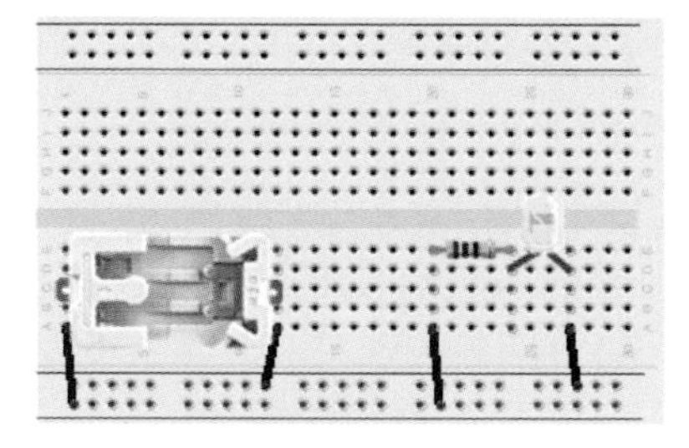

Figure 1-53. *Breadboard circuit*

5. Jameco (for a complete list of supplier abbreviations, see "About Part Numbers" on page xiv).

You'll use two single male headers soldered to the CR2032 battery holder so that it's easy to insert into the breadboard. Figure 1-54 shows the pair of headers you'll need, and Figure 1-55 shows them next to the battery holder.

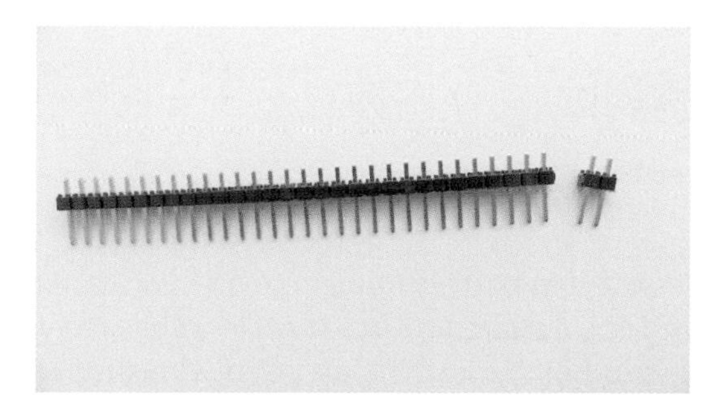

Figure 1-54. *Snap off two male headers for use with the battery holder*

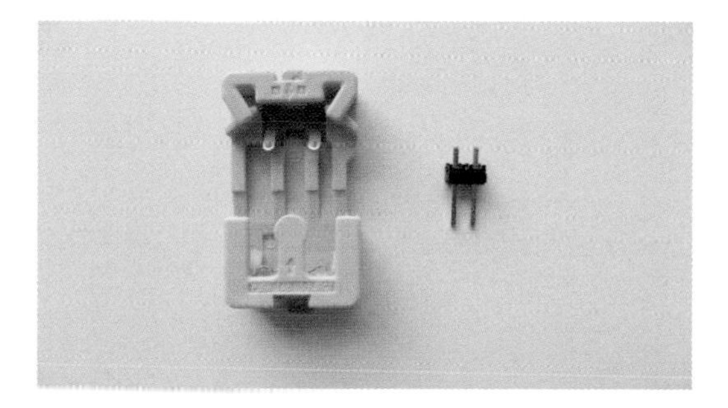

Figure 1-55. *CR2032 battery holder and two male headers*

On either half of the breadboard, all the holes in each row are connected to one another. The gap in the center separates the halves of the row, so the only way to make connections to all the holes in a row is to place a jumped wire connecting the two halves. The leftmost and rightmost columns are connected vertically and are generally used for negative and positive power connections. Figure 1-56 shows how the breadboard is wired underneath.

When looking at the breadboard, you'll notice that there are numbers and letters that can be used to indicate the row and column of each hole. You'll use these markers for reference in your wiring.

Place a single male header in holes E1 and E12, with the longer end of each header pointing down into the breadboard (see Figure 1-57).

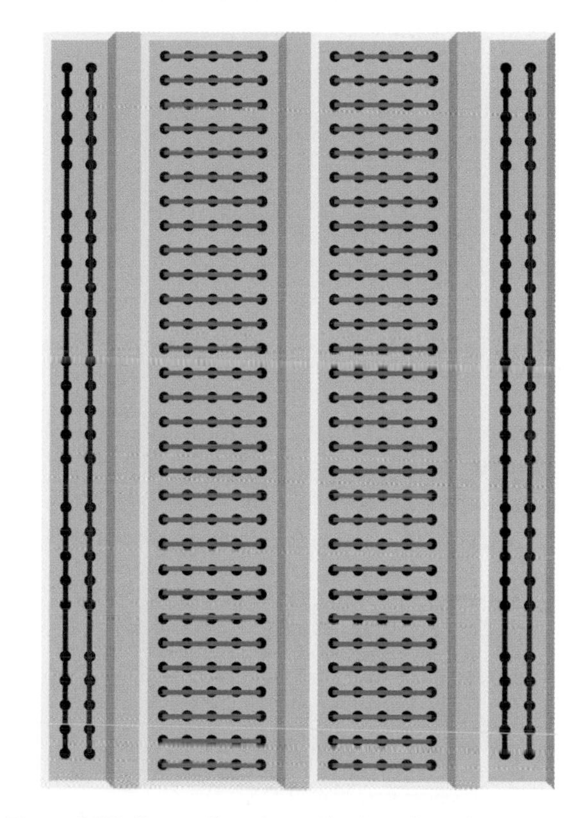

Figure 1-56. *Connections beneath a breadboard*

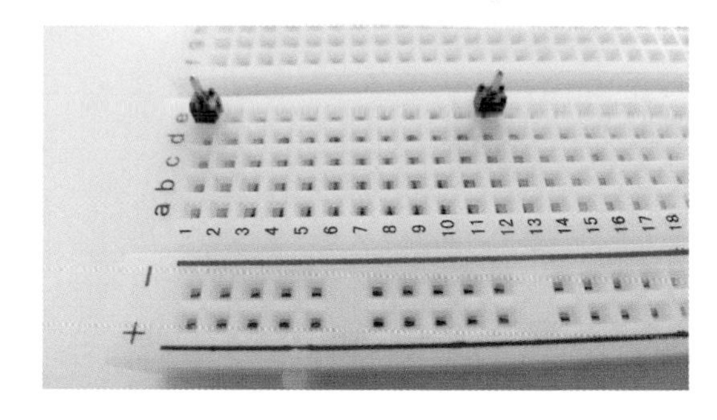

Figure 1-57. *Male headers in E1 & E12*

Place the battery holder on top of the headers so that they are inserted into the holes on either side. Orient the battery holder so that E1 connects to the positive side of the battery holder as shown in Figure 1-58.

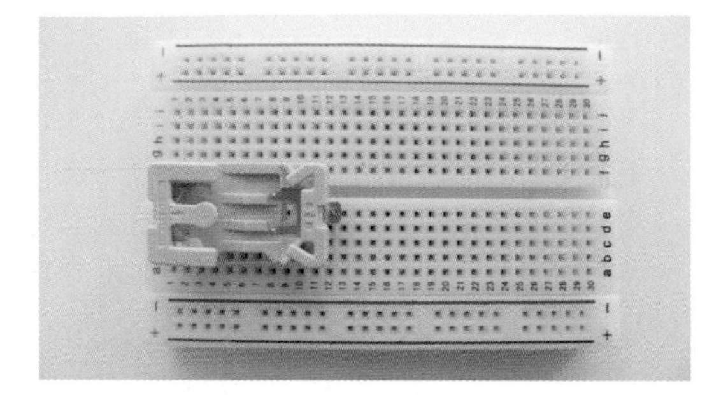

Figure 1-58. *Battery holder placed on top of the male headers*

Solder the connection between each header and battery terminal (see Figures 1-59 and 1-60).

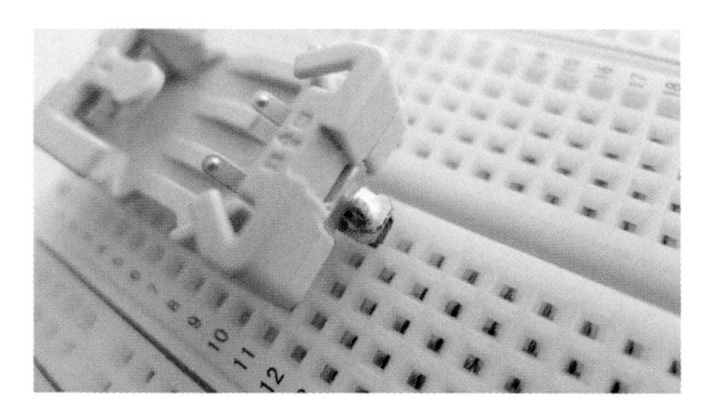

Figure 1-59. *Soldered battery terminal*

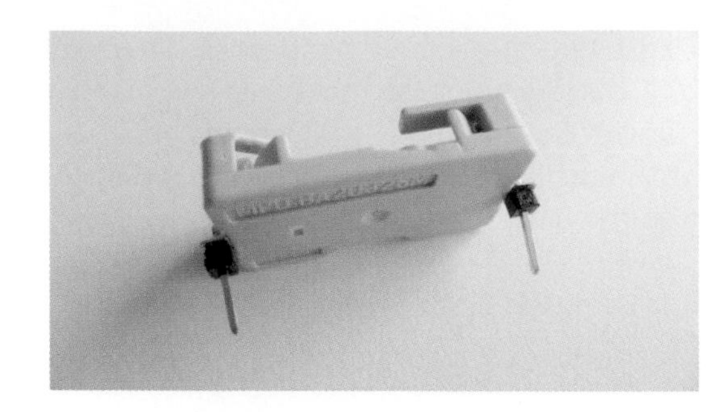

Figure 1-60. *Once the headers are soldered, you will also be able to remove this battery pack for use in other projects*

Now that your battery holder is ready, you can start wiring up your circuit. Cut a short length of red wire and strip both sides. The exposed wire should be long enough so that it can be inserted into the breadboard but not so long that it will leave additional area exposed. Your wire is now ready for use with the breadboard (see Figure 1-61).

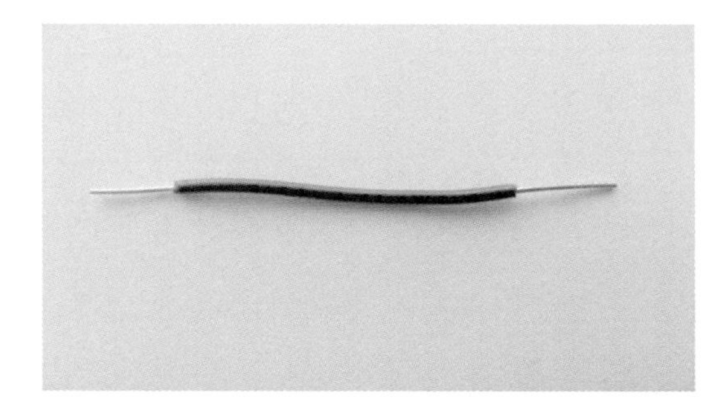

Figure 1-61. *Wire with both sides stripped*

Place one end of the wire into the hole marked A1. Place the other end into a hole in the power bus, marked by the + sign. It should look like Figure 1-62.

Figure 1-62. *Wire connecting A1 and power*

You can leave the wire the length that it is, or trim it a bit to make for a tidier circuit (see Figure 1-63). Just make sure you make the correct connections and that the wire is long enough to be fully inserted into both holes.

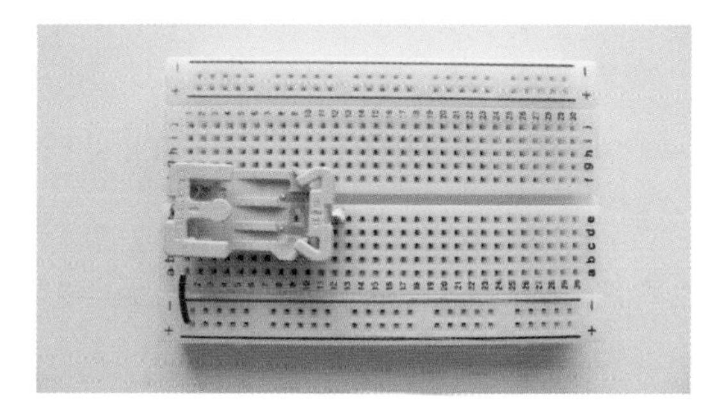

Figure 1-63. *A shorter wire connecting A1 and power*

Do the same using a black wire to connect A12 and the ground bus, marked by the – sign (see Figure 1-64). While any color wire will work, using these colors will help you to more quickly read what connections are being made when you look at the circuit.

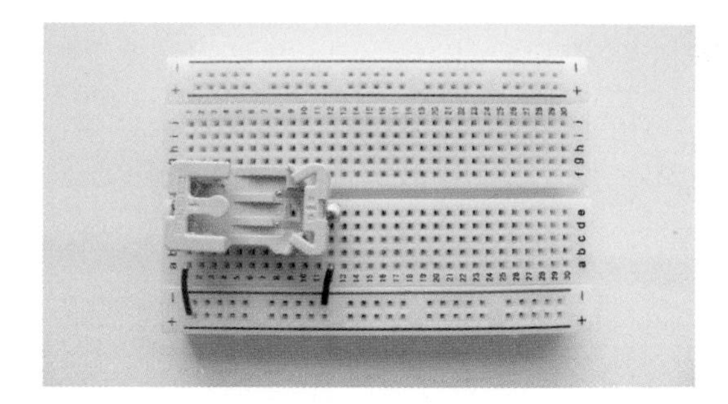

Figure 1-64. *Wire connecting A12 and ground*

The power and ground of the battery are now accessible through any holes in the + and – rails proximate to column A.

Use the following images to complete the assembly of the remainder of the circuit.

Take the 62Ω resistor and bend the legs to a right angle as shown in Figure 1-65. Then trim them to a length slightly longer than the depth of the breadboard.

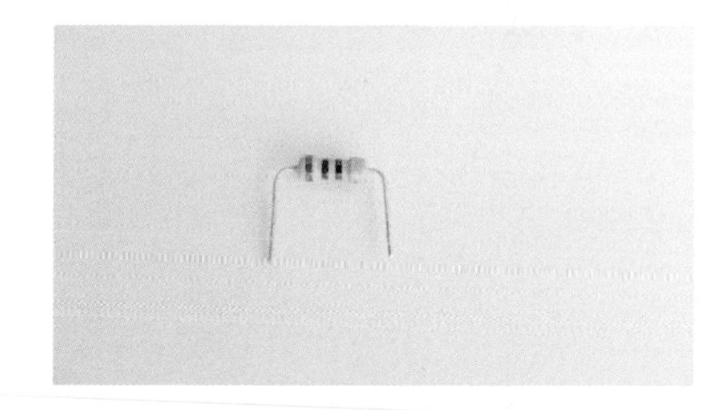

Figure 1-65. *Resistor prepared for use with breadboard*

Place the resistor on the breadboard so that the legs are inserted in E20 and E24, as shown in Figure 1-66.

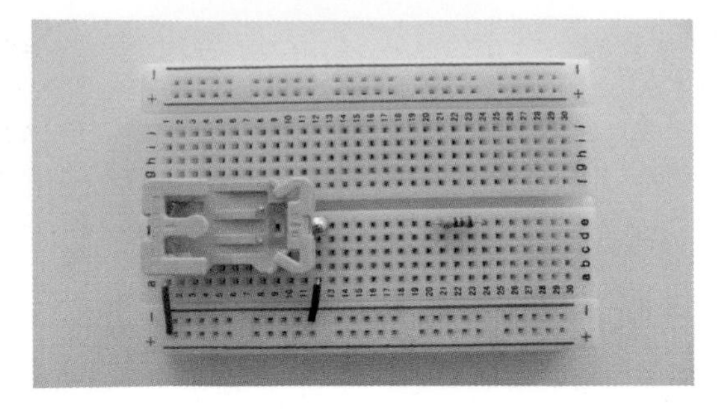

Figure 1-66. *Adding the resistor to the breadboard*

Place the LED so that the positive leg is in D24 and the negative leg is in D27 (see Figure 1-67).

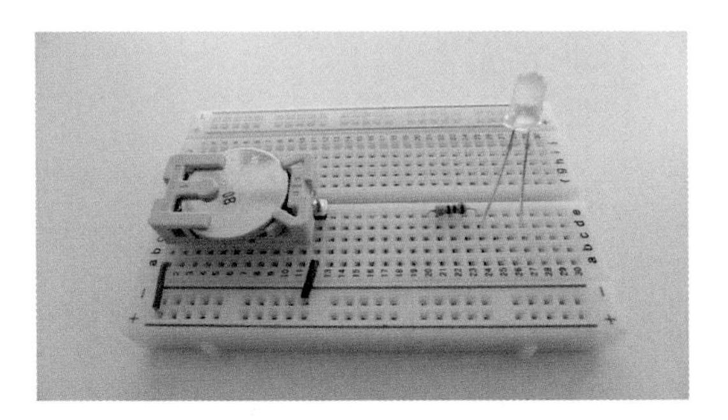

Figure 1-67. *Adding LED to the breadboard*

Use a red wire to connect A20 and + (see Figure 1-68).

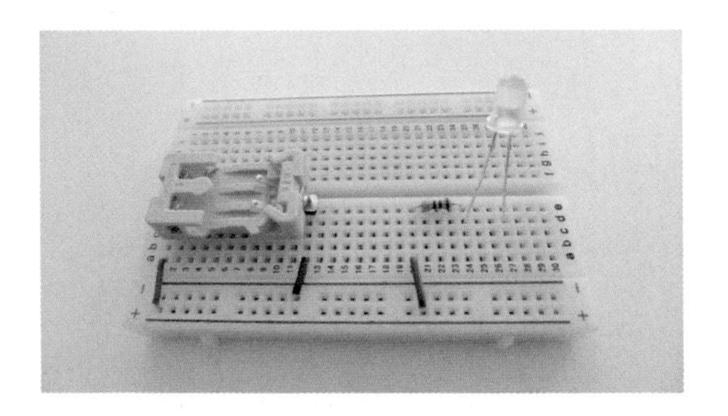

Figure 1-68. *Adding a red wire to connect A20 and +*

Use a black wire to connect A27 and –, as shown in Figure 1-69.

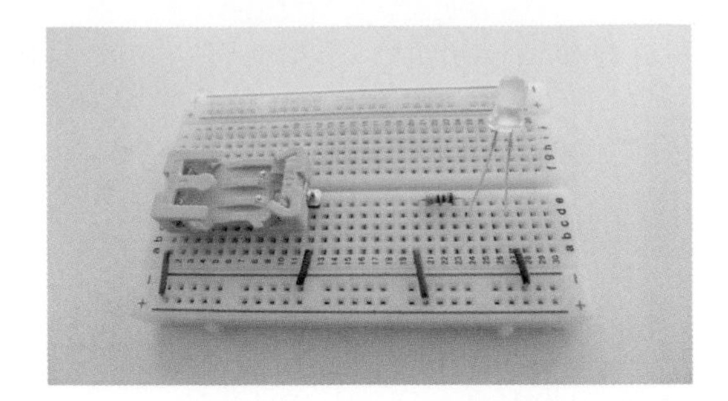

Figure 1-69. *Adding a black wire to connect A27 and –*

Your circuit is now complete! Insert a battery in the battery holder, positive side up (+), to light the LED. It should look like Figure 1-70.

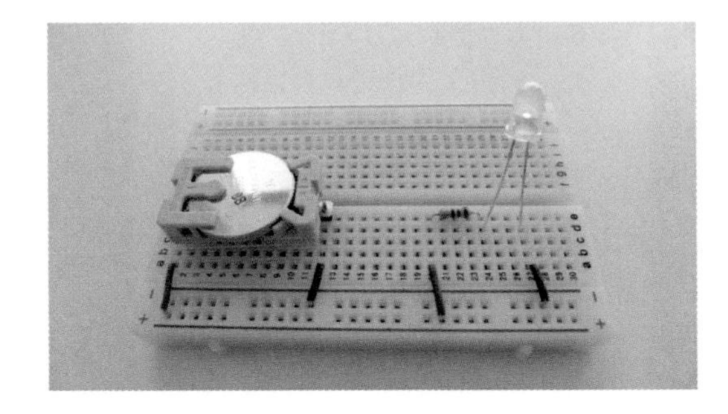

Figure 1-70. *Breadboard circuit with LED lit*

Protoboard Circuit

The protoboard is a logical follow-up to a breadboard. It uses the same components and spacing but makes connections that are far more secure and robust than a breadboard.

Parts and materials, as shown in Figure 1-71:

- (1) CR2032 battery (AF 654, DK P189-ND, SF PRT-00338)
- (1) CR2032 battery holder (AF 653, DK BA2032SM-ND, SF DEV-08822)
- (1) 62Ω through-hole resistor (DK 62QBK-ND)
- (1) 5mm through-hole yellow LED (DK 160-1851-ND, SF COM-09594)

- (2) break-away 0.1" (2.54mm) straight male header pins (AF 392, JC 103369, SF PRT-00116)
- (1) small piece of protoboard without any traces connecting the pads, cut to size (SF PRT-08619)

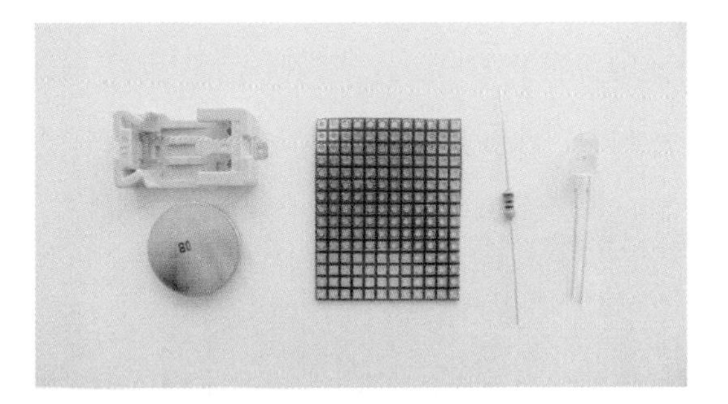

Figure 1-71. *Parts for protoboard circuit*

Tools:

- Soldering iron and solder
- Small snips/flush cutter
- Helping hands

The circuit you are creating will look like Figure 1-72.

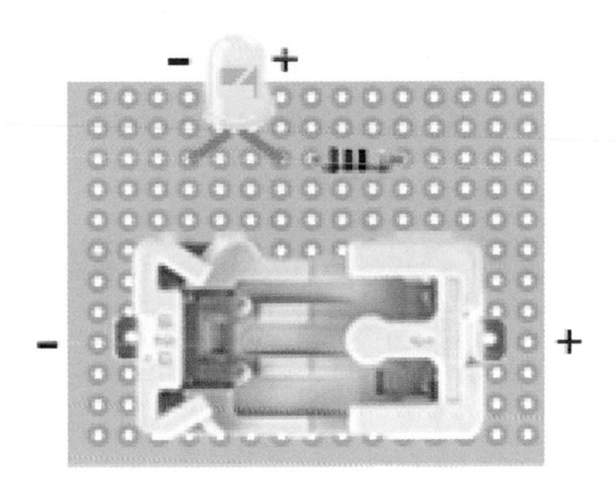

Figure 1-72. *Protoboard circuit*

There are many varieties of protoboard. For this example, you will use the type of board that has no connections between any of the holes (see Figure 1-73). Other types of protoboard will be reviewed in Chapter 2.

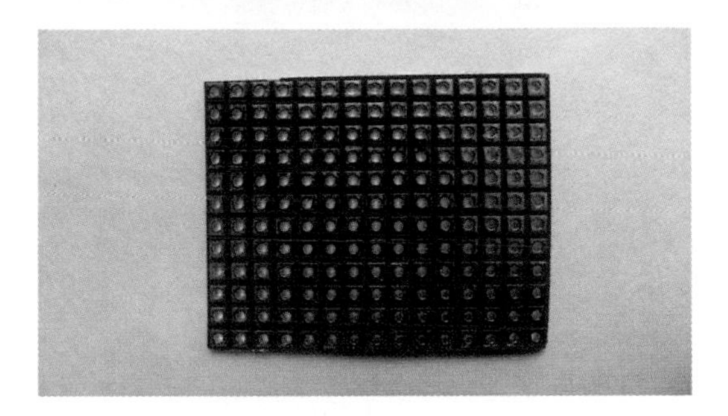

Figure 1-73. *Prototyping board*

First, plan your layout. You can do this by moving components around the board until you've got something that looks right, as shown in Figure 1-74. The shiny silver or copper-colored metal around the holes are called *pads*. If you only have pads on one side, make sure that they are on the bottom where the legs come out.

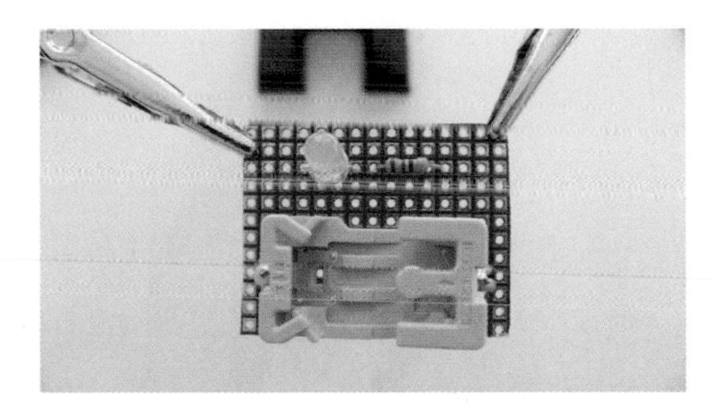

Figure 1-74. *Components placed on the protoboard*

For this circuit, you are placing the negative leg of the LED close to the negative terminal of the battery, the positive leg of the LED close to one side of the resistor, and the other side of the resistor close to the positive terminal of the battery. These locations will allow you to create connections with ease.

Next, turn the board over. For components like LEDs and resistors, you can bend the legs slightly so they don't fall out when you flip the board, as shown in Figure 1-75.

Figure 1-75. *Underside of protoboard with legs*

Helping hands (Figure 1-76) or third hands are a set of clips on an adjustable stand that can hold components in place while you are soldering them. Use some helping hands to stabilize your protoboard. Secure the components in place by soldering a connection between the component legs to the pad on the protoboard surrounding them.

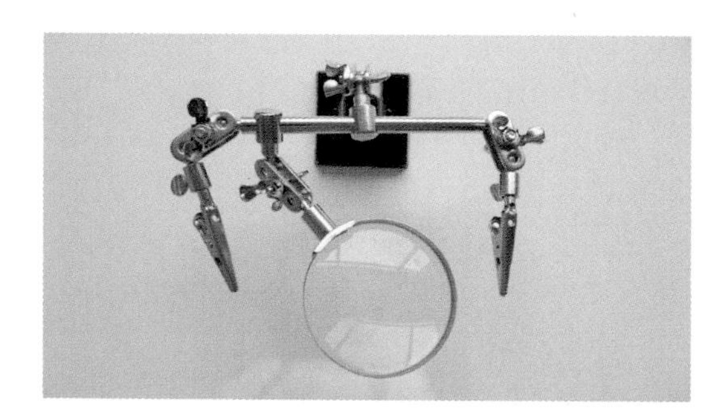

Figure 1-76. *Helping hands*

Once the components are held in place (see Figure 1-77), arrange the legs (and jumper wires if necessary) so they create the necessary connections to complete the circuit, as shown in Figure 1-78. Solder them in place. Snip any excess wire as needed (see Figure 1-79).

Figure 1-77. *Soldering components in place*

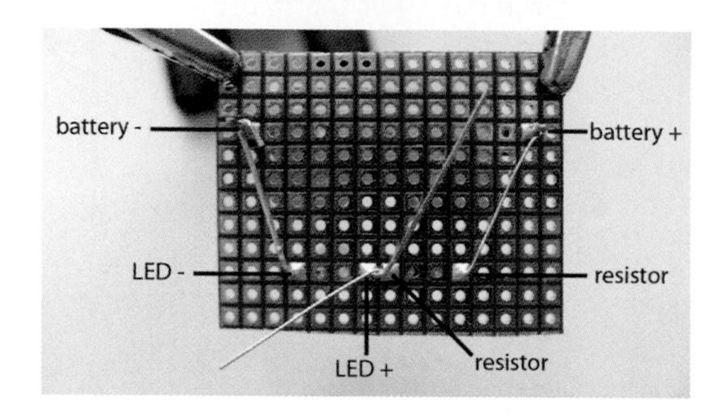

Figure 1-78. *Arranging component legs to create the necessary connections*

Figure 1-79. *Back of board with connections soldered and excess wires snipped*

You now have a very secure circuit. Turn over the board and add the battery to light the LED (see Figure 1-80).

Figure 1-80. *Completed circuit*

This is the same circuit you created with both the alligator clips and the breadboard, just in a slightly more robust implementation.

Conductive Thread Circuit

Now you move into the world of soft circuitry. When using conductive thread, you will sew connections rather than solder them. This allows you to create circuits that are soft and pliable.

Parts and materials, as shown in Figure 1-81:

- (1) CR2032 battery (AF 654, DK P189-ND, SF PRT-00338)
- (1) CR2032 battery holder (AF 653, DK BA2032SM-ND, SF DEV-08822)
- (1) 62Ω through-hole resistor (DK 62QBK-ND)
- (1) 5mm through-hole yellow LED (DK 160-1851-ND, SF COM-09594)
- Conductive thread
- Fabric
- Fabric glue

Tools:

- Needle
- Scissors
- Needle-nose pliers

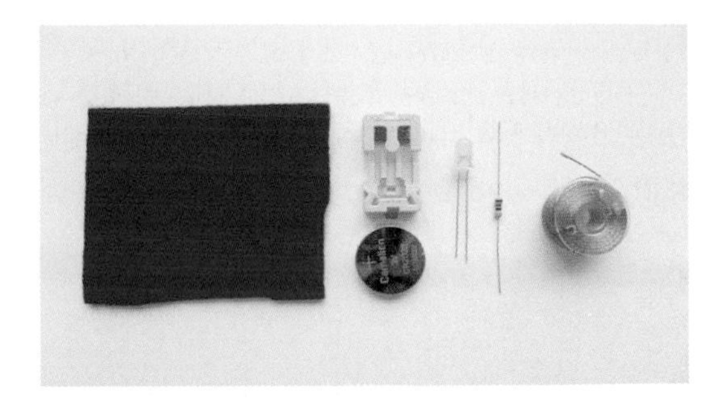

Figure 1-81. *Parts for conductive thread circuit*

Needle Sizes

Choosing a needle for sewing with conductive thread can be a bit tricky. You want one with an eye that is large enough for your conductive thread to pass through but also small enough to pass through the *sewholes* of your components. Figure 1-82 shows a needle that's too big. I usually work with an embroidery needle, but use what is best for you. Be sure to test your needle with all components in your circuit before you begin sewing.

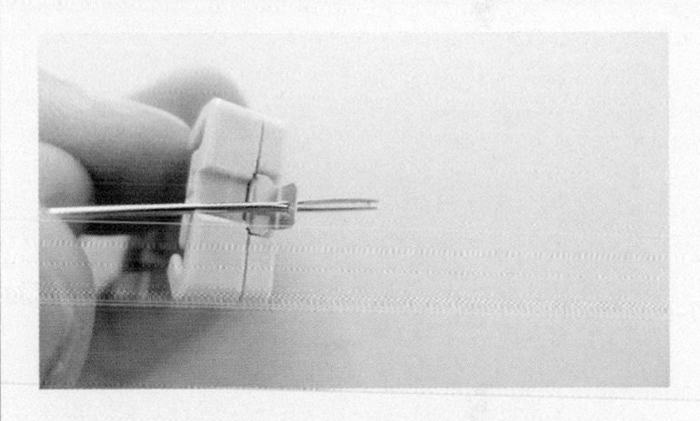

Figure 1-82. *The eye of this needle is too large to pass through the hole of the battery holder*

Before you start assembling the circuit, you need to prepare your parts. In order to sew the LED and resistor in place, you will need to modify their shape a bit.

For this technique, you can use any through-hole component that has long legs. Let's use an LED. You will also need a pair of needle or round-nose pliers.

Take the LED and bend the legs so that they are parallel to the base (Figure 1-83), as if the LED were doing a split. This will allow the LED to sit flat on the fabric.

Figure 1-83. *LED with legs bent*

Grab the end of one leg with the pliers, as shown in Figure 1-84.

Figure 1-84. *LED and pliers*

Twist the pliers so that the leg of the LED rolls around the pliers to create a loop (see Figure 1-85). Repeat on the other side. You've just created loops that can easily be sewn and secured with stitches of conductive thread. Figure 1-86 shows the sewable LED.

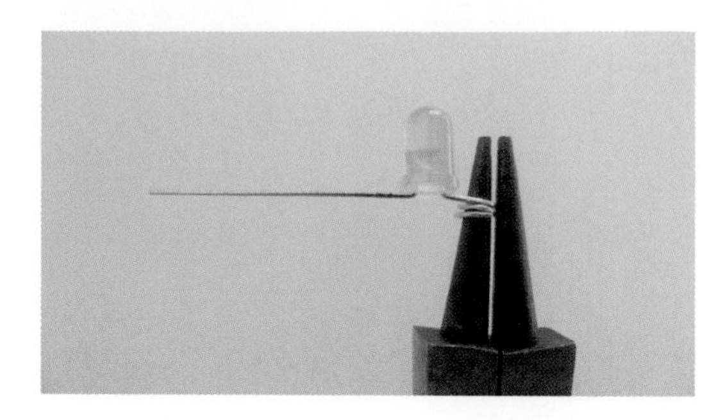

Figure 1-85. *Using the pliers to create a loop*

Figure 1-86. *Sewable LED*

Do the same with the resistor, and you'll be ready to go (Figure 1-87).

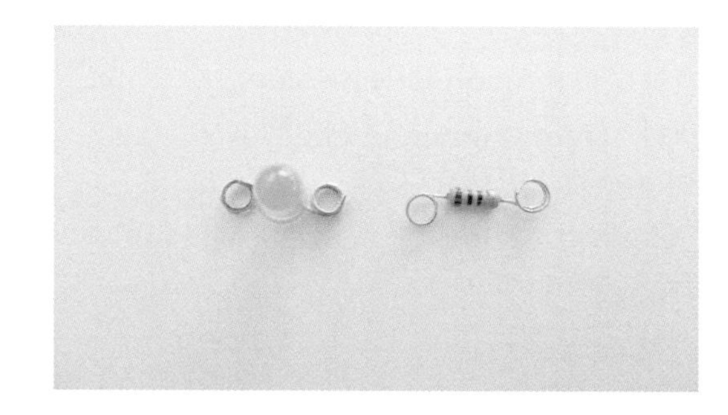

Figure 1-87. *LED and resistor*

Now that the components are ready, you can begin to assemble the circuit. The intended layout will look something like Figure 1-88.

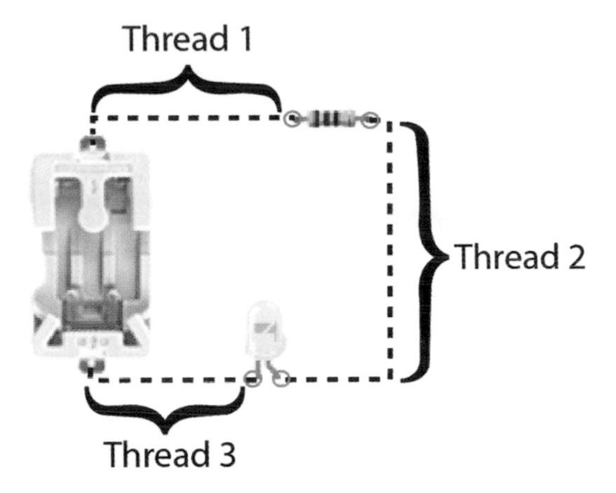

Figure 1-88. *Conductive thread circuit layout*

Let's start with the battery holder. Thread your needle with a piece of conductive thread. If you find that the thread is fraying at the end, you can use a bit of beeswax or even moisture to tame the frays while you are threading it. Pull the two ends of the thread so they are equal lengths and knot them together.

Pass the needle from the back of the fabric up through the hole of the positive terminal of the battery holder (see Figures 1-89 and 1-90). Repeat the stitch around the battery holder several times to make a secure connection (Figure 1-91).

This type of CR2032 battery holder has been adopted by the DIY community because the terminal holes can be used for sewing. Keep in mind that they are not intended for this purpose. The edges of the holes are sharp and conductive thread sewn very tightly can get cut and fray. Conductive thread sewn too loose, on the other hand, will result in an unreliable connection. When you are sewing, try to find the middle ground in the tightness of your stitches.

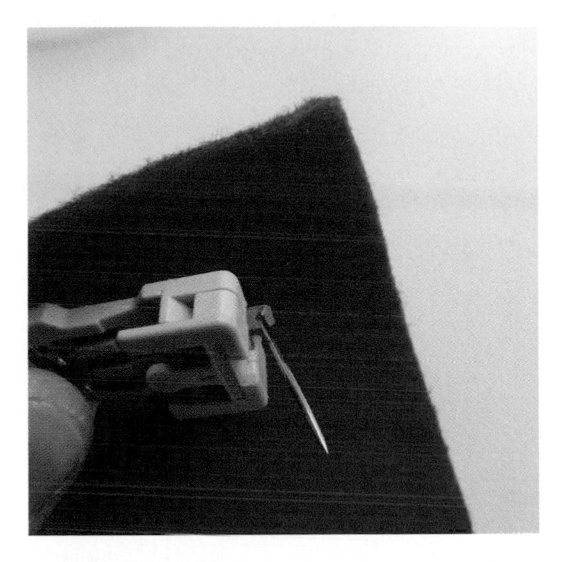

Figure 1-89. *Passing the needle from the back of the fabric through the positive terminal of the LED*

Figure 1-90. *Pulling the thread through*

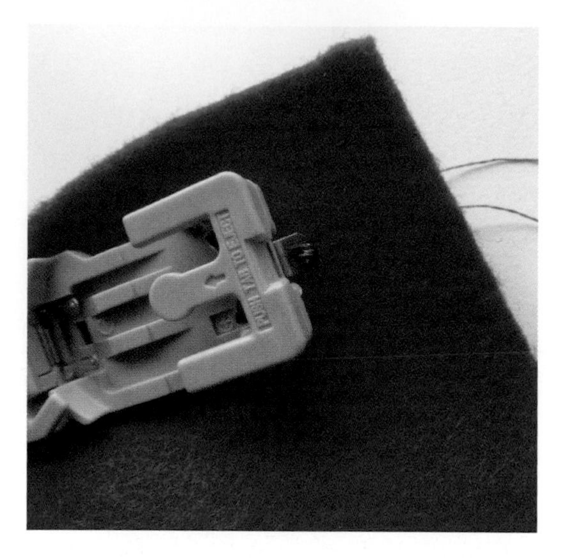

Figure 1-91. *Repeating the stitch to secure the connection*

Conductive thread has a strong tendency to tangle (see Figure 1-92). Take your time when sewing, and keep an eye on the thread as you go!

Figure 1-92. *Tangled conductive thread*

Once the terminal of the battery holder is secure, continue to sew in the direction of the resistor. Once you reach the resistor, make several secure stitches around the loop, tie a knot in the back, and snip off the excess thread. As Figure 1-93 shows, *do not* continue on to the other side of the resistor with the same piece of thread. If you do, that will create a short circuit.

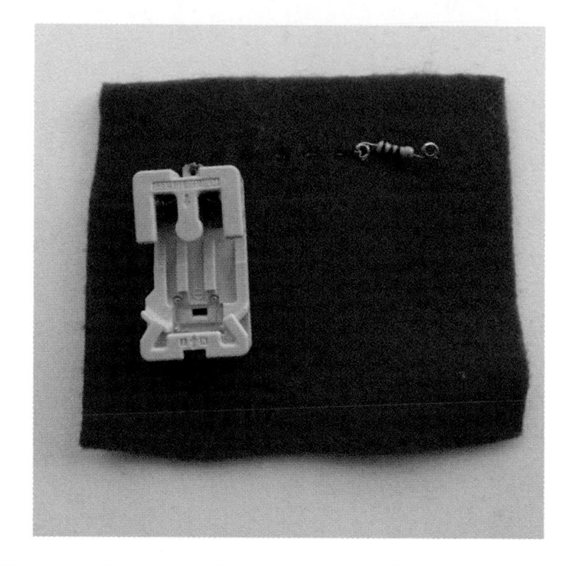

Figure 1-93. *Thread #1 connects the positive terminal of the battery holder to the resistor*

Using a *new piece of thread*, sew the other loop of the resistor, and then sew a line to the LED (see Figure 1-94). Next, sew the positive leg of the LED, knot it in back, and trim it.

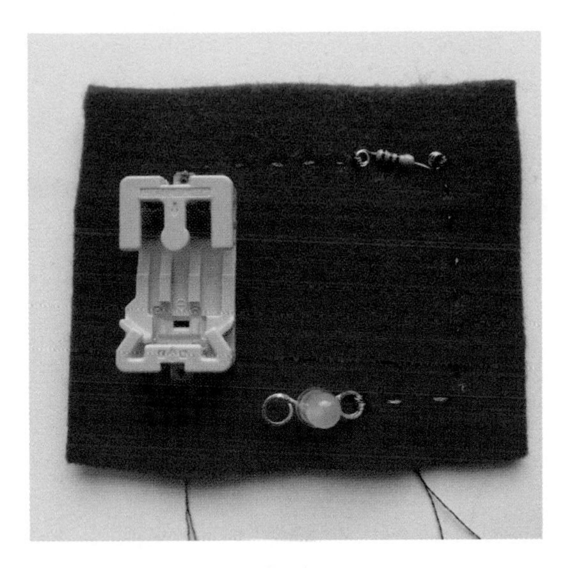

Figure 1-94. *Thread #2 connects the resistor and LED*

Using a *third* piece of thread, sew the negative loop of the LED and connect it to the negative terminal of the battery holder (see Figure 1-95). Tie and trim.

Figure 1-95. *Thread #3 connects the LED and the negative terminal of the battery holder, completing the circuit*

Your circuit is now fully sewn but the job is not quite done. Figure 1-96 shows what the back of the circuit might look like.

Figure 1-96. *Untrimmed tails of conductive thread can lead to short circuits*

These long conductive thread tails create a high probability for short circuits. If one tail accidentally gets brushed into the wrong place, it can create an opportunity for the electricity to sneak straight past the LED directly to ground, creating a short circuit.

When trimming conductive thread, be mindful of the length. Too long means you have the possibility of shorts. Too short and this burly slippery thread might undo its own knot. Shoot for something in the middle (Figure 1-97) and then put a dab of fabric glue, fray stopper, or nail polish on the knot to secure it (Figure 1-98).

Figure 1-97. *Trimmed thread trails*

Figure 1-98. *Securing the knots with fabric glue*

Add the battery and you've got a completed soft circuit!

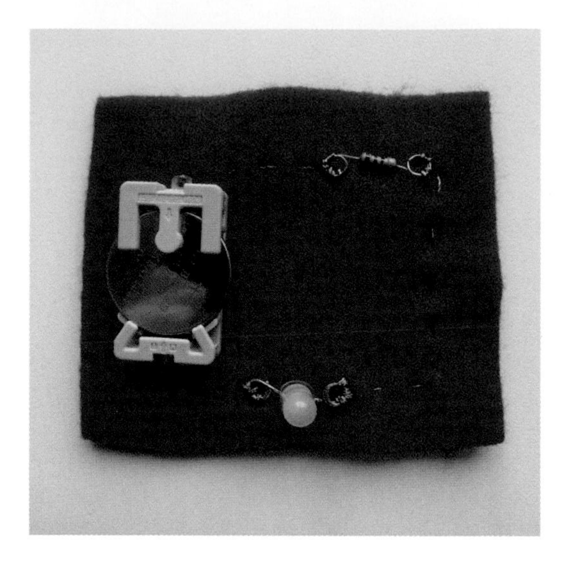

Figure 1-99. *Correct completed circuit*

Conductive Fabric Circuit

Another approach to soft circuitry is using conductive fabric. Iron-on conductive fabric is particularly exciting because a circuit can be created, cut, and adhered quickly. Also, there is the potential for designs to become visually intricate with limited effort.

Parts and materials, as shown in Figure 1-100:

- (1) CR2032 battery (AF 654, DK P189-ND, SF PRT-00338)
- (1) CR2032 battery holder (AF 653, DK BA2032SM-ND, SF DEV-08822)
- (1) 62Ω through-hole resistor (DK 62QBK-ND)
- (1) 5mm through-hole yellow LED (DK 160-1851-ND, SF COM-09594)
- Conductive thread
- Iron-on conductive fabric (LE[6] A1220)
- Fabric
- Fabric glue

6. LessEMF (for a complete list of supplier abbreviations, see "About Part Numbers" on page xiv).

Tools:

- Needle
- Scissors
- Iron
- Ironing board
- Needle-nose pliers

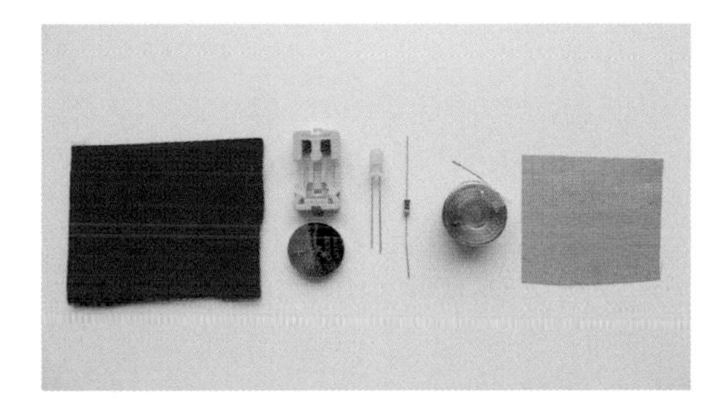

Figure 1-100. *Parts for conductive fabric circuit*

First, sketch and cut your "circuit" design out of the conductive fabric. Make sure there are three openings where you will be able to place the battery holder, resistor, and LED. Arrange the fabric adhesive-side down on the nonconductive fabric. Once your fabric is properly arranged (see Figures 1-101 and 1-102), carefully move it to an ironing board.

Figure 1-101. *Conductive fabric ready to be ironed*

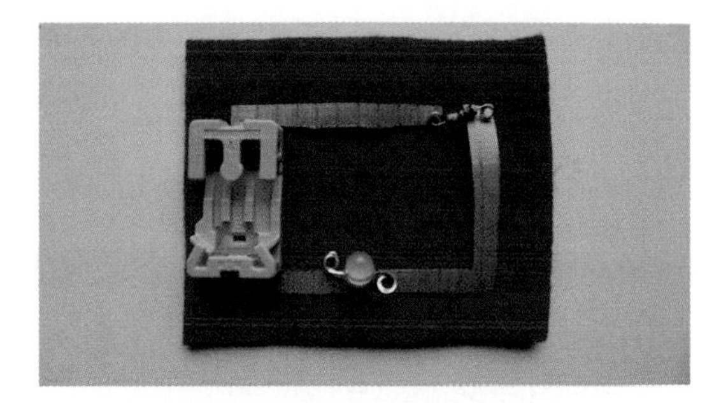

Figure 1-102. *You can use your components to check the spacing of your conductive fabric*

Gently cover the piece to be ironed with a thin piece of cotton or muslin (see Figure 1-103). This will prevent the fabric from getting excessively hot and will protect your iron from getting adhesive gunk on it.

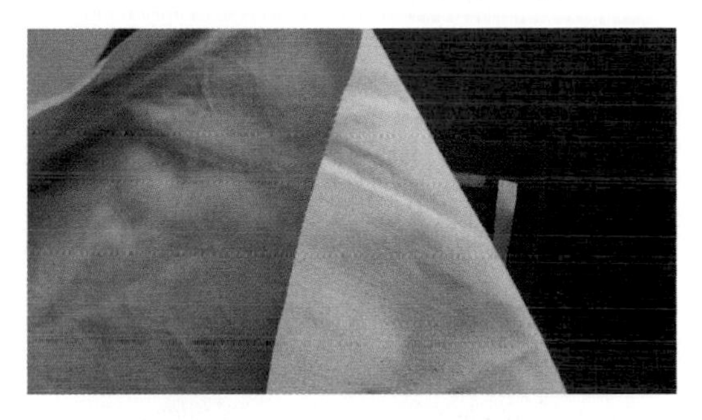

Figure 1-103. *Using a thin piece of fabric to protect your circuit when ironing*

With the iron set to low and the steam turned off, gently iron the circuit until the adhesive melts and the fabric sticks, as shown in Figure 1-104. Lightly press the iron down in different locations rather than sliding it around.

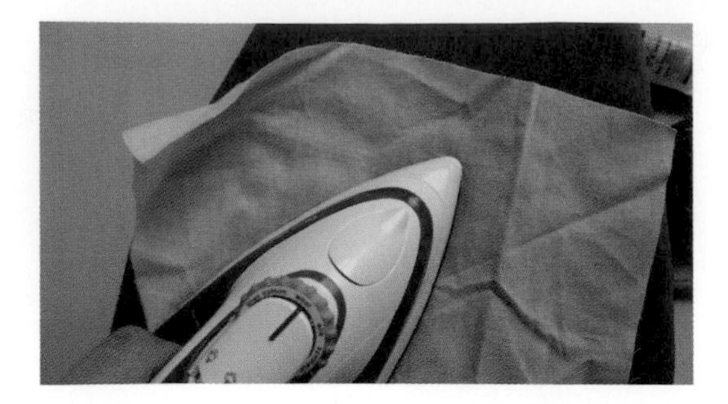

Figure 1-104. *Ironing the conductive fabric*

Because conductive fabric conducts both electricity and heat, it will be quite hot after you iron it! Be sure to let it cool for a few minutes before touching it.

With the conductive fabric secured, use conductive thread to connect the components to the fabric traces, as shown in Figure 1-105. Remember to use separate pieces of thread for each trace.

Figure 1-105. *Creating a connection between the conductive fabric and battery holder terminal using conductive thread*

Finish off your knots in back, and your circuit is complete! Add the battery to light it up.

Figure 1-106. *Completed conductive fabric circuit*

Advantages and Disadvantages

Now that you've built the same circuit six different ways, it's time to look at the advantages and disadvantages of each. Use the following information when considering which method to use when creating a circuit:

Alligator clip circuit

Advantages

Quick; allows integration of nonstandard components (like DIY sensors); plays well e-textile toolkits; useful for testing components

Disadvantages

Bulky; not optimal for complex circuits; unstable; potential for shorts

Wire circuit

Advantages

Extremely flexible and customizable

Disadvantages

Not practical for complex circuits; robustness depends on choice of wire and amount of strain placed on circuit

Breadboard circuit

Advantages

Quick and efficient; easy to modify

Disadvantages

Bulky; not terribly robust; looks weird on a shirt

Protoboard circuit

Advantages

Secure connections; potentially smaller than a breadboard

Disadvantages

Still a bit bulky; difficult to integrate with textiles

Conductive thread circuit

Advantages

Flexible; pliable; customizable; and fashionable

Disadvantages

Potential for shorts; some conductive threads have significant resistance; time consuming

Conductive fabric circuit

Advantages

Fast(er) approach to soft circuits; potential for interesting designs

Disadvantages

Subject to shorts; still requires sewing

Conclusion

This chapter covered what circuits are as well as how to build them using a variety of conductive materials. In the following chapter, you'll develop a more in-depth knowledge of these materials and learn how to decide what to use when.

2 Conductive Materials

When making wearable electronics, you must consider which materials to use in your circuits. Because bodies have a tendency to bend, twist, and shake their booties, the materials used in wearable circuits are subject to a lot of wear and tear. Conductive materials used in wearable circuits need to be durable, flexible, and sometimes even soft.

In Chapter 1, you learned how to assemble circuits using a variety of conductive materials and tools. In this chapter, you will get to know these materials a bit better. You'll also learn about criteria to consider when choosing among them for a wearable electronics project.

Conventional Conductors

Let's start with the conductive materials and tools that are conventionally used for creating circuits and work our way up from there.

Alligator Clips

Alligator (or crocodile) clips, shown in Figure 2-1, are a prototyping tool that consist of a simple insulated wire with a spring-loaded jaw on either end. This means you can quickly and easily clip together a temporary circuit. Alligator clips are often used as a means to prototype connections for a circuit before you make the final commitment of sewing it together with conductive thread. They are also useful when making temporary connections to a part that you are not yet ready to solder.

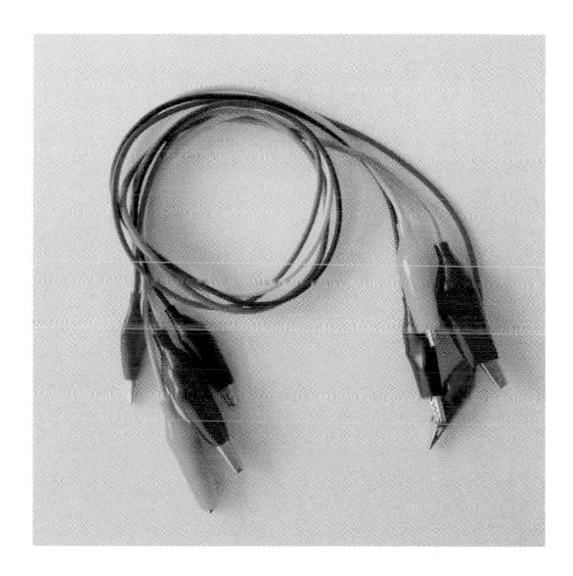

Figure 2-1. *Alligator clips come in a variety of colors*

Wire

Wire (Figure 2-2) is a seemingly ordinary but very useful material. Because of the recent excitement about e-textiles, beginners working on wearables projects often feel there is an expectation to sew all of their circuits using conductive thread, and may overlook wire. Conductive thread is not always required or desirable. Sometimes wire does exactly what you need it to do, particularly if you're working with the right type.

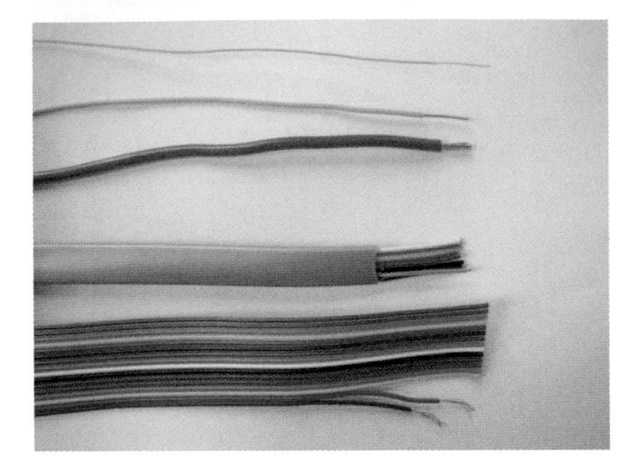

Figure 2-2. *Types of wire (top to bottom): wrapping, solid core, stranded, telephone cable, and ribbon cable*

Solid core wire

Solid core wire is as it sounds—a solid metal core encased in insulation. It is comprised of one solid strand of a particular *gauge* (the diameter of the solid core wire). Due to its stiffness and the fact that it's a single strand of wire, stripped solid core wire is very good for plugging into things like breadboards or female headers.

Some varieties of solid core wire are more flexible than others. 22 AWG (American Wire Gauge) hookup wire is fairly stiff and perfect for prototyping circuits on a breadboard. But if flexed repeatedly, it is likely to snap. This often makes it a poor choice for wearable projects.

Thinner gauge magnet wire and wire wrap wire can actually be nice for certain wearable applications. Because both are so thin, they are extremely flexible and can fit into small places. They can also be added along the seam of a heavier material without adding too much extra bulk.

Stranded wire

Stranded wire contains multiple strands within the insulating sleeve. This causes it to be very flexible and forgiving. If one strand breaks, the other strands maintain the connection, so it is not likely to cause an interruption in the flow of electricity in a circuit. Because it accommodates repeated bending, stranded wire is a good option if you need to use wire around joints like the elbow or the knee.

Grouped wire

In designing wearable electronics, much thought should go into simplifying circuitry to reduce bulk and aid with troubleshooting. Having several loose wires running along the same path can be annoying, bulky, and problematic. Fortunately, there are many types of wire that come in grouped bundles that you can use:

Speaker cable
: Speaker cable (see Figure 2-3) is extremely flexible and widely available. This is a simple and clean way to run two connections over a longer distance.

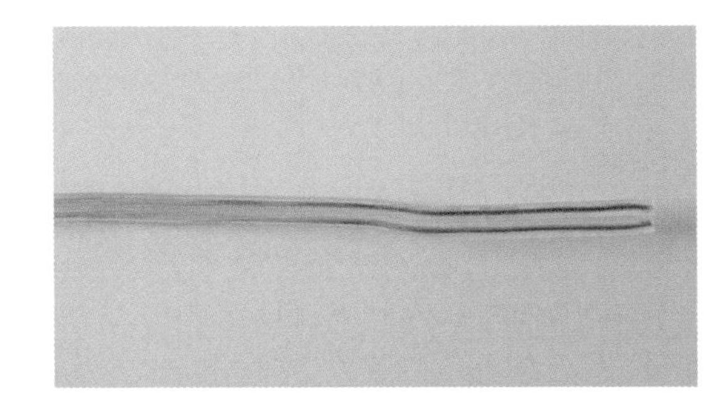

Figure 2-3. *Speaker cable*

Ribbon cable
: Ribbon cable, shown in Figure 2-4, is flat, flexible, and lightweight. It comes with anywhere between 4 to 80 insulated wires running in parallel on a flat plane. You can also peel away a select number of conductors you need as a separate chunk.

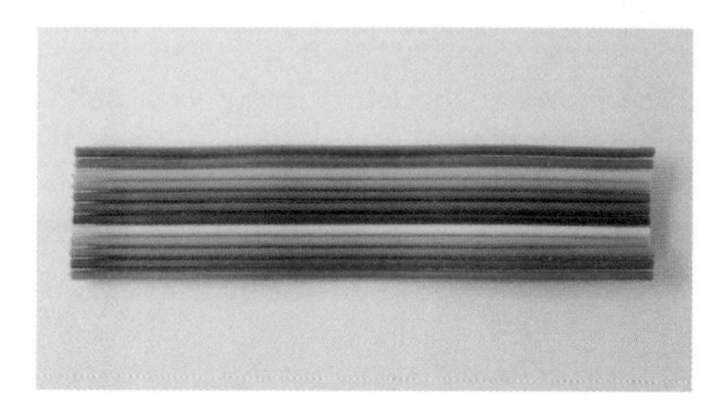

Figure 2-4. *Ribbon cable*

Phone, Ethernet, and other cables
: Though a bit bulkier, these common household cables can offer a spiffy multiconductor solution in a pinch. Just snip off the connector, and you can access the bundle of wires within.

Breadboards

Breadboards are a prototyping tool that allow for quick connections to be made by simply plugging wires into holes. Inside the breadboard (see Figure 2-5), the lower layer contains *buses*, or lengths of a conductor that connect multiple holes. On each long side there are two buses intended for power (+) and ground (–). Perpendicular to those, in the middle, are shorter rows interrupted by the middle ridge that are intended as the canvas on which the majority of the circuit is constructed.

Breadboards offer a fast way to test your circuits but not a robust way to wear them. Protoboard is often a good way to create a more secure version of a circuit prototyped on a breadboard.

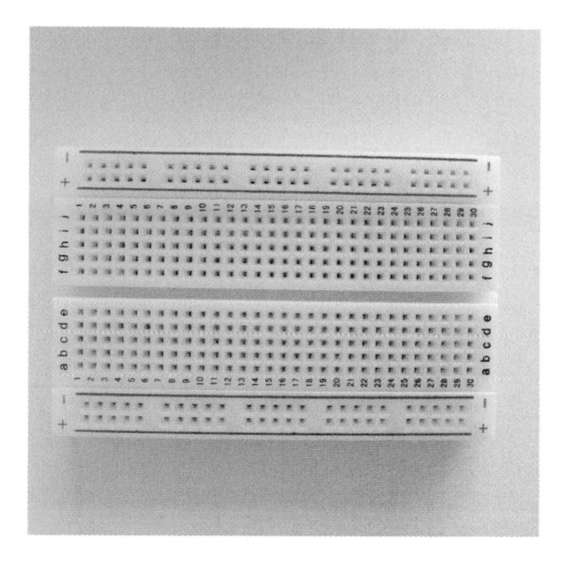

Figure 2-5. *A half-sized breadboard*

Protoboard

A protoboard (or perf board) is a type of circuit board that can serve as a base for more permanent circuits. Shown in Figure 2-6, it contains regularly spaced holes that are lined or ringed with a conductive material such as copper. These conductive areas are called "pads." Less expensive protoboards only have pads on one side. Higher-quality protoboards feature through-plated holes, which means the holes are lined with conductive materials and there are pads on both sides of the boards.

Components are soldered to the protoboard, which makes for stable and sturdy connections, far more secure than those created on a breadboard. Protoboards can also be cut to size, making them easier to fit into small places.

Figure 2-6. *Protoboard*

Electrical connections between components are made using solder, component legs, jumper wires, and at times through connections included in the design of the protoboard itself. Basic protoboards have no connections between the holes, but some protoboards contains strips of holes that are connected by conductive traces. Others, such as Adafruit's Perma-Proto boards (shown in Figure 2-7) mimic the layout of connections found on a breadboards. And on some protoboards, all of the traces are connected. When this is the case, you must cut connections (rather than create them) using a utility knife.

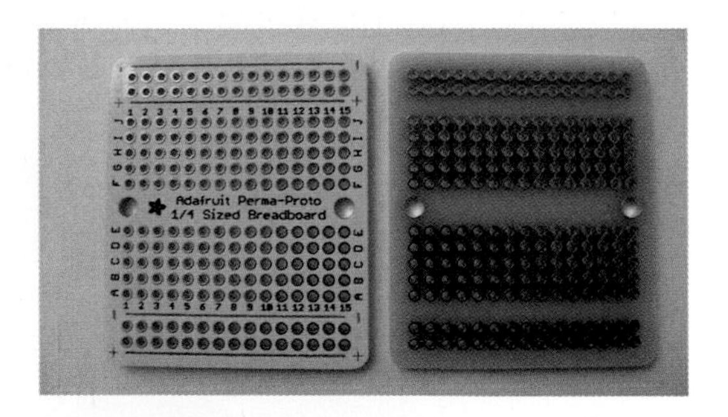

Figure 2-7. *Adafruit produces protoboards laid out like breadboards in full, half, and quarter breadboard sizes*

Conductive Thread

With traditional conductors under your belt, you can now explore textiles! Let's take a look at some softer options.

Conductive thread (Figure 2-8) is thread that contains conductive metals, such as silver or stainless steel. It has been widely adopted by makers and artists as material with which to make soft electrical connections.

Figure 2-8. *Bobbin of conductive thread*

"The Musical Jacket" (see Figure 2-9) is an early example of conductive thread in use. It integrates a wearable MIDI synthesizer with an embroidered keypad that the wearer can use to play notes and create sounds.

Figure 2-9. *"The Musical Jacket," created by Rehmi Post, Maggie Orth, and Emily Cooper*

Conductive thread is a tricky material. When used in the appropriate context by the right person, it can create supple, subtle, and visually stunning circuits. But it also has the potential to be gnarly, knotty, and ineffective. Under the pressure of an impending deadline, I've known it to bring many students to tears and projects to self destruction by way of a seam ripper. Keep in mind that knowing the properties of your materials and how they align with use cases can save you a whole lot of heartache!

If you are new to sewing, take the time to do some stitching with standard cotton thread before moving on to conductive thread. Because of the metallic content, some conductive threads are a bit more difficult to work with. They can be bulkier, quicker to tangle, and also slippery, which means knots will sometimes untie.

There is an extensive array of conductive threads, most of which are sold in large quantities for industrial purposes. Table 2-1 includes a selection of threads that are more readily available in smaller quantities. If you find that these don't meet your needs, keep in mind that there are more options out there—you just have to be a bit more creative in how you look for them. Contact manufacturers for sample requests. Coordinate with classmates, people in your hackerspace, or people on the Internet to share a bulk order. Or suggest to your favorite electronics distributor that it stock the material you want.

Properties of Conductive Thread

Here are the properties you want to consider when choosing a conductive thread to work with:

Thickness
: Two-ply? Four-ply? Others? The thickness of the thread affects how easy it is to sew with, determines which type of needle you'll need, and whether you'll be able to use it in a sewing machine. Also, the more plys, the less resistive the thread will be. For instance, the four-ply version of a conductive thread would be more conductive than the two-ply version.

Resistance
: Resistance is a material's ability to resist the flow of electricity. This is one of the most significant factors to consider when choosing a conductive thread. Some threads have a relatively high resistance, which affects how they can be used in a circuit and what components they can be used with. For instance, motors need lots of current but conductive thread can only deliver a limited amount, so they are not an ideal match. See "Ohm's Law" on page 3 for more information about resistance.

Material
: Different threads contain different conductive materials, and as a result, these threads do not all have the same properties. For instance, stainless steel thread is highly conductive and resistant to corrosion, whereas silver-plated nylon thread has a higher resistance but is much softer and more pliable.

Color
: There aren't any choices to be made in this category at the moment, but there should be. Why would the e-embroiderers of the world want to have their palettes limited to a singular, monochrome silver? Let's hope some vibrantly colored conductive threads will be spinning 'round your bobbins soon…

Insulation
: Insulation is useful for preventing short circuits. Insulated conductive thread does exist, but at the time of this writing there are no insulated conductive threads available in small quantities. See Chapter 5 for details on ways to insulate conductive-thread circuits.

Working with Conductive Thread

Sewing with conductive thread can be a bit more challenging than regular sewing. The best way to get to know a thread and its challenges is to do some tests before you get started.

With hand sewing, keep in mind that you may need a needle with a larger eye. Also, conductive thread can be a bit slippery, so it is advisable to tie your knots well, and at times even reinforce them with a bit of fabric glue.

With a sewing machine, a good rule of thumb is to use the conductive thread in the bobbin rather than for the top stitch. Some two-ply threads will run through the needle OK, but it depends on the thread and on your machine. Industrial or hardier home sewing machines seems to handle conductive thread a bit better. It's smart to keep some spare needles on hand when trying out new threads in case you end up breaking a needle in the process.

Electronic components cannot be directly sewn with a sewing machine. This means that even if you machine stitch your conductive thread traces, you will need to leave long tails of thread on either end (see Figure 2-10) so that you can hand stitch the connection to the electronic component.

One of the biggest challenges in working with conductive thread is preventing shorts. Because conductive thread is uninsulated, there are many opportunities for parts of the circuit to touch that aren't intended to. For instance, if you fold a sewn circuit in half, there's a good chance that traces will touch each other temporarily or permanently, preventing its operation. Similarly, if you have long tails on the back of a piece of conductive thread embroidery, there is the potential for them to move around and come into contact. Keep this in mind when sewing with conductive thread. Keep your circuits neat, trim, and organized, and it will make you much happier in the long run. Strategies for planning and insulating soft circuits are reviewed in Chapter 5.

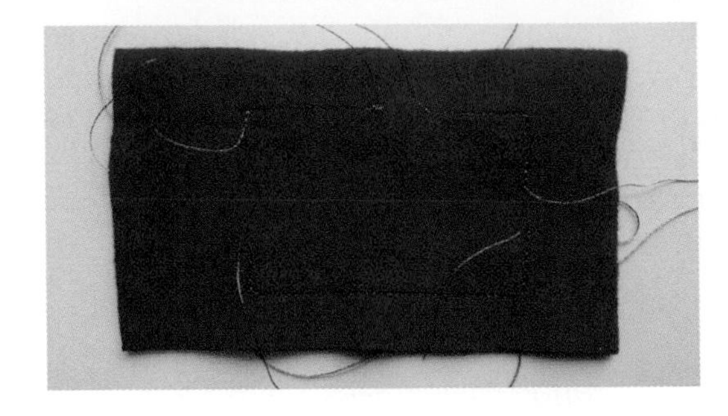

Figure 2-10. *Conductive thread traces created with a sewing machine*

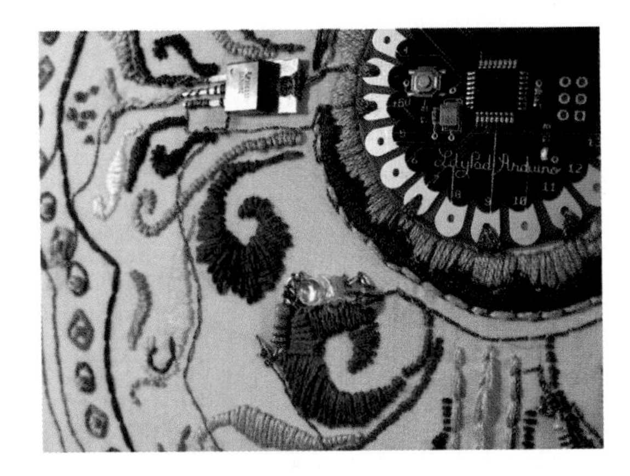

Figure 2-11. *Becky Stern's "A Tribute to Leah Buechley" is an embroidery using conductive and nonconductive thread*

Types of Conductive Thread

Table 2-1 provides a comparison of some conductive threads that are available in small quantities through electronics supply companies.

Conductive Fabric

Conductive fabric (Figure 2-12) is a wondrous material. Whereas conductive thread can present issues with resistance, many conductive fabrics do a

Table 2-1. Comparing conductive threads

Name	Manufacturer	Source	Part number	Ply number	Resistance (Ω/ft)	Material	Notes
Stainless thin conductive thread	n/a	Adafruit	640	2	16	316L stainless steel	Stiff
Stainless medium conductive thread	n/a	Adafruit	641	3	10	316L stainless steel	Stiff
Stainless thin conductive yarn / thick conductive thread	n/a	Adafruit	603	3	12	316L stainless steel	Furry
Conductive thread (thin)	Bekaert	SparkFun	DEV-10118	2	9	Stainless steel	
Conductive thread (thick)	Bekaert	SparkFun	DEV-10120	4	4	Stainless steel	
Conductive thread (extra thick)	Bekaert	SparkFun	DEV-10119	6	1.4	Stainless steel	
Conductive thread (117/17 two-ply)	Shieldex	SparkFun	DEV-08544	2	300	Silver-plated nylon	Likely to oxidize over time, discontinued
Conductive thread (234/34 four-ply)	Shieldex	SparkFun	DEV-08549	4	14	Silver-plated nylon	Likely to oxidize over time, discontinued
Conductive thread (60g)	n/a	SparkFun	DEV-11791		28	Spun stainless steel	Hairy

better job in circuits with higher-current demands. And because this fabric comes in a large sheet, there is more room to play with the visual design of the circuit. Figure 2-13 shows the "IM Blanky," a project that incorporates conductive fabric into both the visual and circuit design.

Figure 2-12. *Conductive fabric comes in a variety of styles*

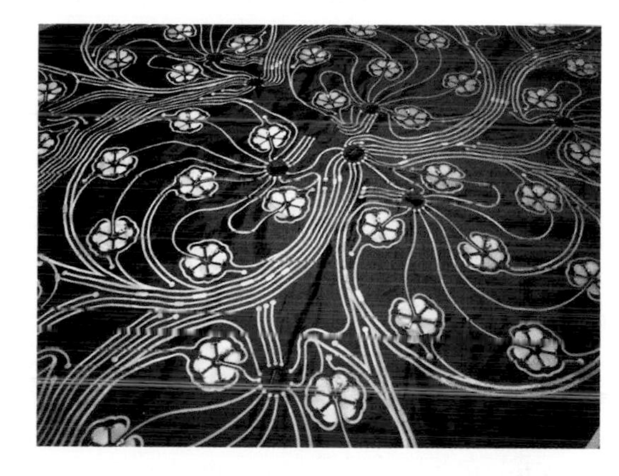

Figure 2-13. *"IM Blanky" by Studio (n-1) (Carol Moukheiber and Christos Marcopoulos with Rodolphe el-Khoury)*

Properties of Conductive Fabric

The considerations when working with conductive fabric are slightly different than that of conductive thread:

Type
: What type of fabric is it? Woven? Ripstop? Knit? Plated? Does it fray or wrinkle? Can it handle the conditions of your intended use?

Stretch
The stretchiness of a fabric relates to its type but is worth special mention. Bodies are bendy. A stretchy fabric can be particularly helpful for parts of a garment that needs to shape-shift or bend frequently. Does the fabric stretch at all? If so, in one direction or two?

Substrate
Conductive fabric is generally composed of several layers. What is the base, nonconductive layer? Is it nylon, polyester, or something else? This will ultimately affect the care and comfort of the garment you are creating.

Plating
The plating of the fabric is the conductive part. As with conductive thread, it's worth considering what the metallic content is and how that affects its performance and longevity.

Weight and thickness
Is this fabric thick, thin, heavy, or light? This is important to consider in the context of what you'll be making. If you're working to create something lightweight like a t-shirt, it might not make sense to use a conductive fabric that is thick and heavy.

Surface resistance
While many conductive fabrics have an extremely low surface resistance, there are some exceptions. Be sure to check this before committing to a type of conductive fabric.

Color
All of the fabrics compared in Table 2-2 are silver in color, but there are many conductive fabrics that are copper as well, and a few that are other colors. If your project is better suited for another color, be sure to do some research to see what else is available.

Working with Conductive Fabric

There are two ways to incorporate conductive fabric: sewing and the use of iron-on adhesive.

Conductive fabric can be sewn just like any other fabric—it's just a question of how you'd like to work it into a circuit. Quilting and appliqué techniques can be used when sewing conductive fabric onto a nonconductive substrate.

The important thing to consider when sewing conductive fabric is that many conductive fabrics tend to fray. Threads from frayed pieces can wander and inadvertently create shorts. This can be mitigated through hemming, serging, or other traditional sewing techniques.

Another approach is to use iron-on adhesive. Conductive fabric can be purchased with iron-on adhesive already applied (e.g., ShieldIt Super) or you can apply your own (e.g., Heat & Bond and others are available at most fabric stores). Keep in mind that iron-on adhesive is not conductive, so connections between separate pieces need to be bridged with conductive thread.

The advantages of this approach are that it makes it extremely easy to cut out traces for a circuit, and it prevents fraying. Conductive fabric can be cut using scissors, an X-acto knife, or even a laser cutter. Laser cutting is an easy way to create detailed and intricate designs. With this approach, once circuit traces are cut, they can simply be ironed on!

Types of Conductive Fabrics

Table 2-2 shows a small selection of conductive fabrics, but there are many more out there. Online retailers such as Fine Silver Products (*http://bit.ly/1wzOicl*) or LessEMF (*http://lessemf.com/*) often sell sample packs, so you can get your hands on a large variety in order to determine which will best suit your project. Similarly, datasheets are available for most of the products and can be consulted when you have questions about their electrical and physical characteristics.

Table 2-2. Comparing conductive fabrics

Name	Source	Part number	Purpose	Type	Stretch	Substrate	Plating	Weight(g/meter squared)	Thickness (in mm)	Surface resistance (in Ω/sq.)
Zell	SparkFun	DEV-10056	RF Shielding	Ripstop	None	Nylon	Tin/nickel over silver	77	0.003 or 0.1	< 0.02 or < 0.1
MedTex 130	SparkFun	DEV-10070	Wound care, antimicrobial	Knit	Two direction (warp and weft)	78% nylon, 22% elastormer	High ionic silver	140	0.45	< 5
MedTex 180	SparkFun	DEV-10055	Wound care, antimicrobial	Knit	One direction	78% nylon, 22% elastormer	High ionic silver	224	0.55	< 5
ShieldIt Super	LessEMF	A1220	RF and microwave shielding	Ripstop with hot melt adhesive	None	Polyester	Nickel and copper (low corrosion)	230	0.17	< 0.5
Woven conductive fabric	Adafruit	1168	Electronics	Woven	None	Silver plated nylon	High ionic silver			
Knit conductive fabric	Adafruit	1167	Electronics	Knit	Two direction	Silver	High ionic silver			< 1
Knit jersey conductive fabric	Adafruit	1364	Electronics	Knit	Two direction	63% cotton, 35% silver yarn, and 2% spandex	High ionic silver			46 ohms per foot across the rows (stretchier direction) and 460 ohms per foot across the columns (less stretchy direction)

Other Conductive Materials

There are many other conductive materials to work with beyond threads and fabrics. Some are more exotic and harder to get, while others are available at your average fabric or hardware store.

Conductive Yarn

Conductive yarn (Figure 2-14) is like conductive thread, except fluffier and a little harder to control. It is excellent for knitting or weaving conductive patches into textiles as well as for creating knitted stretch and pressure sensors.

Figure 2-14. *Conductive yarn*

Conductive Fiber

Conductive fiber (AF 1088, SF DEV-10868), shown in Figure 2-15, is a powerful raw material that can be used in a variety of projects. Soft and lightweight, it can be spun, felted, and more.

Figure 2-15. *Conductive fiber*

Conductive Felt

Conductive felt is not something you can buy off the shelf, but artists and makers have been creating their own. By combining conductive fiber or conductive wool (such as steel or copper wool) and sheep's wool, you can create a tactically pleasing variable resistor or electronic switch! More on this in Chapter 7.

Intended Uses

Conductive fabrics and threads are manufactured primarily for medical and industrial purposes. Certain conductive fabrics have antimicrobial properties, some conductive materials are often used for antistatic purposes, and there is even an industry of electromagnetic field (EMF) safety products. LessEMF, Inc., is a store in Albany, New York, that sells an extensive array of conductive materials for EMF shielding.

The intended applications for these materials provides some indication of why many are not sold in smaller quantities and why aesthetic aspects of the materials have not been well considered. But conductive materials are beginning to appear in more widely available consumer products. The first conductive fabric I saw included in a product in a store was the Echo Touch Gloves at the MOMA Design Store in 2010. Conductive-fabric fingertips have become a common feature found in many gloves for the use of touchscreen devices. Similarly, companies that produce heart rate and brain wave monitors have started using conductive fabric electrodes. Perhaps as the use of conductive materials becomes more commonplace, a wider range of aesthetic options will be available.

Conductive Ribbon

There are many types of conductive ribbon (Figures 2-16, 2-17, and 2-18), but only a few that are available in small quantities. Ribbons can be entirely conductive, or they may include multiple conductors separated by nonconductive material such as nylon or polyester. See Table 2-3 for details.

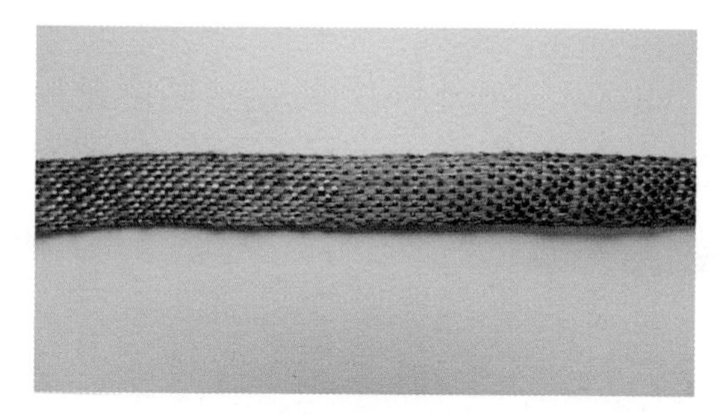

Figure 2-16. *Single-conductor ribbon*

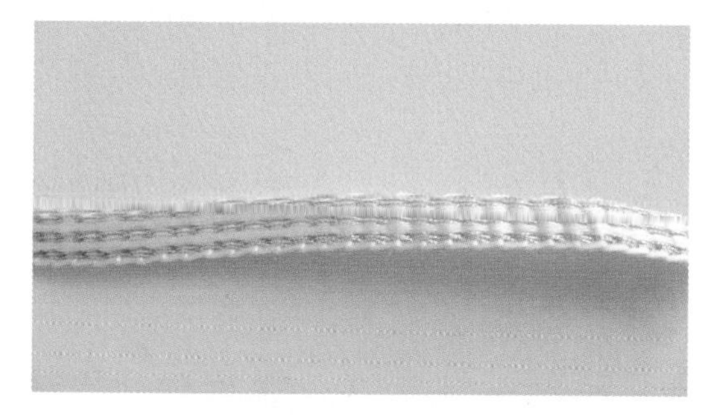

Figure 2-17. *Three-conductor conductive ribbon, uninsulated*

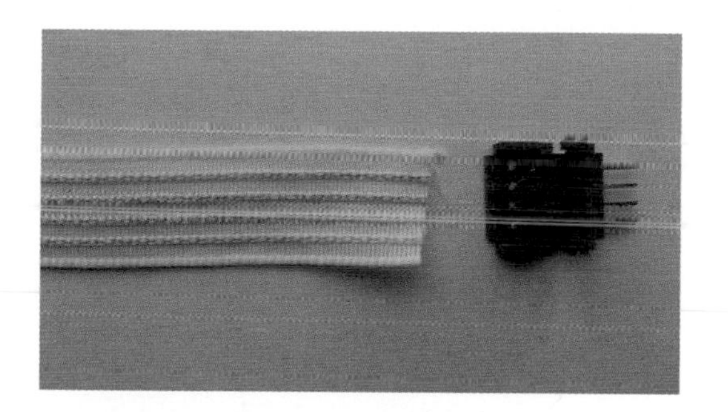

Figure 2-18. *Four-conductor conductive ribbon with crimp connector; this ribbon is lighter weight and has the advantage of the conductors being insulated, meaning you can fold the ribbon against itself without creating shorts*

Table 2-3. Comparing conductive ribbons

Name	Source	Part number	Material(s)	Surface resistance (in Ω/f)
Three-conductor conductive ribbon	SparkFun	DEV-10172	68% stranded tinsel wire 32% polyester	0.1
Four-conductor conductive ribbon (insulated)	SparkFun	DEV-11680	100% Polyester; grosgrain weave, with silver-plated nylon	~16
Four-conductor conductive ribbon (insulated)	SparkFun	DEV-11680	100% Polyester; grosgrain weave, with silver-plated nylon	~16
Fabric ribbon (four-channel wire)	Adafruit	1373	100% Nylon with wires	~0.1
Stainless-steel conductive ribbon (17mm)	Adafruit	1243	316L stainless steel	1.2
Stainless-steel conductive ribbon (5mm)	Adafruit	1243	316L stainless steel	1.2

Conductive Fabric Tape

Conductive fabric tape (LE A225), shown in Figure 2-19, comes on a roll with a peel-off backing. What's neat is that the adhesive is also conductive. So unlike iron-on conductive fabric, which has nonconductive adhesive, you can place one piece of conductive fabric tape on top of another to create a solid electrical connection.

Figure 2-19. *Fabric tape has conductive adhesive*

Conductive Hook and Loop

Conductive hook and loop (Figure 2-20) is like a conductive version of Velcro. What's brilliant about it is that it can act as a secure and sensible electronic switch for clothing.

Figure 2-20. *Conductive hook and loop (AF 1324)*

Conductive Paint

Conductive paints such as Bare Conductive Paint (Figures 2-21 and 2-22) or CuPro-Cote paint (LE A292-4) can be used to paint, draw, or silk-screen circuits. It tends to work best on a nonporous substrate.

Figure 2-21. *Bare Conductive Paint (SF COM-10994)*

Figure 2-22. *Bare Conductive Paint Pen (SF COM-115210)*

Everyday Stuff

There are many commonly available conductive materials that can be repurposed in circuits. When shopping for unconventional conductive materials, it is helpful to bring a multimeter along so you can test for conductivity.

For instance, some organzas have metallic thread that runs in one direction of the weave. You can see the multimeter tests of some in Figures 2-23 and 2-24. Sew a line of thread perpendicular to that, and boom! You've got cheap conductive fabric, available from your local fabric store.

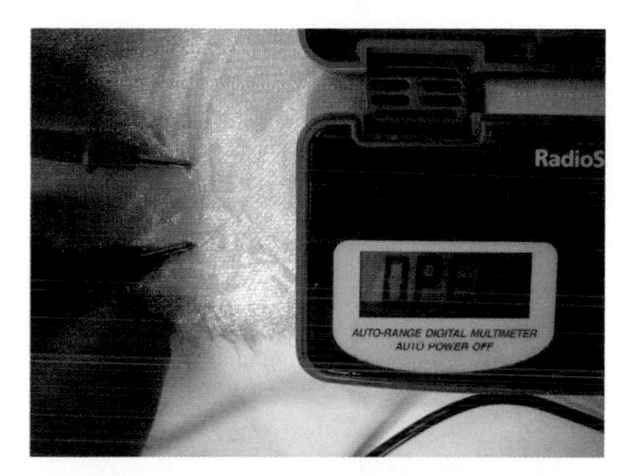

Figure 2-23. *Organza (nonconductive weave)*

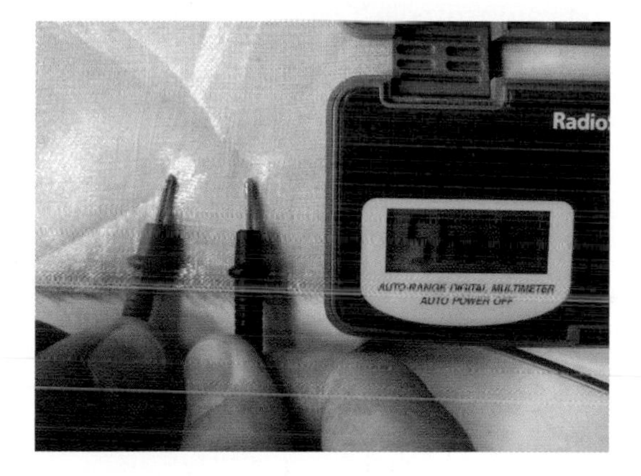

Figure 2-24. *Organza (conductive weave)*

There are also nice conductive materials available in the sculpture section of your local art store. For example, some malleable meshes (Figure 2-25) intended for sculpture are also conductive. These are handy because they can be shaped, cut to size, and soldered to.

Figure 2-25. *Armature meshes are available at most art supply stores*

Keep your eyes open when you're out in the world. There may be more interesting conductive materials than you'd think!

Choosing Conductive Materials

Now that you know a variety of methods for constructing circuits, the question will be what to use and when. Along the way, advantages and disadvantages have been highlighted for each method, but here are a few more things to think about:

Ease of use
: It is important to consider your own abilities. Which method is most comfortable for you? What do you prototype fastest with? If you've racked up years of experience with a breadboard, it may be more comfortable to do a first prototype that way than to start by whipping out the conductive thread. On the other hand, if you're an expert seamstress, sewing up a quick circuit may seem far less daunting than circuit boards and wires. What is most comfortable for you will probably work best, at least for the first time around.

Cost

Cost should be considered in relation to what it is you're trying to make. Often with a prototype, the inclination is to work with what's cheapest and most expendable. But there are times when prototypes turn into showpieces and the choice of materials matters more. You may also be telling a story with your choice of materials, imagining a future time and scenario where they are more ubiquitous and less expensive. On the other hand, you may be working on a prototype for something that is coming to market where the cost of every component and material is crucial. Or perhaps you're just teaching an underfunded workshop where you need to be clever but thrifty with your choice of materials. No matter what, cost is always a factor worth considering.

Insulation

Most wires are insulated. Nearly all soft conductive materials are uninsulated. Will this work for your design? Or are there insulation strategies that could meet your needs? I take a closer look at this in future chapters, but it's worth keeping in mind from the get-go.

Resistance

Resistance is a consideration more likely to come up in the realm of softer materials, particularly conductive thread. Low-powered LEDs and thread play nicely together; motors and thread do not. Think about how much current your components require and whether the resistance of your materials will cause any problems in the delivery of the current they need.

Flexibility

Large pointy circuit boards and crevices and curves of the body don't often work well together. What flexibility and form does your circuit need to take?

Experiment: Wearable Circuits

Now that you know about the various ways to construct a circuit, it is time to create your own. Make two circuits: one hard and one soft. You can use the circuit you created in Chapter 1 or another one of your choosing. Once you're done, take notes about the strengths and weaknesses of each as well as in what context you could see each being used.

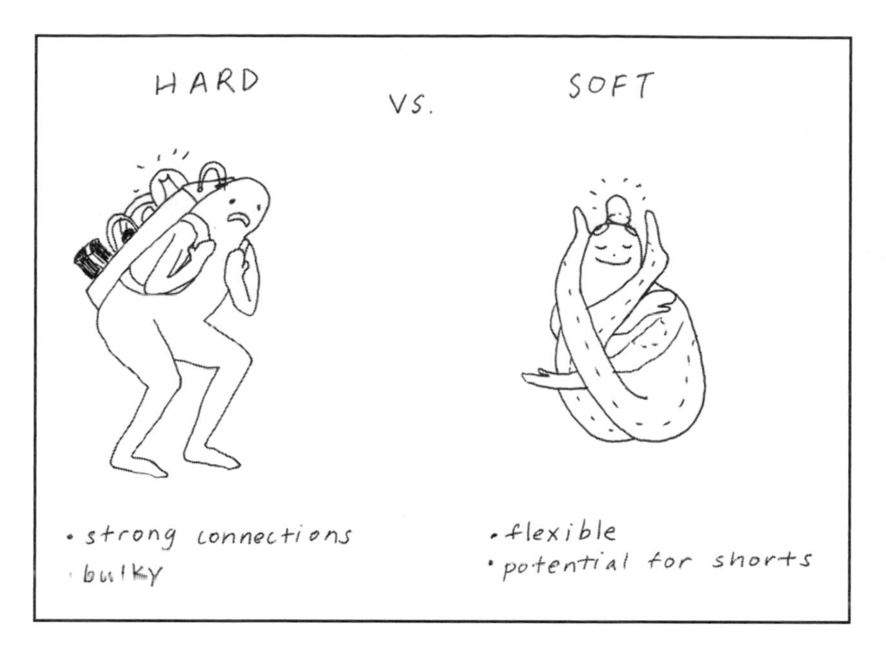

Figure 2-26. *Hard circuit versus soft circuit (illustration by Jen Liu)*

What's Next

As you can see, there are many options available in terms of conductive materials that can be used to build a circuit. In Chapters 3 and 4, you'll learn about components that you can use to create more exciting and complex circuits. In Chapter 5, you'll learn more about how to put these components and conductive materials to work when constructing wearable circuits.

Switches 3

Now that you understand how to design and construct a basic circuit, you can begin to think about how to control it.

A switch is something that enables, prevents, or diverts the flow of electricity. It creates or breaks the physical connection of two conductors. Familiar switches include a standard toggle switch that controls the lighting of a room, the slide switch on the barrel of a classic flashlight, or even the blinking red, "DO NOT TOUCH" button on the control panel of a spaceship. And when you put your imagination to use, you can create switches in forms that you wouldn't expect.

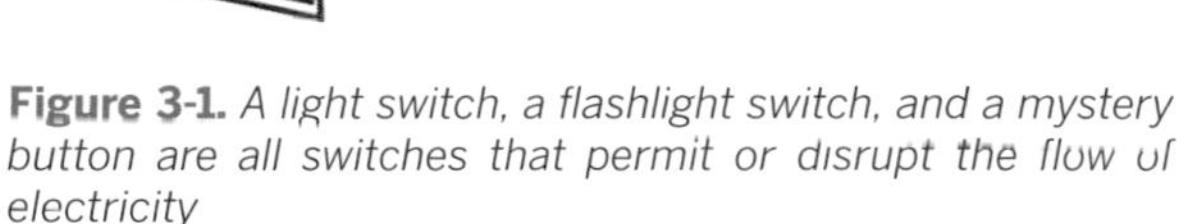

Figure 3-1. *A light switch, a flashlight switch, and a mystery button are all switches that permit or disrupt the flow of electricity*

Switches are awesome because they can act as either an intentional input or a passive sensor. You'll see that users can activate a circuit by doing something as deliberate as pushing a button on a circuit board or as subtle and intuitive as standing up, blinking their eyes, or even giving someone a hug.

In this chapter, you will learn how switches work, how to use them in simple circuits, what types of switches are available, and how you can make your own.

Understanding Switches

The circuit symbol for a basic switch looks like Figure 3-2.

Figure 3-2. *Circuit symbol for a switch*

If you were to integrate a switch into the basic circuit you built in Chapter 1, the circuit diagram would look like Figure 3-3.

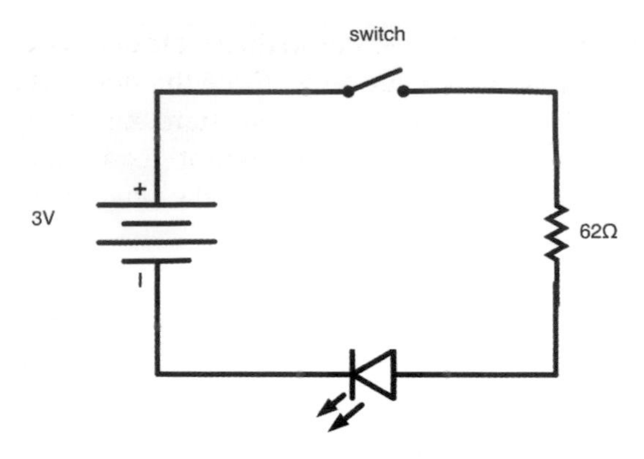

Figure 3-3. *Circuit diagram for simple LED circuit with a switch*

When the switch is *closed*, the two contact points will be connected, and electric current will be able to flow. When the switch is closed, the LED will light up.

When the switch is *open*, the two contacts are not connected, so the circuit is interrupted. Electric current is unable to flow through the circuit, so the LED will not light.

Poles and Throws

The switch represented in Figures 3-2 and 3-3 is called a *single pole single throw* switch, or SPST switch:

Pole
: Refers to the number of separate circuits controlled by the switch

Throw
: Refers to the number of positions each pole can be connected to

SPST switches have two terminals. When the switch is in the "off" position, the connection is open. When it is moved to the "on" position, it closes the connection, and electricity is able to flow. An example is a simple rocker switch like the one pictured in Figure 3-4.

Figure 3-4. *A SPST rocker switch*

The symbol for a single pole *double* throw (SPDT) switch looks like Figure 3-5.

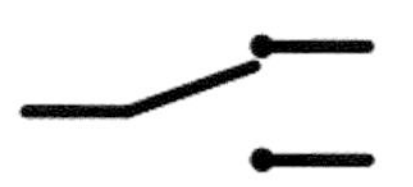

Figure 3-5. *Circuit symbol for a SPDT switch*

This type of switch can be used to switch between two different circuits. Figure 3-6 shows a circuit where only one of the two LEDs will light depending on the position of the switch.

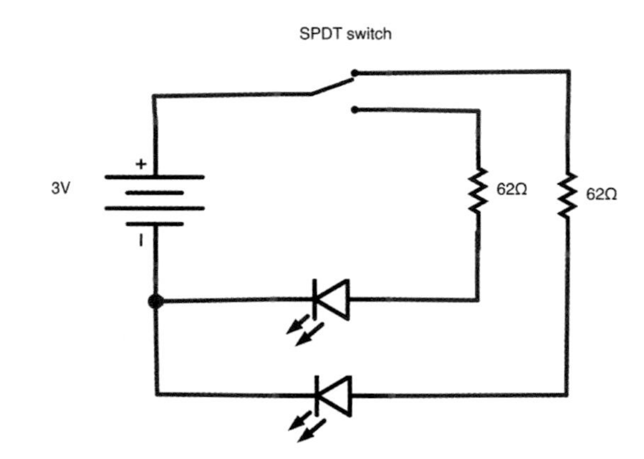

Figure 3-6. *In this circuit, a SPDT switch is used to switch between which LED is lit*

The toggle switch in Figure 3-7 is an example of a SPDT switch. Note that it has three terminals on the bottom for the three connections you saw in the diagram.

Figure 3-7. *SPDT toggle switch*

If needed, a SPDT switch can also operate as a SPST. Just don't connect anything to the second throw, as shown in Figure 3-8.

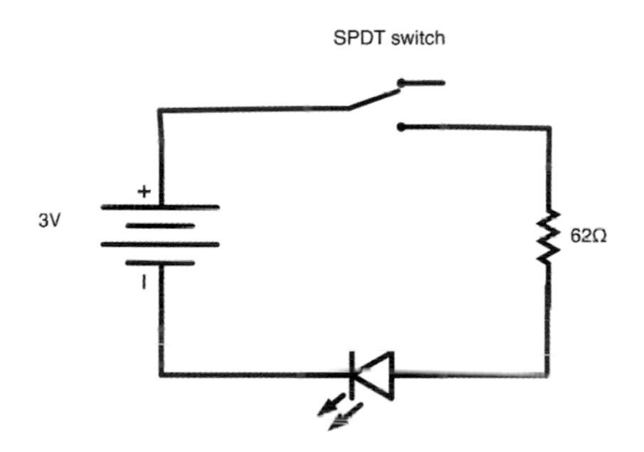

Figure 3-8. *In this circuit, the SPDT switch is used as a SPST*

In the world of electronic switches, there are far more complex combinations of poles and throws, but it's not likely you'll be using them in basic wearable electronics projects. The most important thing to understand is what a switch is and how it works in your circuit.

Types of Switches

There are two categories of switches that you'll encounter: momentary and maintained.

Momentary switches stay in their state only as long as they are being activated. Once the switch is released, it returns to its previous state. An example would be the buttons on a remote control. Momentary switches can be normally open or normally closed (see Figure 3-9):

Normally open (NO)
: Refers to a momentary switch whose default state is for the contacts to be open. If it is activated by the user, the switch will close.

Normally closed (NC)
: Refers to a momentary switch whose default state is for the contacts to be closed. When the switch is activated, the switch will open, deactivating the attached circuit.

Figure 3-9. *Circuit symbols for normally open and normally closed buttons*

Maintained switches are the opposite of momentary switches. They will stay in whatever state you put them in. Think of a light switch in a room. When you turn the lights on, the switch doesn't spring back to off when you let go of it.

Off-the-Shelf Switches

Now that you understand some basic switch terminology and how to use a switch in a circuit, you can look at what sort of electronic switches are out there.

Within the world of electronics, there is a wide range of switches available. Which of these switches are best suited for wearables? For wearables, you generally want either discrete switches to turn things on or off or to change between modes, or switches that work with the body's natural movements. Here are some useful options.

Tactile Buttons

Tactile buttons are momentary buttons that provide light tactile feedback. Because they are momentary, they can be normally open (NO) or normally closed (NC). Their packaging is flat and slim, which enables them to sit well within the profile of garments. For wearables, a broader button face tends to be better than a smaller one, as it provides a larger landing pad for an incoming fingertip. Figure 3-10 shows an assortment of momentary buttons.

Figure 3-10. *Large and small tactile switches*

Some tactile switches have four legs. Their circuit symbol is shown in Figure 3-11.

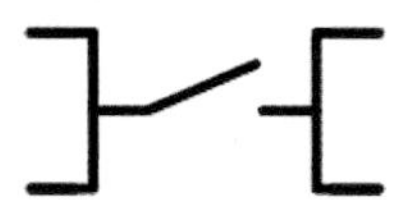

Figure 3-11. *Circuit symbol for tactile switch*

Figure 3-12 shows how to use a tactile button as a simple SPST switch.

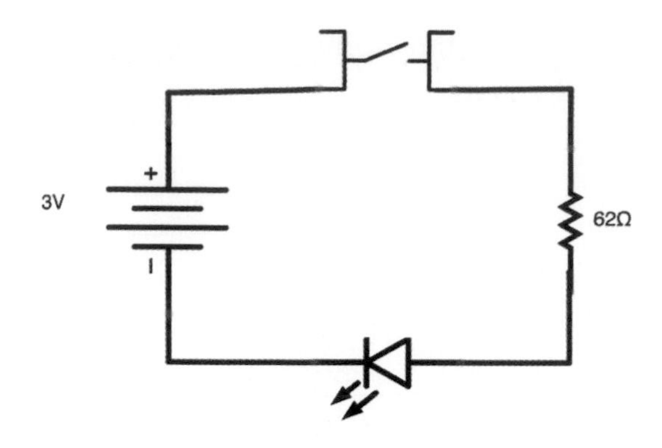

Figure 3-12. *Using a tactile switch as a SPST*

Tactile buttons work best when mounted on a circuit board.

Latching Buttons

Latching buttons (sometimes called tactile on/off buttons) are maintained switches that respond to being pressed. Press it once and the button will stay closed. Press it again and the button will stay open. They are an excellent option for integrating into the seam of a piece of clothing.

The latching button shown in Figure 3-13 would work well as an on/off button that could be subtly situated in a collar, cuff, or sleeve.

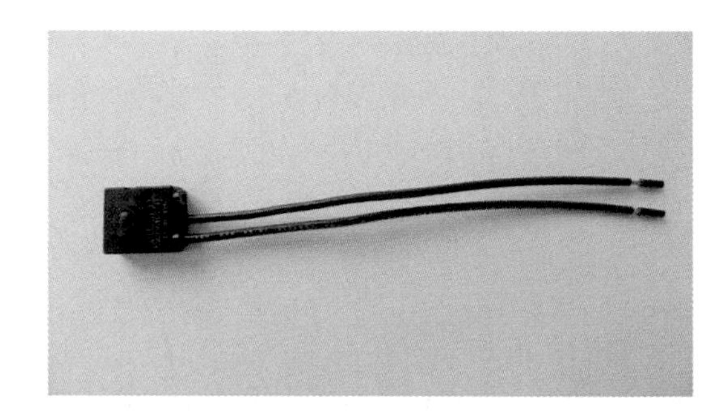

Figure 3-13. *A latching button (AF 1092)*

Toggle Switches

A toggle switch (see Figure 3-14) is a maintained switch that is operated by a lever. Toggle switches are not often used in wearables because the protruding lever is not the most comfortable thing to wear. They tend to work best when mounted on a control panel.

Figure 3-14. *Toggle switch*

Slide Switches

Slide switches (Figure 3-15) are a type of maintained switch that is controlled by moving the switch back and forth in a linear motion. They are small and stable, useful for functions like turning a project on and off. Many of these mount well on a circuit board.

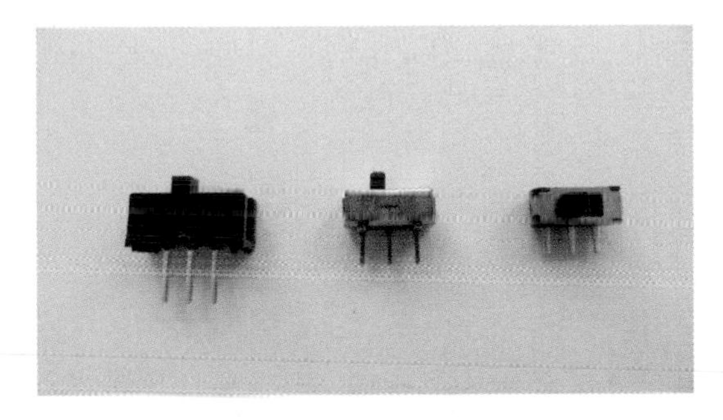

Figure 3-15. *Slide switches*

Microswitches

A microswitch is a highly sensitive switch that can be operated by a small movement. It is a type of momentary switch. A microswitch can come as a wire, lever (Figure 3-16), or roller, which can make it more easily triggered by a certain type of mechanical action. When positioned well, they can be used to detect subtle body movements.

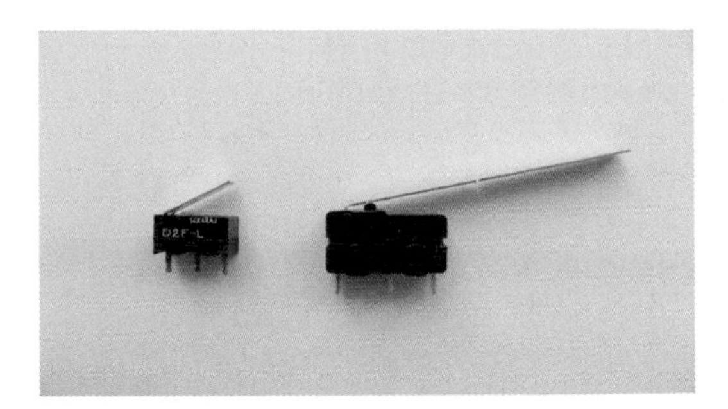

Figure 3-16. *Microswitches with levers*

Tilt Switches

Tilt switches (see Figure 3-17) open or close based on the orientation of the switch. In the past, tilt switches contained mercury. These days, they usually contain a small conductive ball. When the switch is upright, the ball will sit on top of the two contacts, bridging them and closing the circuit. When the switch is inverted, the ball moves and breaks the connection.

Tilt switches work well with wearables because they can disappear into a garment and respond to the movements of the body without the wearer even thinking about it. Raising your hand or touching your toes can suddenly become a way to activate a circuit.

Figure 3-17. *A tilt switch*

There are also multiaxis tilt switches (Figure 3-18). These contain multiple sets of contacts. This is useful when you are looking to sense multiple orientations, such as whether you are lying on your front, back, left side, or right side.

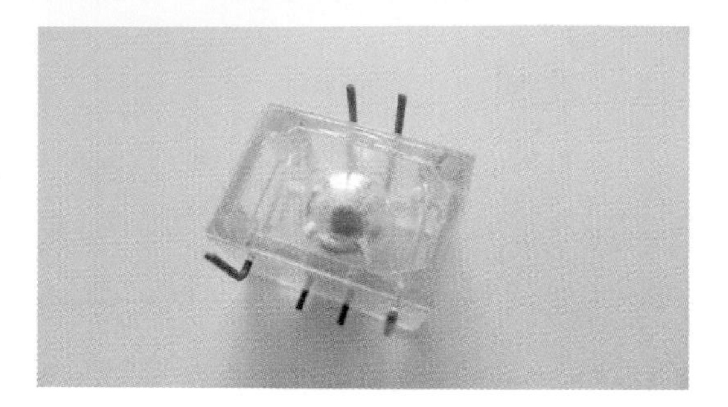

Figure 3-18. *A multiaxis tilt switch*

DIY Switches

If the range of off-the-shelf switches don't meet your needs, you can always make your own. Arrange two conductors in such a way that they will sometimes touch and sometimes not, and you've got yourself a switch! Generally the aim is to come up with a switch that behaves reliably—that is, unless random activation has worked itself into the concept for your project.

DIY switches can be fully integrated with the materials used in your own project so they almost become invisible to the user.

Contact points for switches can be made with any conductive material you can find (see Chapter 2 for inspiration). These examples use iron-on conductive fabric (LE A1220), but you can substitute conductive thread, yarn, paint, mesh, or any other conductive material of your choosing.

Sandwich Switch

A sandwich switch is the DIY version of a normally open momentary switch. Press it to close the circuit, and release to open it. I call this a sandwich switch because it is comprised of layers. You could also think of it as a lasagna or layer cake, but at some point this would turn into a buffet instead of a circuit-making exercise.

Consider the bread of your sandwich to be a nonconductive material, preferably a stiff but squishy one like felt, foam, neoprene, or fleece. You want it to yield to the touch but spring back afterward. This material will give the switch its physical form and also insulate the conductive materials that lie within. You will also need a third piece of this material to use later for the Swiss cheese of the sandwich (see Figure 3-19).

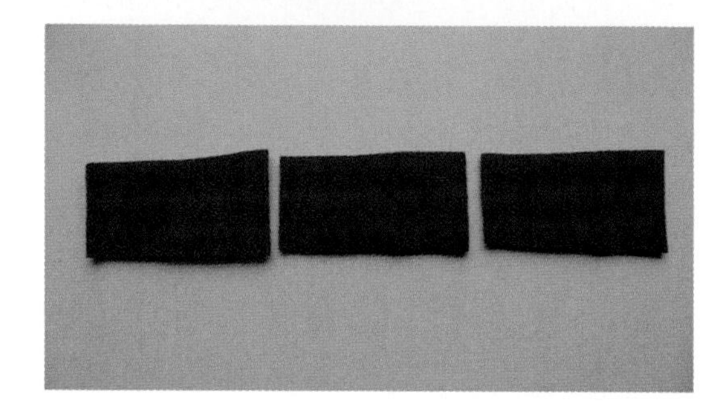

Figure 3-19. *Cut three pieces of a nonconductive material in the same size*

Next is the mustard, mayo, or pesto of your sandwich (Figure 3-20). This is the conductive material. Line two of the nonconductive pieces with a conductive surface, as shown in Figures 3-21 and 3-22. You can see the combined pieces in Figure 3-23.

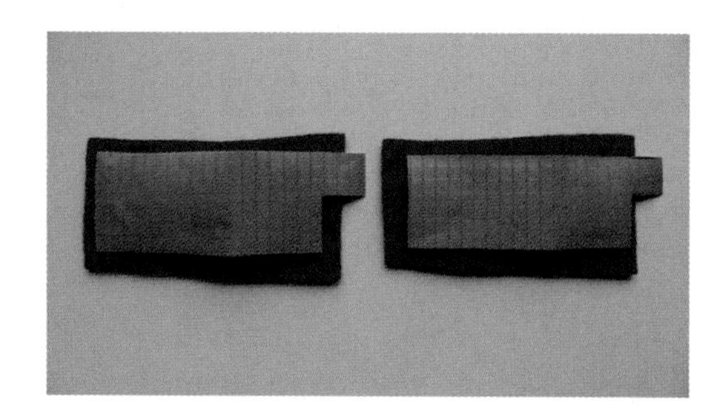

Figure 3-20. *Cut two patches of conductive fabric that are slightly smaller than the pieces of nonconductive fabric; include a tab long enough so that it can wrap around an edge*

Figure 3-21. *With each piece iron down the conductive fabric except for the tab*

Figure 3-22. *Flip each piece over, fold the tab over, and iron it down*

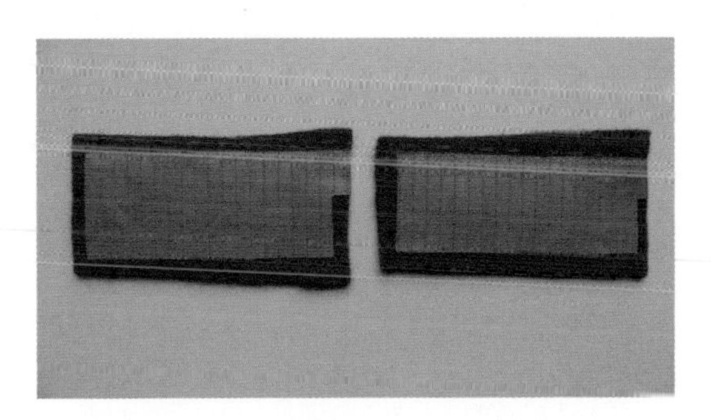

Figure 3-23. *Your resulting pieces will look like this*

Finally, there is the Swiss cheese, which goes in the middle. This should be a nonconductive material that contains some holes. The idea is that when the switch is at rest, this insulating material is enough to prevent the two conductors from touching; but when the switch is pressed, the pressure forces the conductive materials to connect. The holes can be inherent to the material, like netting, or you can cut them yourself. Experiment with different densities of holes and different thicknesses of materials to achieve the sensitivity you desire. Figure 3-24 shows one possible choice.

Figure 3-24. *Cut a hole in the middle layer*

Now it is time to complete the switch. Assemble the layers so that the conductive surfaces face inward with the perforated layer in the middle, as shown in Figures 3-25 and 3-26. Then sew or glue the sandwich layers together (see Figure 3-27).

Figure 3-25. *Place one end piece conductive side up; then place the middle piece with the hole on top of it*

Figure 3-26. *Place the other end layer conductive side down*

Figure 3-27. *Sew the switch together*

Keep in mind that your switch will perform differently when sewn or glued together than when the layers are simply placed on top of each other. Securing the materials together helps to reinforce the overall structure.

Figure 3-28 shows how to connect alligator clips to the switch; you can see the switch in action in Figure 3-29.

Figure 3-28. *Connect alligator clips*

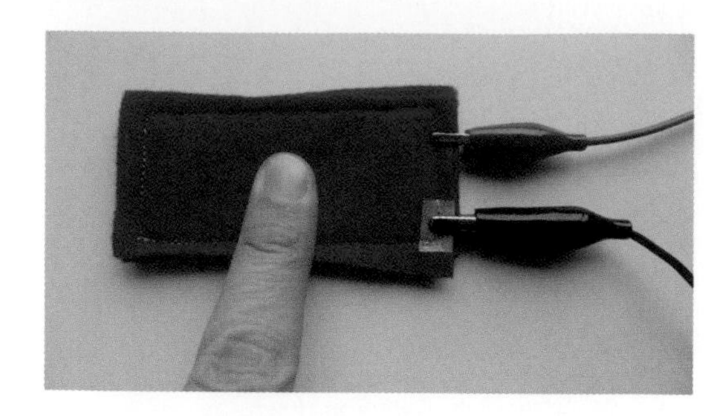

Figure 3-29. *Pressing the switch*

Contact Switch

A contact switch contains two conductive surfaces that will at some point make contact with each other. This is fun to play with in wearables because you can consider the way the body moves. Different body parts make contact when your arms are by your side, your heels click together, or your head is in your hands.

The contact switch in Figure 3-30 is constructed with two conductive patches on a nonconductive surface. When the material is at rest, the switch is open, as shown in Figure 3-30. When the material is folded in half, the two conductors touch each other and the switch is closed (see Figure 3-31).

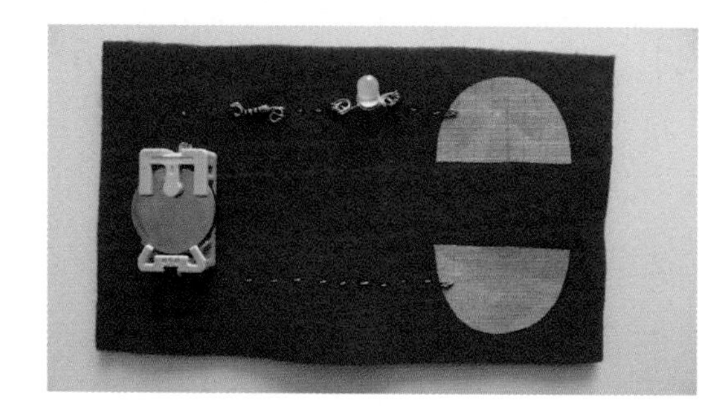

Figure 3-30. *Contact switch open*

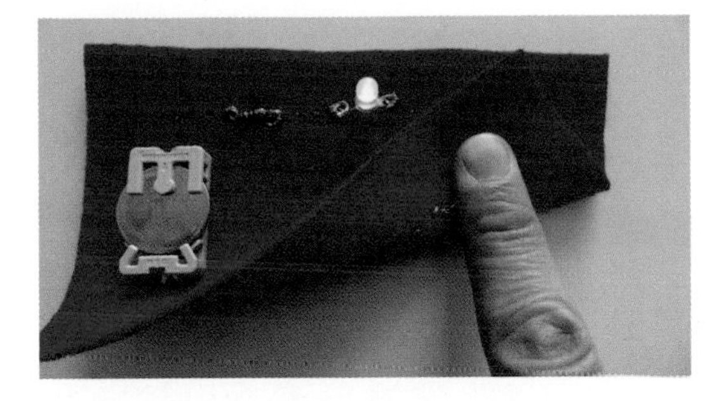

Figure 3-31. *Contact switch closed*

These days, some gloves come with conductive fingertips intended for use with smartphone screens (Figure 3-32). These can be modified so the fingertips act as a contact switch. By using the same circuit as with the contact switch, you can create a glove that lights up when the index finger is touched to the thumb (Figure 3-33) and turns off when the connection is released (Figure 3-34).

Testing Switches

When creating switches from scratch, it is important to test that they work reliably. Here are two ways you can do this.

The first is with a multimeter. Use two alligator clips to connect each side of the switch to the probes of the multimeter:

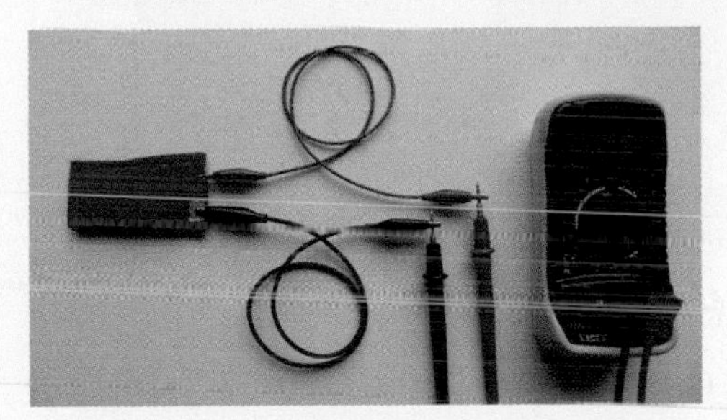

Set the dial of the meter to the continuity setting. Assuming your switch is normally open when it is at rest, you will hear nothing. When you activate it, the meter will beep:

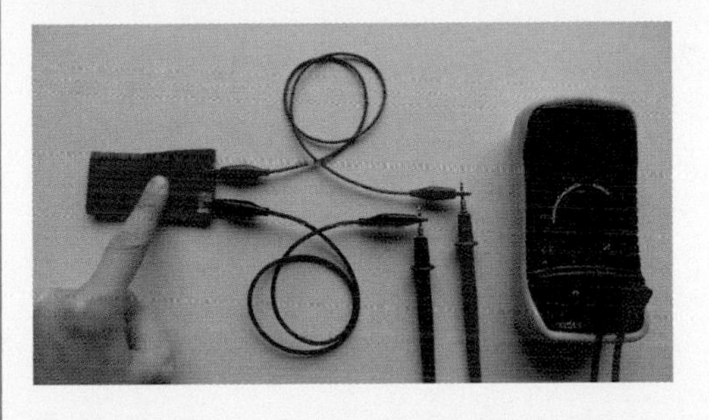

You can also test your switch using a simple alligator clip circuit. Just add the switch in series:

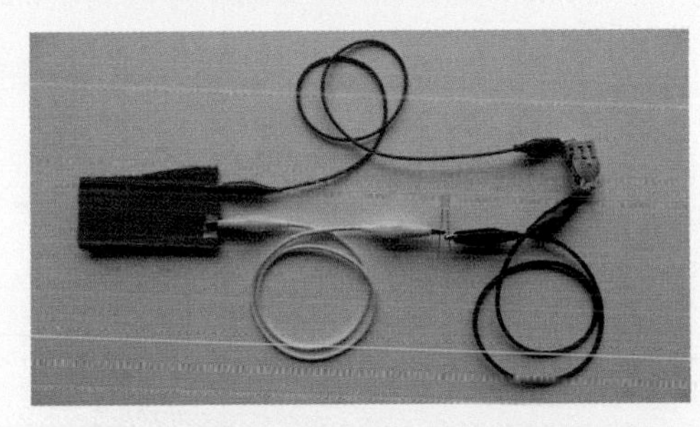

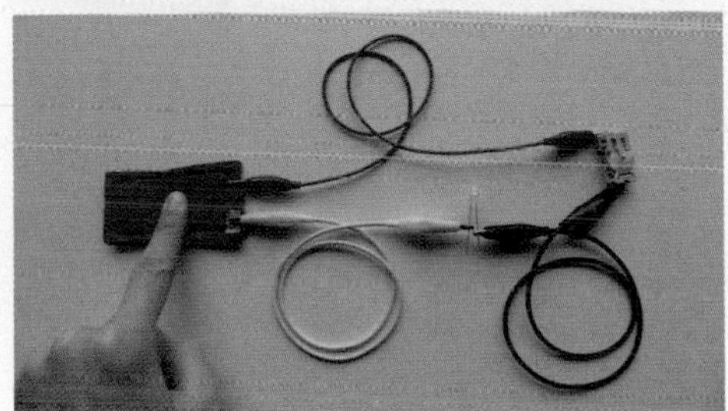

Figure 3-32. *Gloves with conductive fingertips*

Figure 3-33. *Glove switch closed*

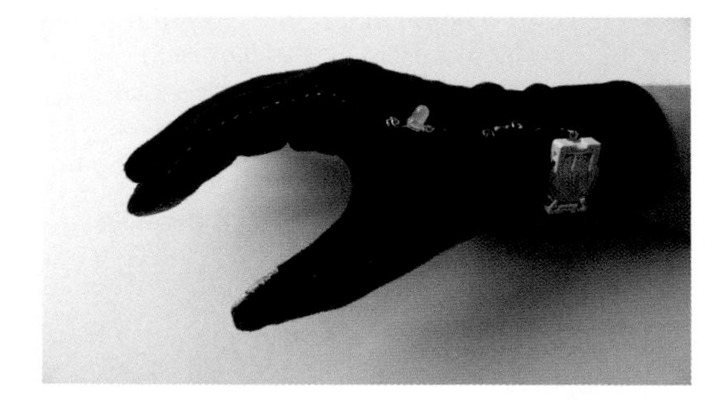

Figure 3-34. *Glove switch open*

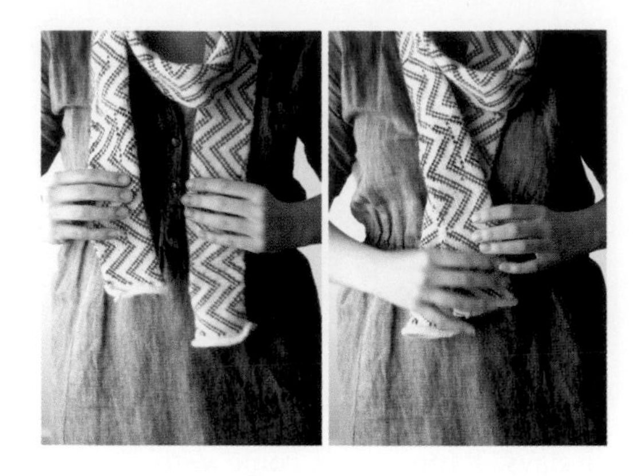

Figure 3-35. *This contact switch created by Erin Lewis integrates conductive thread with knitted yarn: when the pink patches of the scarf are held together, the switch is closed*

Figure 3-36. *In "Embrace Me" by Studio 5050, conductive fabric on the chest of the hoodie closes a circuit when two wearers hug, causing a pattern on the back of the hoodies to light*

Figure 3-37. *"BlinkCam" by Andrew Schneider uses conductive fabric eyelash switches to control a modified Polaroid camera; every time the wearer blinks, the eyelash switches close the circuit, causing the camera to take a picture*

Bridge Switch

A bridge switch leaves a small break in the circuit that can be bridged by any piece of conductive material (see Figure 3-38). Create the entire circuit on one surface and leave an exposed break in one of the conductive traces. When another object or garment that is conductive is put in contact with the break (Figure 3-39), it bridges the connection and closes the circuit.

Figure 3-38. *Bridge switch open*

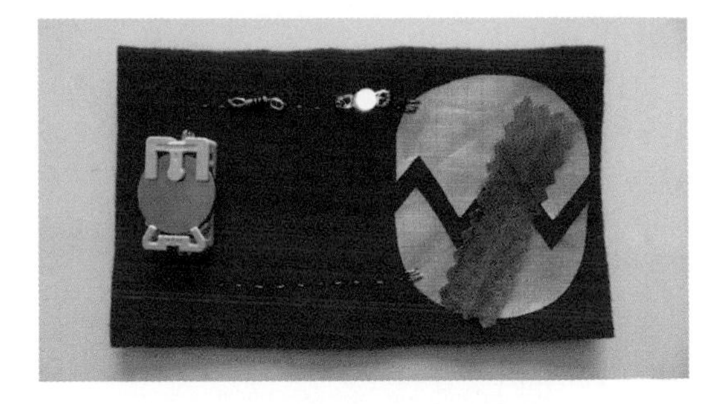

Figure 3-39. *Bridge switch closed with a piece of copper mesh*

Figure 3-40. *This bridge switch designed by Jackson McConnell uses conductive fabric on the back of a pendant to close the connection on the chest of the t-shirt*

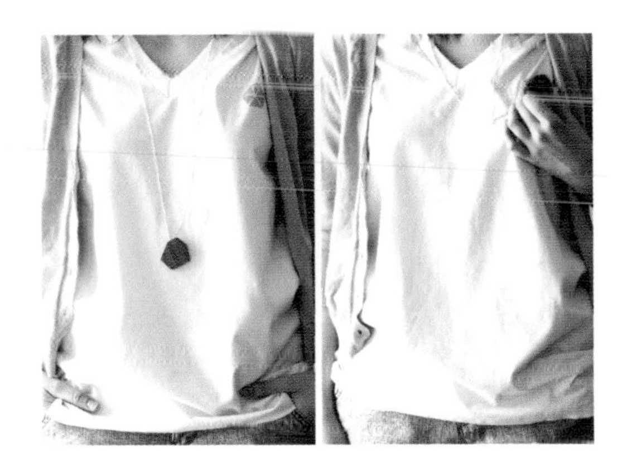

Figure 3-41. *Jackson's switch in use*

Pinch Switch

A pinch switch leverages the foldability of fabric. Take a stiff strip of material and line it with two pieces of conductive material. Affix it to a base so that it sits like an open loop (Figure 3-42). When you pinch it together, the two conductive pieces will touch and close the connection (Figure 3-43).

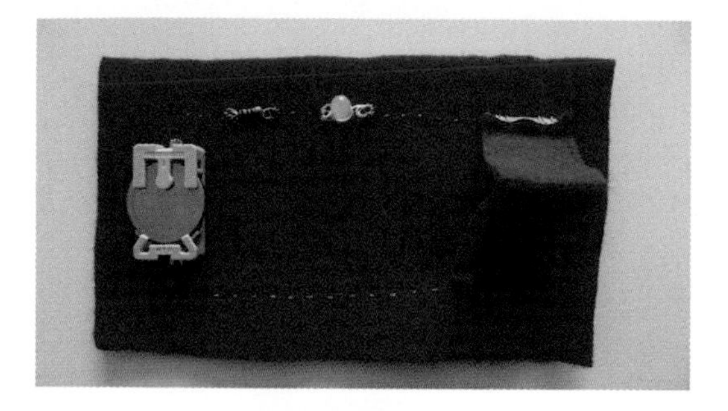

Figure 3-42. *Pinch switch open*

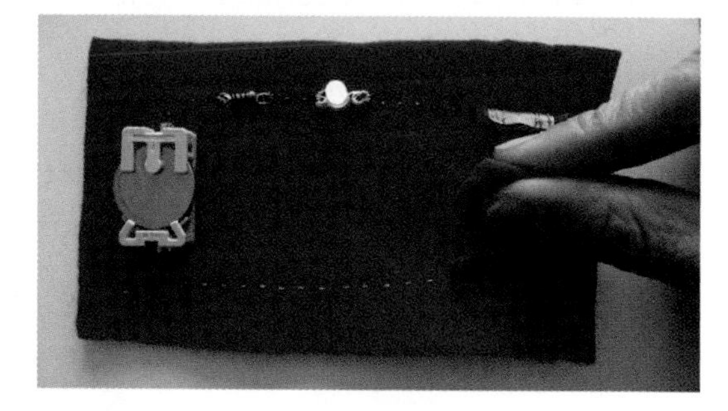

Figure 3-43. *Pinch switch closed*

Figure 3-44 shows an example of a pinch switch embedded in a wearable.

Figure 3-44. *This pinch switch design by Hazel Meyer triggers a wireless signal to be sent to a nearby wearable*

Other DIY Switches

Those were just some examples to get you started, but you can probably imagine many more ways that conductive materials can be used to create a switch.

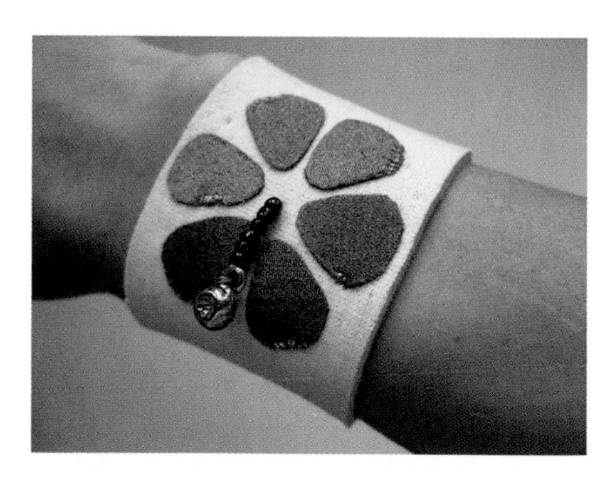

Figure 3-45. *A DIY tilt-sensing bracelet by Hannah Perner-Wilson*

Experiment: Social Switches

When properly planned, a switch can create the opportunity for a social interaction. Using simple switches, try to create a wearable circuit that lives on multiple bodies and responds to a social interaction between two or more people.

Here are some possible ways to break up your circuit:

"Power Me Up"
: One wearable contains the battery, and one wearable contains the rest of the circuit. The wearables are connected using two contact switches. Figure 3-46 shows the circuit diagram, and Figure 3-47 shows a project that implements this design.

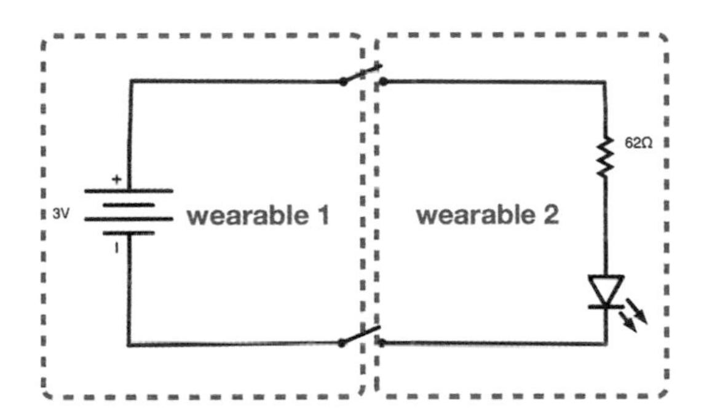

Figure 3-46. *"Power Me Up" circuit*

Figure 3-47. *These antenna use the "Power Me Up" model (prototypes created at the Digital Futures Playshop)*

"You Complete Me"
: There is a break in the circuit on one wearable that is completed by a piece of conductive material on the other wearable. The example shown in Figure 3-48 uses a bridge switch.

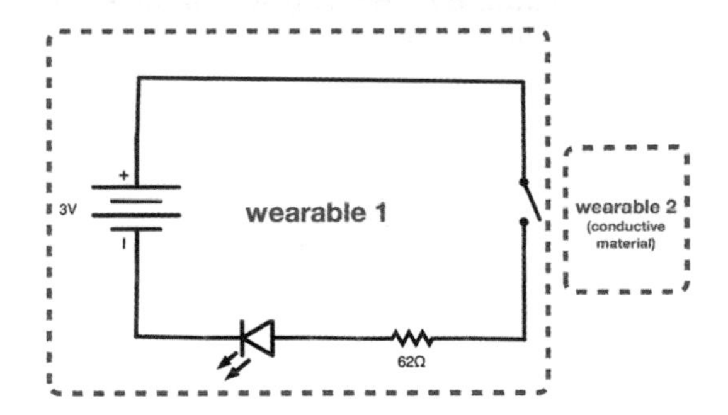

Figure 3-48. *"You Complete Me" circuit*

"We're All In This Together"
: Each circuit component is on a different wearable, and wearers must connect in the appropriate way in order to complete the circuit. The connection between wearables is made with contact switches. Figure 3-49 shows the circuit diagram, and Figure 3-50 shows an implementation of it.

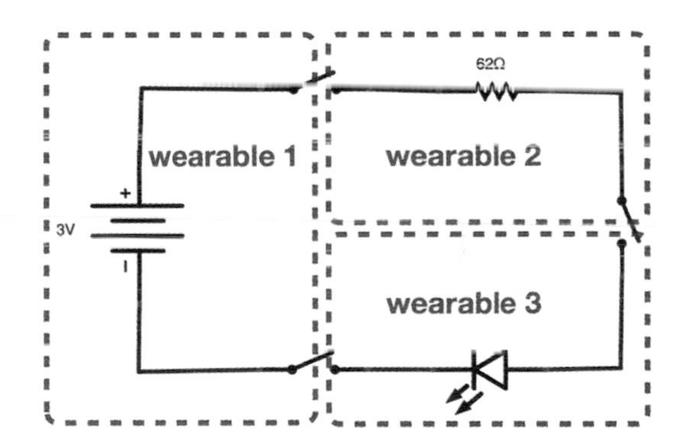

Figure 3-49. *"We're All In This Together" circuit*

Figure 3-50. *These hand coverings use the "We're All In This Together" circuit design (prototypes created at the Digital Futures Playshop)*

Conclusion

Switches are your first opportunity to create interactive circuits as well as interfaces that live on the body. Later on, you will learn about more complex inputs, but even through the use of a simple switch and thoughtful design, it is possible to create highly engaging body-based interactions.

E-Textile Toolkits

4

In the previous chapters, you learned about the diverse range of conductive materials that can be used to construct circuits. But as you saw when constructing soft circuits in Chapter 1, standard electronic components need to be modified if they are to be used with soft conductive materials.

E-textile toolkits are families of modules that are designed specifically for use with nontraditional conductive materials such as conductive thread or conductive ribbon. In this chapter, you will learn about several types of e-textile toolkits, the modules they contain, and how to get started using them in a circuit.

Because these toolkits are emerging platforms, there is no standard vocabulary for these systems or their parts. Before you get started looking at them in detail, let's define some of the terms we'll be using:

E-textile
: Stands for "electronic textile." This refers to a category of electronic parts that can be used in combination with textiles and other soft materials.

Toolkit
: Refers to a set of tools that can be used for any number of purposes. This is different from a "kit," as a kit's parts are usually meant to be used together to assemble a single thing. Other terms you might run into that mean roughly the same thing are "platform" or "system."

Module
: Refers to a discrete unit that is part of a toolkit. Modules are printed circuit boards that contain electronic components and their necessary connections. Modules can be connected together using conductive materials to create a complete circuit.

The toolkits that will be reviewed in this chapter are designed primarily for use with conductive threads or ribbons, but they can ultimately be used with whatever conductive materials you like.

LilyPad

The LilyPad (Figure 4-1) was the first set of widely available electronic components specifically intended for integration with nontraditional conductive materials. First released by SparkFun Electronics in 2007, the LilyPad was based off of years of research by Leah Buechley during her time at the University of Colorado (see Figure 4-2).

Figure 4-1. *LilyPad Arduino 328*

Figure 4-2. *Early LilyPad prototype by Leah Buechley*

There are several characteristics that set LilyPad modules apart from more traditional circuit boards:

Sewable
: LilyPad modules are designed specifically to enable electrical connections made with hand-sewn conductive thread. Connection points are situated around the perimeter of the boards so that they can be easily accessed. These areas are referred to as "sew tabs," "sew holes," or "petals."

Rounded edges
: All LilyPad boards have rounded edges. This works well in the context of floppy flexible fabric substrates. If you bend a body part while wearing a LilyPad component, you won't get jabbed by a sharp corner.

Thin
: LilyPad boards are a bit thinner than a traditional circuit board, which means they are less bulky and easier to incorporate into a lining, pocket, or seam.

Purple
: LilyPad modules are purple. This is meant to make them friendlier, a bit more attractive, and it also makes them easier to pick out from a sea of parts.

The LilyPad has set a precedent for many e-textile toolkits that have been developed since. You will see that many of the design choices—such as the round boards and sew tabs—are echoed in the toolkits that follow.

Modules

There are a variety of modules within the LilyPad toolkit. Some of these are similar to components that you used in Chapter 1 and others you will learn to use in chapters that follow.

Microcontrollers are the brains of your circuit. They are tiny computers that will live in your jacket lining, pocket, or under your hat. The LilyPad toolkit contains multiple options for microcontroller boards (see Figure 4-3). You'll learn more about microcontrollers in Chapter 6.

Figure 4-3. *LilyPad microcontroller options: LilyPad Arduino 328, LilyPad Arduino Simple, LilyPad Arduino Simple Snap, LilyPad Arduino USB, LilyPad Twinkle, and LilyPad Tiny*

There is also a useful selection of switches and sensors available as LilyPad parts, including a slide switch, a push button, a light sensor, a temperature sensor, and an accelerometer (Figure 4-4). These modules contain all of the necessary resistors and connections so their pins can be connected directly to a microcontroller module without the need for any additional circuitry.

Figure 4-4. *LilyPad Light Sensor, Temperature Sensor, Accelerometer, Button*

There are several types of actuators in the LilyPad toolkit that can be used for decoration, feedback, or display, including a vibrating motor, a buzzer, and several types of LEDS. See Figures 4-5 and 4-6.

Figure 4-5. *LilyPad Buzzer and LilyPad Vibe Board*

Figure 4-6. *LilyPad MicroLED, LED, Tri-Color LED, and the Lily Pixel*

LilyPad protoboards (Figure 4-7) enable you to assemble and solder together denser and more complex portions of your circuit. This is a great way to create stable connections and reduce the likelihood of shorts. Learn more about how to work with LilyPad protoboards in Chapter 8.

Figure 4-7. *LilyPad protoboards*

Battery holder and connector modules can provide power to your circuit (Figure 4-8). The LilyPad Power Supply steps up a AAA battery to 5V. The LilyPad Coincell Battery Holder is available both with and without a switch. The LilyPad LiPower and LIlyPad Simple Power offer connections for rechargeable lithium polymer batteries. See Appendix B for more on batteries.

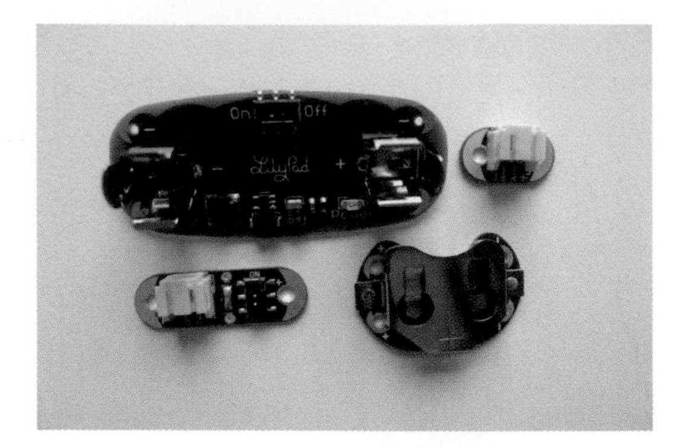

Figure 4-8. *LilyPad battery holders and connectors*

Finally, wearability and wireless go hand in hand. You would not want your clothing attached to a long communications cord. The LilyPad XBee (Figure 4-9) is a useful wireless solution. You'll learn how to work with XBee radio transceivers in Chapter 9.

Figure 4-9. *LilyPad XBee*

Experiment: Let's Get Twinkly

To get comfortable working with LilyPad components, let's make a simple circuit with the LilyTwinkle. This microcontroller module comes preprogrammed to "twinkle" LEDs attached to the four output pins. Let's see how the circuit is assembled.

Figure 4-10. *LilyTwinkle*

Figure 4-11. *The LilyTwinkle (right) is much smaller than a LilyPad Arduino 328 (left)*

Parts and materials:

- (1) LilyTwinkle (SF DEV-11364)
- (1) LilyPad Coin Cell Battery Holder (SF DEV-10730)
- (1) CR2032 battery (AF 654, DK P189-ND, SF PRT-00338)
- (4) LilyPad LEDs (SF DEV-10081)
- (7) alligator clip test leads (AF 1008, RS 278-1156, SF PRT-11037)
- Conductive thread (if you would like to sew your circuit)

Make the connections shown in Figure 4-12 using either alligator clips or conductive thread (Figure 4-13). If you need a refresher on how to sew circuits with conductive thread, see Chapter 2.

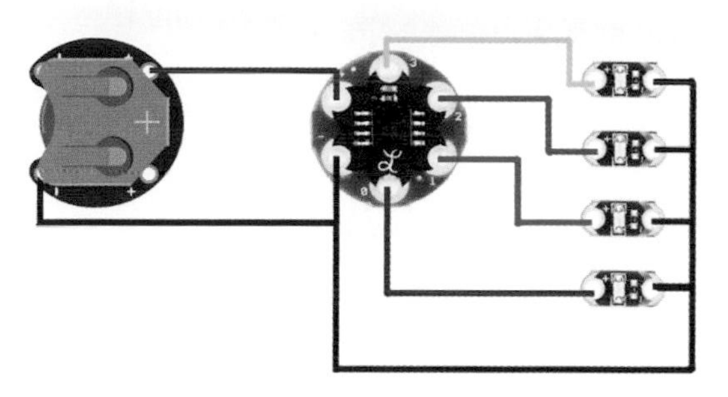

Figure 4-12. *Connections diagram for LilyTwinkle*

Once your circuit is complete, insert the battery in the battery holder. Your LEDs should twinkle away!

Figure 4-13. *LilyTwinkle circuit made with conductive thread*

The LilyTwinkle program lights LEDs at random, very much like fireflies. Remember that if you only have one LED connected, there may be times when there is a long pause before that LED lights up again. If an LED doesn't light immediately, give it a bit of time or connect the other LEDs to make sure that the LilyTwinkle is working properly. Also check the polarity of the LED to make sure it is connected correctly.

Experiment: Let's Get Tiny

Parts and materials:

- (1) LilyTiny (SF DEV-10899)
- (1) LilyPad Coin Cell Battery Holder (SF DEV-10730)
- (1) CR2032 battery (AF 654, DK P189-ND, SF PRT-00338)
- (1–4) LilyPad LEDs (SF DEV-10081)
- (4–7) alligator clip test leads (AF 1008, RS 278-1156, SF PRT-11037)
- Conductive thread (if you would like to sew your circuit)

While the LilyTwinkle works well for ambient lights, you may want a more consistent behavior. The LilyTiny is the exact same hardware as the LilyTwinkle but comes loaded with a different program. It can be used to blink, beat, fade, or breathe LEDs attached to allocated pins.

Table 4-1 shows the breakdown of which pin does what.

Table 4-1. LilyTiny pin functions

Pin	Function
0	"Breathing" fade
1	Heartbeat pattern
2	Steady blink
3	Random fade

If you're just after one of these behaviors, it'll make sense to use just one of the pins. Figure 4-14 shows how to make those connections.

Figure 4-14. *LilyTiny circuit made with alligator clips*

If you'd like to have multiple LEDs performing the same behavior, you can add them in parallel to the first (Figure 4-15). The total number of LEDs that you can light will depend on the manufacturer of the battery, but usually you can get at least eight or more to work.

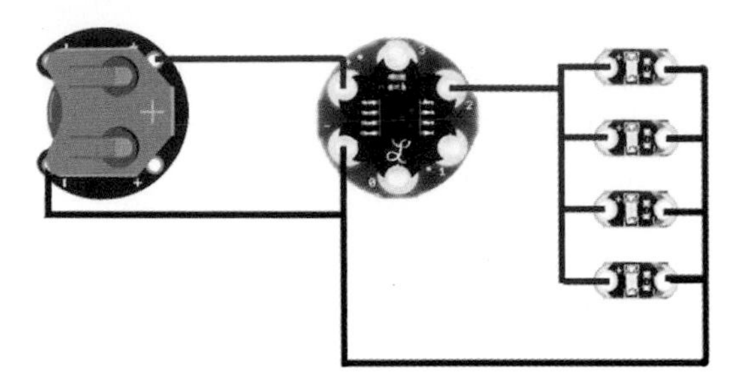

Figure 4-15. *LilyTiny LEDs in parallel*

As you can see, these circuits are simple and quick to construct. This is an excellent choice for someone who wants more dynamic behavior for LEDs but doesn't want to fuss with programming.

Flora

Created by Adafruit Industries in New York, the Flora is the newest e-textiles toolkit to hit the wearables scene. Working off of many of the design choices first introduced by the LilyPad, the Flora makes some significant leaps forward from an engineering perspective and offers an exciting selection of new modules.

Modules

Though not official Arduino boards, the Flora microcontroller modules (see Figure 4-16) are able to be programmed with a version of the Arduino IDE. The Flora main board is slightly smaller than a LilyPad Arduino and features built-in USB support as well as robust power management. The Gemma is an even smaller board that offers some of the functionality of the main board.

Figure 4-16. *Flora Main Board and Gemma*

The Flora toolkit offers slightly different sensors than the LilyPad (Figure 4-17). The Flora Lux Sensor is a multifaceted light sensor that senses infrared, full-spectrum, and human-visible light. The Flora Accelerometer has the bonus feature of being a compass as well. The toolkit also includes a color sensor, which is excellent for color-matching programs.

Figure 4-17. *Flora Lux Sensor, Flora Color Sensor, Flora Accelerometer/Compass Sensor*

One of the Flora's featured modules are the Smart NeoPixels (Figure 4-18). These are by far the most heavy-duty LEDs you've seen in the context of e-textiles toolkits. They are very bright, chainable, and individually addressable, which means that if you have a strand of 10 pixels, you can set the color and brightness of each one individually.

Figure 4-18. *Flora Smart NeoPixel*

Finally, there is a GPS module (Figure 4-19), which is excellent for outdoor, location-aware projects.

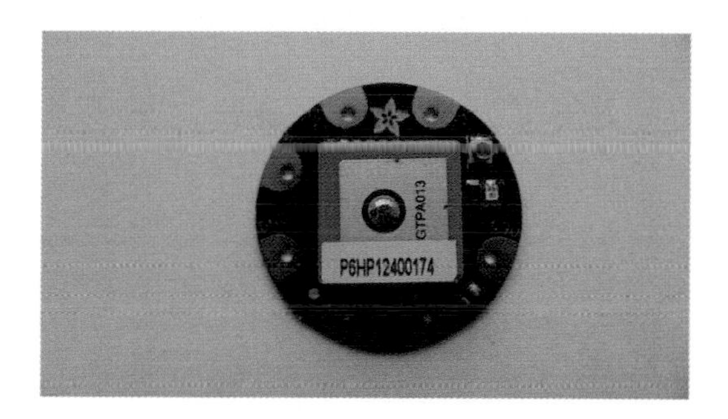

Figure 4-19. *Flora GPS*

You'll learn about some of the Flora modules in Chapters 6, 7, and 8.

Aniomagic

Aniomagic (Figure 4-20) is an e-textile toolkit that can be used to create dynamic lighting patterns. It is similar to the LilyPad and the Flora in that connections are made using conductive thread and that sew tabs provide access to connections on the printed circuit boards.

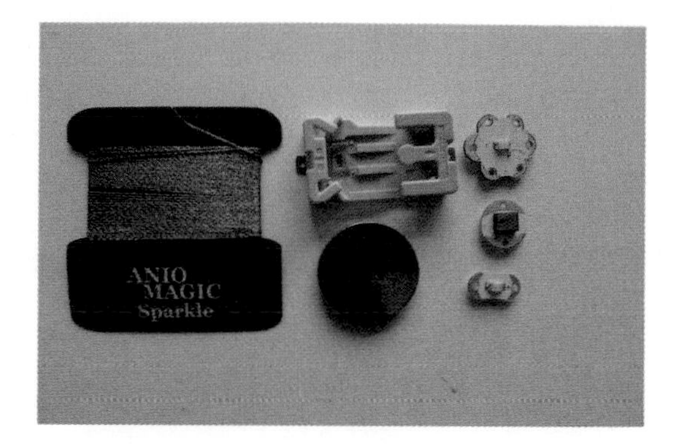

Figure 4-20. *Aniomagic modules*

There are two main features that distinguish Aniomagic from other e-textile toolkits:

- It has a graphical programming interface. This means that you can determine the behavior of the LEDs with drop-down menus and sliders rather than writing code.
- The Aniomagic Sparkle Board is programmed optically. This means that the program is transmitted onscreen through the display of a rapid sequence of shapes. No need for a USB cable.

The advantage of these features is that it is very easy to create programs, and an Aniomagic circuit can be reprogrammed by any device that has a web browser. This means that when you are on the bus, at a party, or in a meeting, you can reprogram your wearable circuit on the fly.

Modules

The Aniomagic toolkit includes two microcontroller modules—the Sparkle (Figure 4-21) and the Chiclet. The Sparkle is meant for use with LEDs only. The Chiclet is for use with sensors.

Figure 4-21. *Aniomagic Sparkle Board*

Actuators include both basic LEDs as well as Lightboards (see Figure 4-22) that contain an onboard controller for more complex operations. Both are available in a variety of gem-inspired colors, including amethyst, emerald, diamond, quartz, sapphire, and more.

Figure 4-22. *Aniomagic Light Boards*

The Aniomagic toolkit also includes a variety of sensors, including sound, light, and touch.

Figure 4-23. *Aniomagic Sound Sensor Board*

Experiment: Let's Get Sparkly

In this experiment, you will use a Sparkle circuit to achieve some dynamic behaviors using standard LEDs. While an Aniomagic Sparkle Kit will come with its own LEDs, you can also substitute LilyPad or other types of LEDs.

The parts and materials (Figure 4-24) needed for this experiment are all included in the Aniomagic Sparkle Kit (MS[1] MKAN2, SF KIT-12729). It contains the following:

- Sparkle Board
- (4) LEDs
- CR2032 battery holder
- CR2032 battery
- Conductive thread

When looking at the Sparkle Board, you will see that there are six pins available: power (+), ground (–), and four pins intended for use with LEDs. The Sparkle Board itself has an onboard LED whose behavior is controlled by the program.

The circuit layout is shown in Figure 4-25.

1. Maker Shed (for a complete list of supplier abbreviations, see "About Part Numbers" on page xiv).

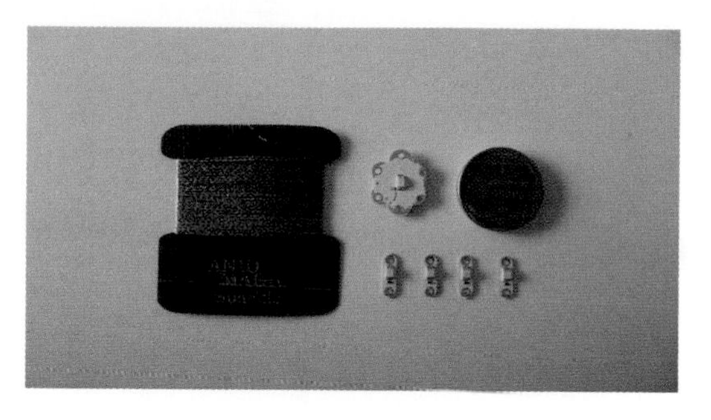

Figure 4-24. *The Aniomagic Sparkle Kit*

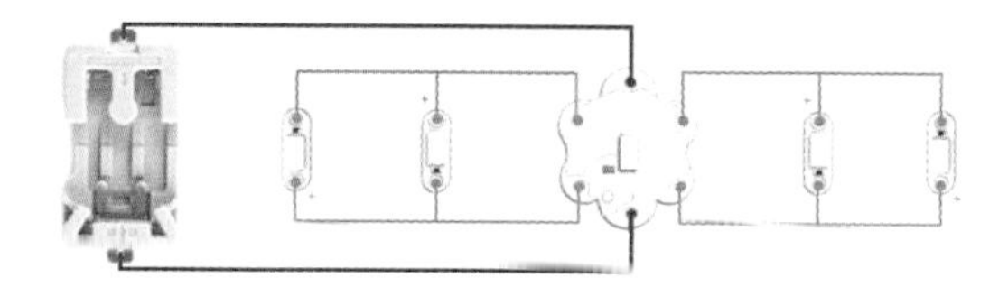

Figure 4-25. *Aniomagic Basic Circuit layout*

Note the polarity of the LEDs in the wiring diagram.

Use either alligator clips or conductive thread to make the necessary connections. Figure 4-26 shows how the circuit looks with conductive thread.

Figure 4-26. *Aniomagic Basic Circuit with conductive thread*

Insert a battery to power the circuit. Now you're ready to program your Sparkle Board. To create a program for your Sparkle board, go to the "Program" page on the Aniomagic website (*http://www.aniomagic.com/program/*).

What you encounter will look something like Figure 4-27.

On the left side of the screen is an illustration of the Sparkle Board. This is the area of the screen that will eventually deliver the program to your Sparkle Board. In the middle is a column titled "Normal." These controls you will use to create the program for your Sparkle circuit. On the right is a column titled "Special." Just ignore this one for now—it contains controls related to sensor circuits.

Review the options in the "Normal" column and make your selections. Once you've configured your settings, you're ready to load the program onto the Sparkle board. There is a thin curved line located below the red LED (Figure 4-28). This is used to switch the Sparkle board into programming mode. Press this "button," as shown in Figure 4-29. The red LED should start blinking rapidly.

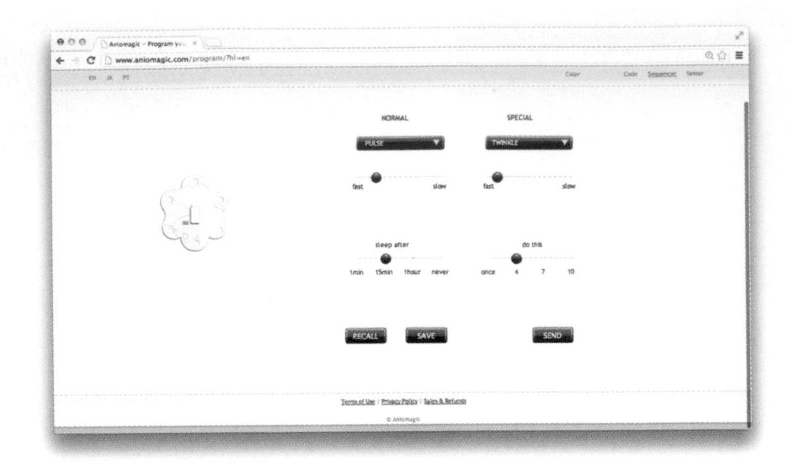

Figure 4-27. *Programming Aniomagic*

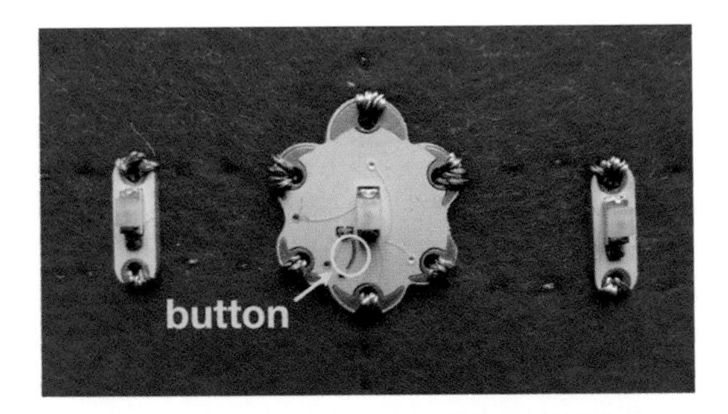

Figure 4-28. *Aniomagic programming mode "button"*

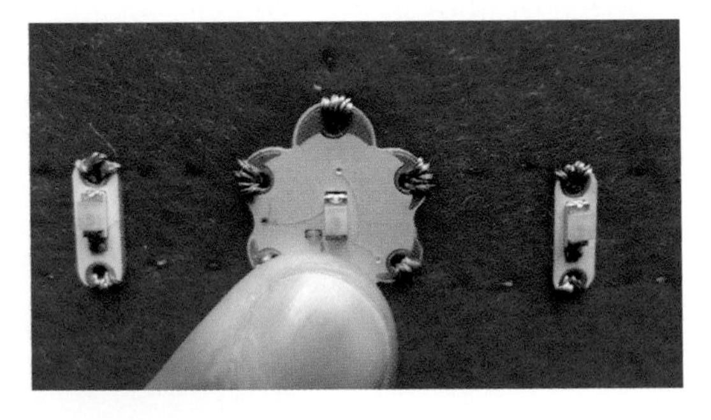

Figure 4-29. *Aniomagic—pressing "button" to enter programming mode*

Hold the Sparkle Board up to the screen so it is facing the illustration of the Sparkle Board (see Figure 4-30). It's helpful if it is quite close to the screen and if the room you are in is not too bright.

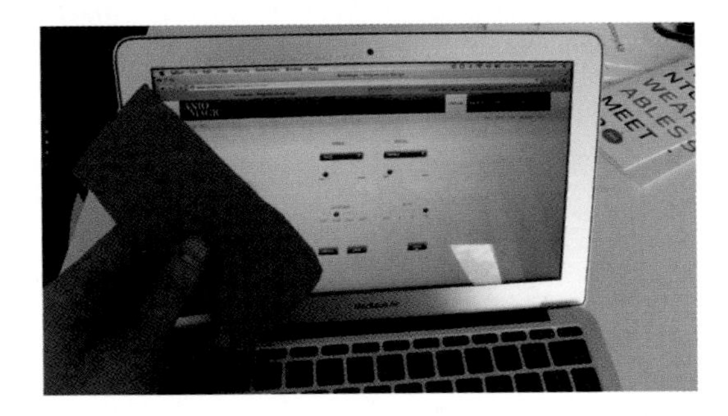

Figure 4-30. *Aniomagic—holding circuit up to screen for reprogramming*

Press the "Send" button. The illustration of the Sparkle Board should transform into a series of rapidly flashing shapes. Once the shapes stop flashing, take the Sparkle Board away from the screen. The red LED should now be steadily lit (Figure 4-31) and the program should be running!

Figure 4-31. *Aniomagic "Program is running" LED*

Figure 4-32. *Aniomagic Sparkle circuit in action*

If your Sparkle Board does not respond to programming, you can try resetting it. Use a pair of tweezers or some wire to bridge the two pins shown in Figure 4-33.

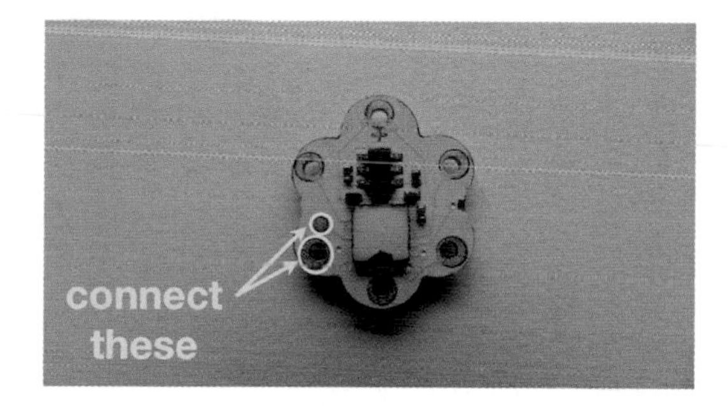

Figure 4-33. *Resetting Aniomagic—use a pair of tweezers to bridge these two connections*

You can also use the small red LED to find out the status of the Sparkle Board, as shown in Table 4-2.

Table 4-2. Sparkle Board modes

Red LED behavior	Board mode
Always on	Program is running
Flashing quickly	Waiting for a new program
Flashing once a second	Sleeping

Figure 4-34. *Aniomagic bracelet (image courtesy of Aniomagic)*

Thinking Beyond

E-textile toolkits expand your options for integrating electronics into soft and wearable projects. But it is also important to think beyond these existing toolkits. If you're just getting started, grabbing a toolset like the LilyPad or the Flora can be really handy to get a project going, particularly if you start with something like the LilyPad Protosnap Kit. In the long run, there's probably not a wearables or e-textile toolkit that precisely meets the needs of your vibrantly unique idea. As you move through iterations of a project, don't be afraid to move between platforms and beyond them. The more knowledgeable and adaptable you become with your tools, parts, and materials, the more you will be able to mold your project to fit the curves and nuances of the human form and create entirely novel wearable solutions!

Making Electronics Wearable

5

There are many challenges you don't anticipate in designing wearable electronics, which is why it is so important to actually wear your prototypes—early and often. In this chapter, I explore what makes something wearable and how to incorporate that into your own designs.

Why Wear It

When embarking on a wearable electronics project, the first and most important question you should ask yourself is this:

"Why does this need to be wearable?"

There are many possible answers to this question, but it's important that you have at least one in mind. Here are a few:

- You are sensing the body in such a way that the sensor needs to be placed on the body.
- You have a display or feedback mechanism that needs to stay with a person at all times.
- Your project needs to travel with the user and not stay in one place.
- You want to create a particularly intimate or immersive experience for the wearer.
- Your project is specifically clothing-oriented, such as a costume or fashion piece.
- You're interested in the future of wearable electronics and want to use making as a way to think about what's next.

Once you have a reason in mind, you can use it to guide the decisions that follow as you design your project.

What Makes Something Wearable

There are many factors to consider when converting an electronic circuit to a wearable form. What works on a breadboard or in a project box doesn't always translate so well to the dynamic, unpredictable, and rugged context of the human body.

Figure 5-1. *"Untitled Wearable 1" by Alex Beriault (photographed by Dax Varona)*

Let's take a look at what you should be thinking about when designing electronics that live in clothing.

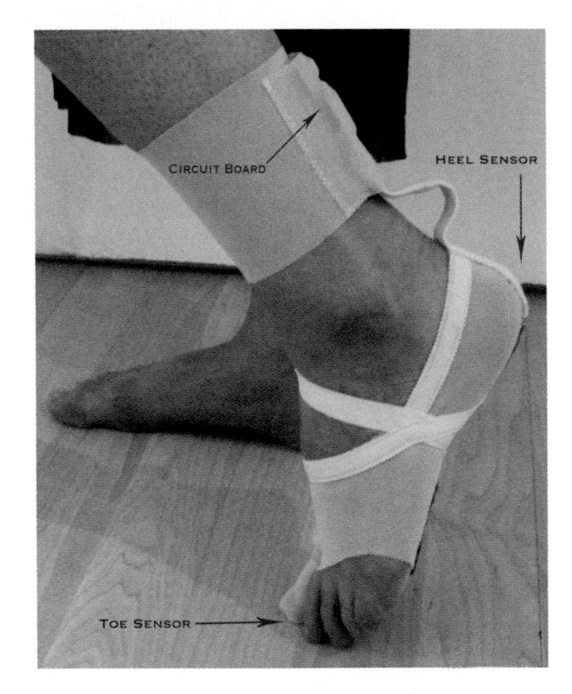

Figure 5-2. *Loretta Faveri's "SoMo" is a wireless, wearable sensor that generates sound through movement; the design of this wearable takes into account size, weight, placement, and strain relief*

Comfort

Here are a few aspects of your circuit to consider when striving to make your wearable electronics comfortable.

Size, weight, and shape

How large, heavy, and bulky are the electronic components that are included in your connections? How much surface area do they take up? How much do they protrude from the body? Do they conform to the body's natural shape? And are they able to move with or accommodate the movements of the body? These are good questions to ask with your prototype.

Generally speaking, small and lightweight packages with curved shapes work best in the body context. This is why many wearables-oriented circuit boards are designed with rounded edges and corners. Curves are more comfortable to wear.

Placement

Comfort can be greatly influenced by where components and connections are placed in a wearable. When placing components, consider how the body moves. Are components in a place that is unobtrusive or is it directly in the line of fire? Are they likely to be protected or subject to constant abuse?

Here are some base guidelines to consider when placing components:

- Keep heavier items close to the core (i.e., the torso, shoulders, or thighs). If a heavier item needs to go out on a limb (no pun intended) make sure it is secured directly to the body (like a wrist watch) rather than flapping around on a piece of cloth.
- When possible, run connections along seams and edges (see Figures 5-3 and 5-4). These areas very naturally accommodate some extra material.
- Pockets offer excellent support and protection for electronic components. You can either work with existing pockets in a garment or create your own.
- Linings also provide lots of opportunities. They can either hide what is going on underneath, or you can use them as a substrate for your electronics so the circuit is removable from the garment.

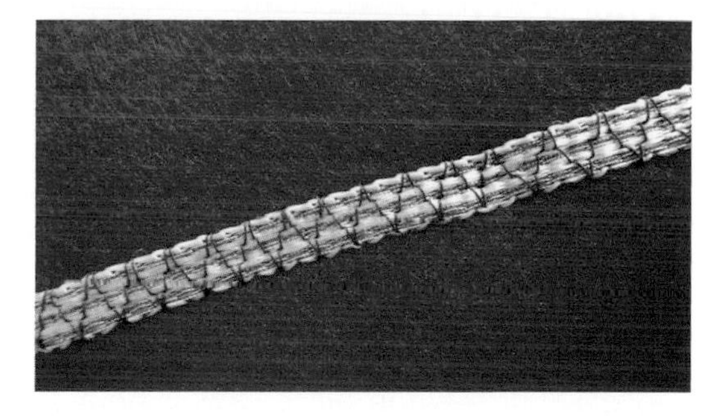

Figure 5-3. *Conductive ribbon secured with a zigzag stitch; this can work well along a seam or edge*

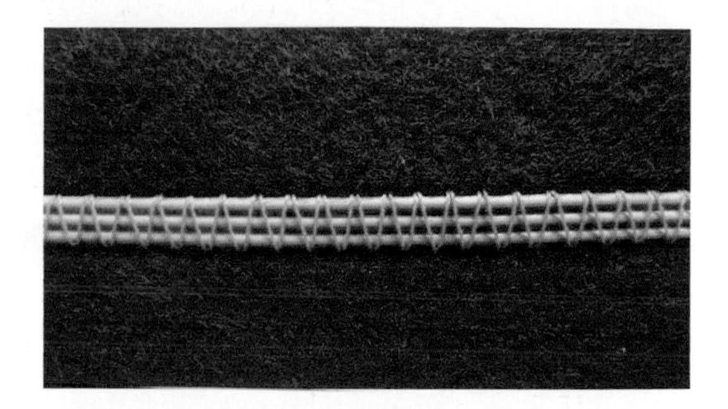

Figure 5-4. *Ribbon cable secured with a zigzag stitch*

Durability

Circuits that live in the body space need to be pretty tough. Bodies bend, squish, bang, tug, and stretch, which is a lot to ask of a circuit. When you make a circuit that's meant to be worn, you need to ensure that it can stand up to wearing, washing, and repairs.

Strain relief

A connection that is continually tugged is likely to eventually break. The way to prevent this is with strain relief.

First, make sure there is ample material to accommodate the full expected range of motion. If you're running a wire along an arm, make sure there's enough to cover the distance when the arm is flexed, not just when the arm is straight.

You can also take measures to relieve the connection of the strain. With wire, this can be accomplished by making a small loop close to the connection and sewing it in place (Figure 5-5). This way any strain is put on the wire, not on the solder joint.

Figure 5-5. *This secured loop of wire provides strain relief for the solder joints of the flex sensor*

Insulation

When creating soft circuits, you have to be vigilant about insulation, as many of these materials are not insulated by default.

Your circuit design should, of course, lay out the conductors so that they do not touch each other. You also need to consider how this layout will perform in the context of the intended use. Clothing is inherently floppy. If you take the garment off and toss it in a pile, could it inadvertently create a short circuit?

There are many material approaches that can reduce the likelihood of short circuits:

Layout
: The physical layout of the circuit itself can prevent conductors from touching.

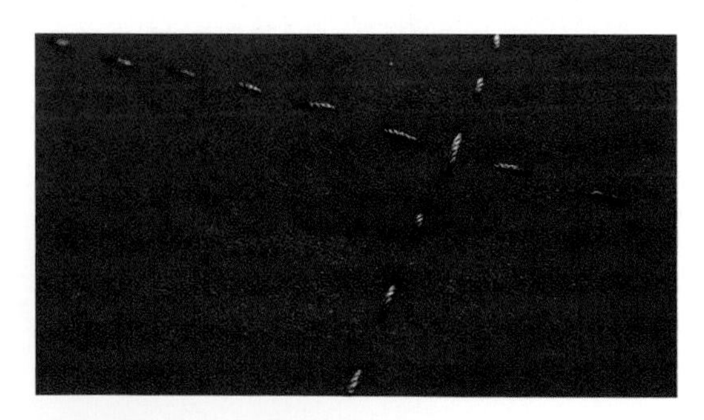

Figure 5-6. *When crossing conductive thread stitches, going on top of the fabric with one thread and below it with the other enables the nonconductive substrate to act as an insulator between the two conductors*

Stitching
: Nonconductive thread can provide a light layer of insulation in key spots. A satin embroidery stitch or zigzag machine stitch can cover a line of conductive thread and shield it from contact with others. Check out Instructables (*http://www.instructables.com/*) for tutorials on how to get up and running with these stitches.

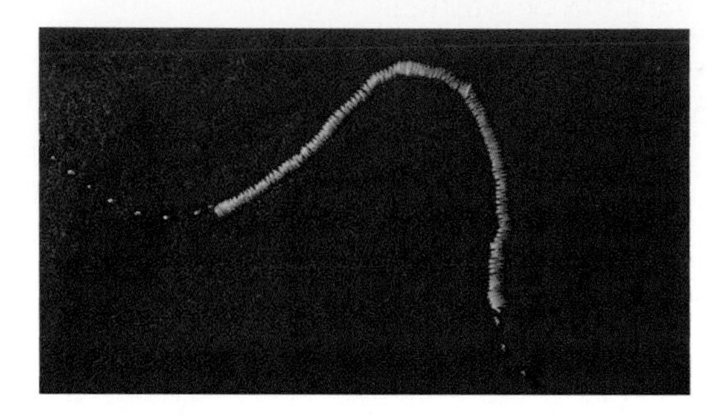

Figure 5-7. *Zigzag stitch used to insulate conductive thread*

Coatings
: Fabric glue, fabric paint, nail polish, and other coatings can provide spot insulation where it's needed.

Figure 5-8. *Fabric paint used to insulate conductive thread*

Layers and linings
: Nonconductive fabric can protect patches or traces of conductive fabrics with an overall layer of insulation. It can be secured with iron-on adhesive or sewn in place. This is helpful for when the garment is folded, crumpled, or handled in an unexpected way.

Modularity

Designing a circuit that contains removable modules can help extend the life and increase the durability and practicality of a piece of wearable electronics. Being able to easily replace certain components means you can both customize and repair a circuit. The ability to remove sensitive components like batteries also makes it easier for the wearable to be washed.

In order to make something modular, you need to work with good connectors so that pieces of the circuit can be easily added and removed.

The worlds of electronics and sewing are full of potential connectors. Take some time to pore through the Digi-Key or Molex catalog or look up a connector that you like on an existing board. Take your multimeter to the fabric store and spend some time in the notions section looking for conductive connectors. Depending on what you need connected and where, there's likely a variety of options available.

Protection

Are there any parts of your circuit that require protection? A layer of foam, batting, or felt could protect exposed header pins. A waterproof case or coating can protect your circuit from the elements. A lining can protect an exposed circuit from bare skin.

Usability

For a wearable electronics project to be considered truly "wearable" it should not only be comfortable to wear: it should also be comfortable to *use*. But what does that mean? You can break it down into the following questions:

- Does it function well as the wearable it is intended to be?
- Do the electronics function as expected?
- Does it make for a "good" or "satisfying" or "successful" experience for the wearer?

The reason that these factors are important is because wearables inhabit an intimate space. Wearables that work well work with you. They start to feel like a part of you. Wearables that don't work well can feel like a invasion of personal space.

These questions can potentially be answered both by wearing the project yourself and by user testing with others. Wearing a project yourself can provide you with some quick and easy answers and also gives you the benefit of firsthand experience. User testing is a great way to get feedback from others as well as to identify and resolve any of your own biases and assumptions that have made their way into your design. You will likely learn things about your project that you never expected.

Figure 5-9. *User testing the "Telepathic Motion-Sensitive Cat Vest" by Calliope Gazetas*

Aesthetics

How wearables look matters. They are objects that occupy your most intimate spaces. They are an extension of both your embodied experience and your sense of self. How they look influences how you use them, when you wear them, how you relate to them, and what kind of emotional attachment you have to them.

Think about how you would like your wearable to look and feel. You may want it to be fuzzy and cozy, sleek and fashionable, or techy and sci-fi—the choice is yours.

When embedding electronics in wearables, you are often faced with the question of whether to hide or reveal them. Hiding increases the opportunity for these technogarments to be seamlessly integrated into your existing habits and styles. But revealing has its own stylistic and functional advantages. It reminds both the wearer and the viewer that there is more going on. And depending on how the integration is handled, it has the potential to add a wow factor.

Figure 5-10. *"The Vega One Jacket" by Angella Mackey (also featured on the cover of this book) gives no indication that there are electronics present in the jacket until they are turned on (photographed by Henrik Bengtsson)*

Know Your Wearer

Is your wearer a grandmother, a chihuahua, a seven-year-old, or a CEO? What makes something wearable will also depend on who is wearing it. Always have the needs of your particular wearer in mind when making design decisions about a new wearable electronics project.

Designing a Wearable

Now that you know what makes something wearable, you can start designing your own. Here are aspects to consider as part of your process.

Choosing a Form

Wearable electronic projects can take on a variety of forms, including jumpsuits, wristbands, glove projects, hats, scarves, socks, jewelry and even singing underpants.

That's all well and good, but how do you choose a wearable form to work with? If you have a circuit that you want to wear, you may have a sense of where on the body you want to wear it but may not have a sense of *how*.

Choosing how to house your electronics is a crucial part of the wearable electronics design process. It provides the canvas on which you can start to plan your circuit. And as your designs become more sophisticated, it might even mean incorporating the design of your circuit into the materials themselves.

Hacking wearables

Seams, pockets, linings, oh my! The way clothing is traditionally designed actually offers a lot of opportunities for wearable technology. Take a look at the clothing you are wearing right now. Assess the nooks, crannies, and pieces of real estate that might be available for electronic components (Figure 5-11).

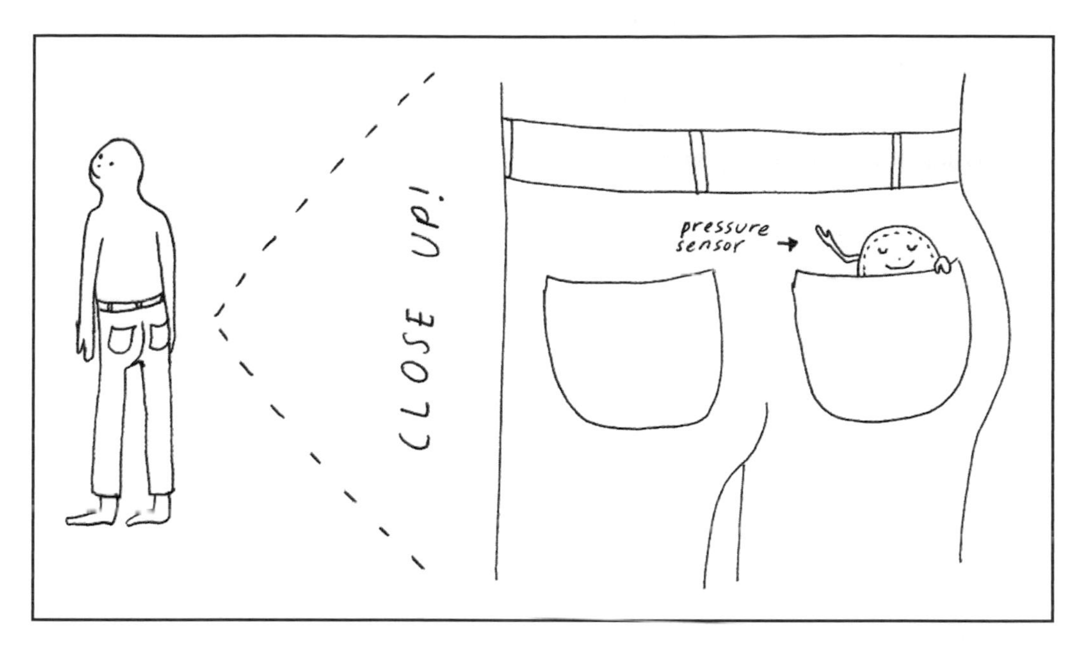

Figure 5-11. *The back pocket in a pair of jeans is an excellent place to put a pressure sensor meant to detect sitting (illustration by Jen Liu)*

Hacking existing clothing and wearable forms is a great way to get a prototype up and running. Thrift stores, discount clothing stores, or even "give away" piles can be great places to start when you need a base for the prototype that you're working through.

A hoodie is an example of a garment that is fairly easy to modify. You can add on to it, creating additional pockets, or a lining to better accommodate or incorporate circuitry. Or you can use a seam ripper to open a seam and elegantly modify the existing design (see Figures 5-12 and 5-13).

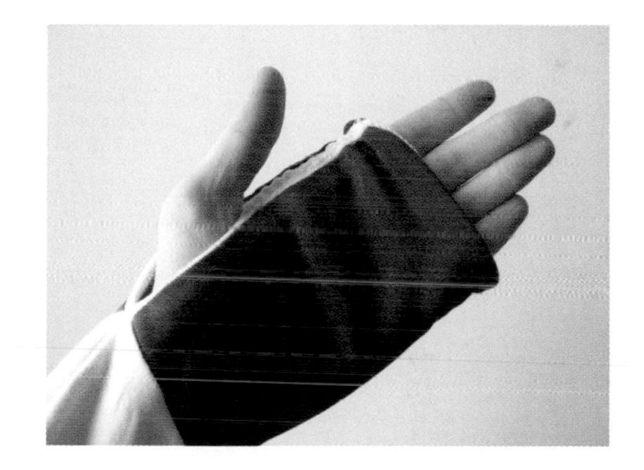

Figure 5-12. *In this example, the cuff of the hoodie sleeve has been extended and modified to include a conductive fabric contact point enabling the hand to act as one half of a soft switch*

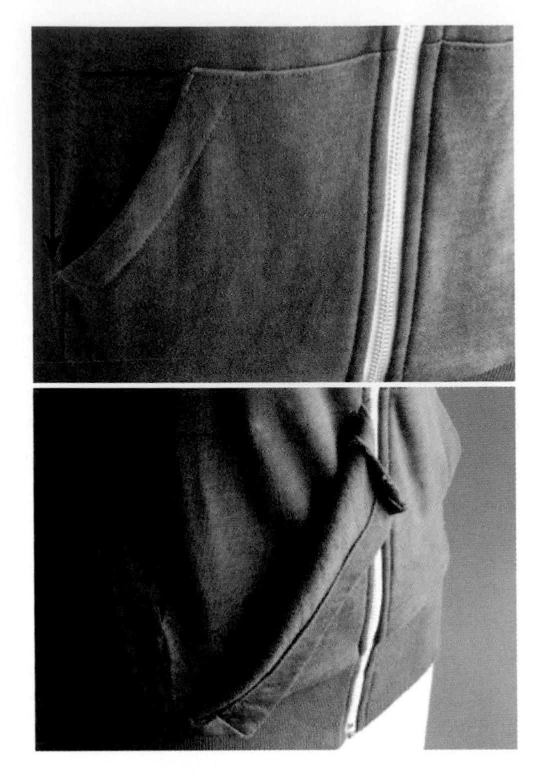

Figure 5-13. *The seams on a hoodie pocket can be opened to provide easy access to the pocket area*

Figure 5-14. *In "Energy Harvesting Dérive," Christian Croft and I hacked a pair of roller shoes so that the turning of the wheels would spin a generator*

Making wearables

If you are a seamstress, fashion designer, leather worker, industrial designer, or jewelry maker, you may be comfortable creating garments or wearable accessories from scratch. This provides an excellent opportunity to incorporate electronics into the design of the wearable itself.

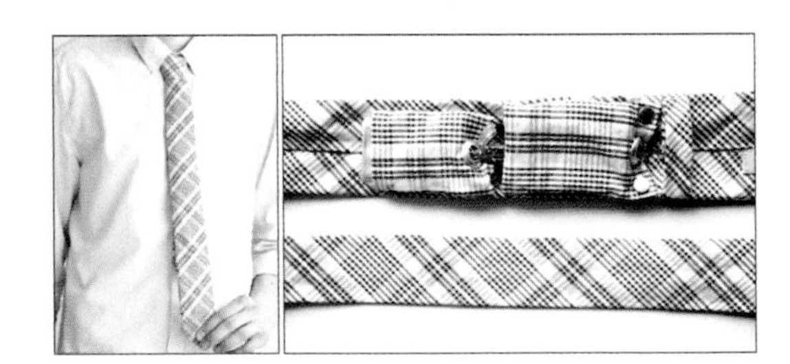

Figure 5-15. *This custom necktie by fashion designer Mystica Cooper seamlessly integrates electronics into the design of the tie*

Figure 5-16. *This bracelet by Leah Buechley incorporates LEDs through the use of beadwork*

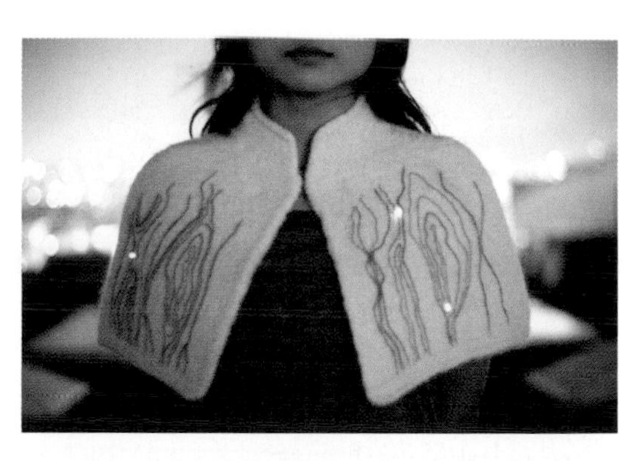

Figure 5-18. *"Soft Electric" by Grace Kim reveals the circuit but elegantly incorporates the conductive thread traces into the aesthetic of the felted cape (photographed by Jeannie Choe)*

Figure 5-17. *"Transformative Textiles" by Oldouz Moslemian uses weaving to integrate fiber optics into the custom-designed material for a pleated dress; fiber optics are only present in the interior of the pleats (photographed by Peter Hoiss)*

Figure 5-19. *"Bubble Pop Electric" by Joanne Jin uses a machine embroidery technique to integrate LEDs into a necklace*

Figure 5-20. *"Muse," the brain-sensing headband by Interaxon, is an example of a newly invented wearable form; this EEG headset is meant to be stylish enough to wear on the go*

Collaboration

One of the most enticing aspects of working with wearable electronics is that it is such an interdisciplinary practice. While you may have a wide range of talents, there is still a strong likelihood that you don't have all of the skills you need to produce your ideal wearable electronics project.

Find someone in your community to work with who has the skills you lack. Interdisciplinary teams are the strongest and can significantly contribute to the long-term success of a project. Plus you'll probably learn a lot along the way!

Choosing Materials

It is often useful to construct your circuit with both hard and soft materials. Wearables usually introduce a variety of design constraints, and using a hybrid approach can help you meet all of your needs.

You can see this strategy reflected in various e-textile kits that have been developed. With the LilyPad Arduino, there isn't a sewn connection between the microcontroller and every resistor, capacitor, and LED. But stitching three traces between the LilyPad Arduino board and a LilyPad Light Sensor is a lot more reasonable and makes more sense if the LilyPad Arduino lives on the shoulder and the light sensor on the cuff of the sleeve.

Revisit "Constructing Circuits" on page 11 to refresh your memory on the advantages and disadvantages of various circuit construction methods. Overall, hard circuits are excellent for creating small, complex, and robust circuits. Soft materials are advantageous when you need circuits that are simple, pliable, flexible, and comfortable.

Choosing Components

The materials you choose will help to determine the types of components you want to work with and vice versa. When it comes to wearables, circuit boards that are relatively small, flat, and smooth tend to be the easiest to integrate. But look at your wearable as well as the space on the body where different parts of your circuit will live. Some areas have more real estate than others.

Printed circuit board design is a highly useful skill for building wearable electronics. However, it is a more advanced skill, so if you're just getting started, simply being thoughtful about your choice of components can get you a long way.

For instance, if you are using a circuit that includes both an Arduino and an XBee radio, you could use a LilyPad Arduino and a LilyPad XBee. But if you're creating something meant to live on the wrist, that solution takes up an awful lot of space. Using an Arduino Fio will save you some room, reduce the number of connections, and streamline your prototype (see "Hello XBees" on page 200).

Creating a Layout

Designing circuits on a circuit board and designing circuits for wearables are two entirely different practices. With wearables, it is essential to really start thinking about circuits in a three-dimensional way. For this reason, it is important to *plan* the layout of your circuit before your start incorporating it into your garment.

There are a few ways to do this. You can lay your circuit out on paper or on screen. Those with a fashion or design background might enjoy sketching the garment and how the components and conductors will be organized on it. Keep in mind that you will definitely need multiple views and will likely need to think about layers of materials.

You can also work things out physically. Take the garment itself (preferably on a dress form, mannequin, or fellow human) and lay out the circuit with tailor's chalk or some paper mockups of components and straight pins. You can even use stickers (Figure 5-21).

Figure 5-21. *SparkFun sells handy LilyPad stickers so you can stick components in different places to try different layouts*

Keep in mind that with soft circuits you really need to consider the layout of your conductive traces. Will they need to cross? If so, what is your plan for insulating them? You can layout your circuit on paper or on screen (Figures 5-22 and 5-23).

You also want to consider how you'd like to group your components. What needs to live where? Items like sensors or actuators may require very specific placement, whereas items like microcontrollers and batteries may be able to be hidden in more spacious or discrete areas of the garment.

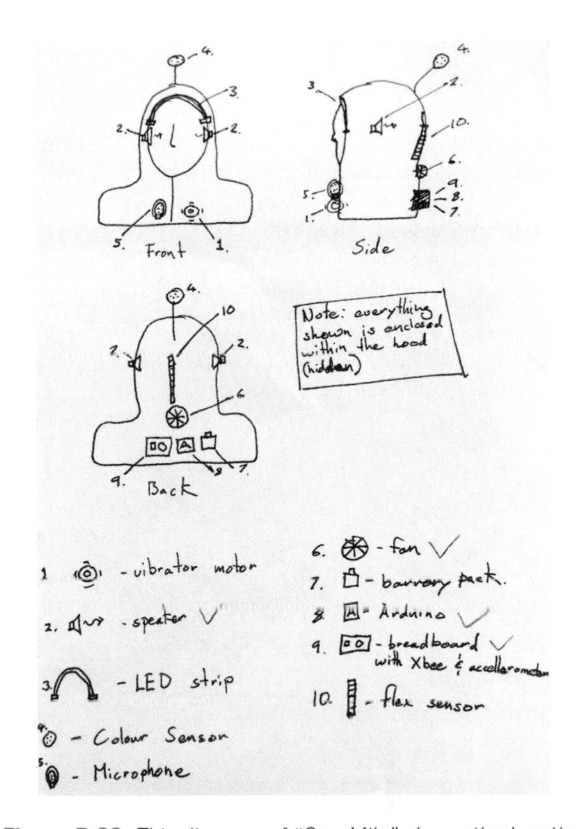

Figure 5-22. *This diagram of "One Mile" shows the location of various components; "One Mile" is a project by Hudson Pridham, Maziar Ghaderi, and Yuxi Wang*

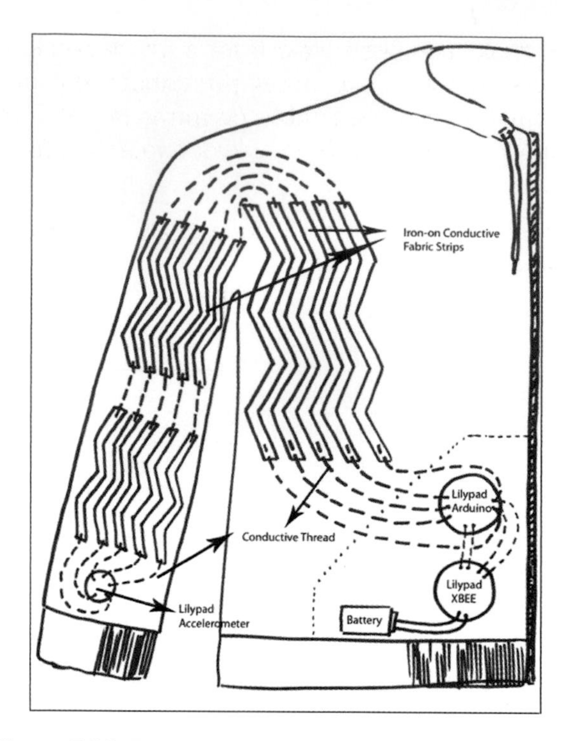

Figure 5-23. *This diagram of the "Audience Jacket" by the Social Body Lab includes both the components and the layout of the connections between them*

Iterative Design

You're never going to get it totally right the first time. Once you create a first prototype of your wearable, it's essential that you wear it, or that you have someone else wear it. Some of your design choices will likely work quite well, but it's also quite likely that there will be things you didn't expect. Be sure to take notes—there's a lot to learn from seeing the way something performs with actual use.

Once you've had a chance to observe the design in use, make some revisions and create a second prototype. And a third. Committing yourself to multiple iterations will line you up for a much stronger, well-informed, and robust project in the long run.

Maintaining Access

Don't forget to leave a backdoor! When incorporating your circuit into your garment, be sure not to enclose your circuit completely. You will need access in order to replace the battery, make adjustments, or make repairs. For example, you can use pockets to allow for access to circuitry:

Experiment: Eight-Hour Wearable

The easiest way to learn how to make something wearable is to wear it. In Chapter 1, you constructed a circuit at least two different ways. Now you're going to take the same circuit and make it wearable.

Select a way in which you would like to wear the circuit. Design a layout for the circuit, construct the wearable, and then wear it for a full day—eight hours straight. Take notes throughout the day about how your wearable performs in different contexts of your life. Then use your notes to inform the next iteration.

Once the second version is done, go ahead and take it out for another spin!

Figure 5-24. *An eight-hour wearable test (illustration by Jen Liu)*

Microcontrollers

6

Hello, microcontrollers! In this chapter, you'll begin to explore your options for microcontrollers that can be embedded in clothing. I cover both how to build the circuits as well as how to create programs that bring the circuits to life.

Here are the parts you will be be using in this chapter (Figure 6-1):

- LilyPad Arduino Simple (SF DEV-10274)
- LilyPad LED (SF DEV-10081)
- LilyPad Button (SF DEV-08776)
- LilyPad Light Sensor (SF DEV-08464)
- FTDI Board (AF 284, SF DEV-10275)
- USB mini-B cable (AF 899, DK WM5163-ND, RS 55010682, SF CAB-11301)
- Alligator clip test leads (AF 1008, RS 278-1156, SF PRT-11037)

You may want to also check out the following optional parts:

- Through-hole LED (DK 160-1703-ND, SF COM-09594)
- 220Ω resistor (DK 220QBK-ND, RS 271-1313)
- 10KΩ resistors (DK 10KQBK-ND, RS 271-1335, SF COM-08374)

Figure 6-1. *LilyPad Arduino Simple, LilyPad LED, LilyPad Button, LilyPad Light Sensor, LilyPad FTDI Board*

A microcontroller is basically a tiny computer. You can think of it as the brain of your project. You may be more familiar with computers that come in the form of a desktop or laptop device. Within the last few years, you've even become accustomed to smaller computers that take the form of smartphones and tablets. But what if a computer could live in your clothing or other things that you wear on your body?

While wearable computing has been an area of research for many years, it's only just recently that it has entered the realm of consumer products.

Microcontrollers have also become much more popular with hobbyists and makers due to their decrease in price and increasing availability and accessibility.

Microcontrollers are computers in their most basic form. This makes them an excellent tool with which to get started exploring how computation can live in the body space.

In this book, you will work with Arduino and Arduino-compatible products to meet your microcontroller needs. Arduino is an open source electronics prototyping platform intended to be used by artists, designers, educators, hobbyists, and basically anyone who wants to make a physical interactive project but isn't an electrical engineer. The name "Arduino" refers to both the hardware and the software. Let's start with the hardware.

Hardware

Arduino boards are printed circuit boards that contain a microcontroller and its related components and circuits. This includes pin breakouts, status LEDs, a reset button, and more. This makes it easy to get microcontroller circuits up and running quickly without the fuss of building out these nitty-gritty aspects of the circuit yourself.

There is a wealth of Arduino boards in a variety of configurations. You can see the range of what's currently available on Arduino's "Products" page (*http://arduino.cc/en/Main/Products*) (Figure 6-2). The Arduino Specs Comparison (*http://arduino.cc/en/Products.Compare*) page provides detailed information about the differences between Arduino Boards (see Figure 6-3).

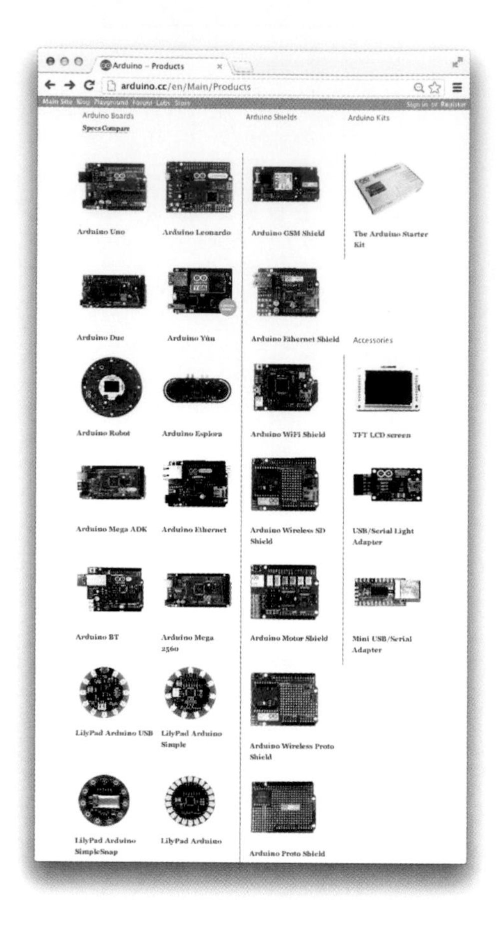

Figure 6-2. *The Arduino "Products" page*

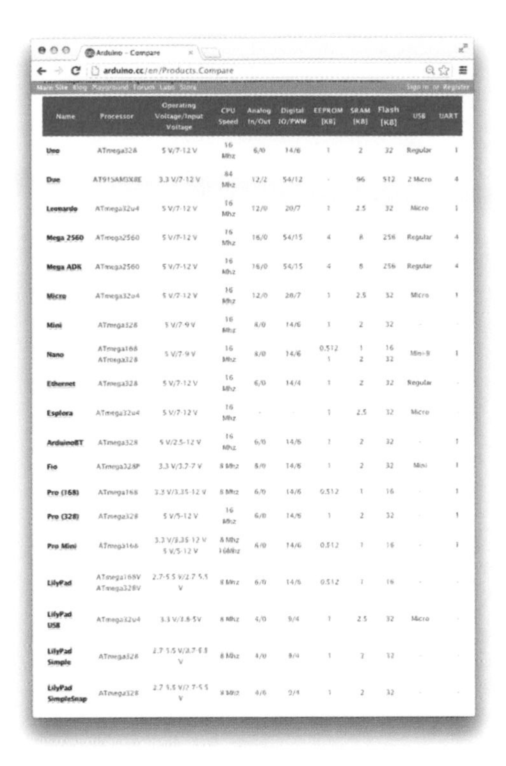

Arduino - Compare

arduino.cc/en/Products.Compare

Name	Processor	Operating Voltage/Input Voltage	CPU Speed	Analog In/Out	Digital IO/PWM	EEPROM [KB]	SRAM [KB]	Flash [KB]	USB	UART
Uno	ATmega328	5 V/7-12 V	16 Mhz	6/0	14/6	1	2	32	Regular	1
Due	AT91SAM3X8E	3.3 V/7-12 V	84 Mhz	12/2	54/12	-	96	512	2 Micro	4
Leonardo	ATmega32u4	5 V/7-12 V	16 Mhz	12/0	20/7	1	2.5	32	Micro	1
Mega 2560	ATmega2560	5 V/7-12 V	16 Mhz	16/0	54/15	4	8	256	Regular	4
Mega ADK	ATmega2560	5 V/7-12 V	16 Mhz	16/0	54/15	4	8	256	Regular	4
Micro	ATmega32u4	5 V/7-12 V	16 Mhz	12/0	20/7	1	2.5	32	Micro	1
Mini	ATmega328	5 V/7-9 V	16 Mhz	8/0	14/6	1	2	32	-	-
Nano	ATmega168 ATmega328	5 V/7-9 V	16 Mhz	8/0	14/6	0.512 1	1 2	16 32	Mini-B	1
Ethernet	ATmega328	5 V/7-12 V	16 Mhz	6/0	14/4	1	2	32	Regular	-
Esplora	ATmega32u4	5 V/7-12 V	16 Mhz	-	-	1	2.5	32	Micro	-
ArduinoBT	ATmega328	5 V/2.5-12 V	16 Mhz	6/0	14/6	1	2	32	-	1
Fio	ATmega328P	3.3 V/3.7-7 V	8 Mhz	8/0	14/6	1	2	32	Mini	1
Pro (168)	ATmega168	3.3 V/3.35-12 V	8 Mhz	6/0	14/6	0.512	1	16	-	1
Pro (328)	ATmega328	5 V/5-12 V	16 Mhz	6/0	14/6	1	2	32	-	1
Pro Mini	ATmega168	3.3 V/3.35-12 V 5 V/5-12 V	8 Mhz 16Mhz	8/0	14/6	0.512	1	16	-	1
LilyPad	ATmega168V ATmega328V	2.7-5.5 V/2.7-5.5 V	8 Mhz	6/0	14/6	0.512	1	16	-	-
LilyPad USB	ATmega32u4	3.3 V/3.8-5V	8 Mhz	4/0	9/4	1	2.5	32	Micro	-
LilyPad Simple	ATmega328	2.7-5.5 V/2.7-5.5 V	8 Mhz	4/0	9/4	1	2	32	-	-
LilyPad SimpleSnap	ATmega328	2.7-5.5 V/2.7-5.5 V	8 Mhz	4/6	9/4	1	2	32	-	-

Figure 6-3. *The Arduino "Specs Comparison" page*

The most common Arduino that beginners work with is the Arduino Uno. This is a basic Arduino with a reasonable amount of functionality (not too much, not too little) all in an accessible package. The only problem is that from a wearable electronics perspective, the Arduino Uno is quite bulky (see Figure 6-4).

Figure 6-4. *Arduino Uno (left) and LilyPad Arduino (right)*

The LilyPad Arduino is an Arduino in a LilyPad package. Like other LilyPad products, it has "petals" or "sewtabs" placed around the edge of the circuit board to facilitate electrical connections made using conductive thread. It uses the same microcontroller as the Arduino Uno and has the same number of inputs and outputs. It is intended for use in electronic textile and wearable electronics applications.

For the examples in this chapter, you'll be using the LilyPad Arduino Simple (see Figure 6-5). This is a simplified version of the LilyPad Arduino. Some pins have been removed to make it easier to create connections. Also, a JST connector has been added so that you can easily plug in a battery. An on/off switch has been added as well.

Figure 6-5. *LilyPad Arduino (left) and LilyPad Arduino Simple (right)*

There is a newer and older version of the LilyPad Arduino Simple. The newer version (DEV-10274) eliminates the ISP header (which is not normally needed) and adds a charging circuit for lithium polymer batteries. There is no marking to distinguish the older board from the new, but you can tell by the location of the on/off switch. In the old version, it sits to the left, just below the JST connector. In the new version (Figure 6-6), it is in the center directly below the microcontroller.

Figure 6-6. *The LilyPad Arduino Simple*

Figure 6-7. *Connectors on the LilyPad Arduino Simple*

Looking at the board, you can spot two connectors that you will use quite frequently, as shown in Figure 6-7. One is the set of FTDI headers. These are male headers that correspond to the female headers of the removable FTDI board. This board is what enables the LilyPad Arduino to communicate with your computer via USB. The second is the JST connector. This offers a quick and secure way to connect a battery to your Arduino circuit.

Around the perimeter are the "sew tabs" that I mentioned earlier. These are labeled and correspond to various pins on the microcontroller. You can see a breakdown of the accessible pins and their functions in Figure 6-8 and Table 6-1.

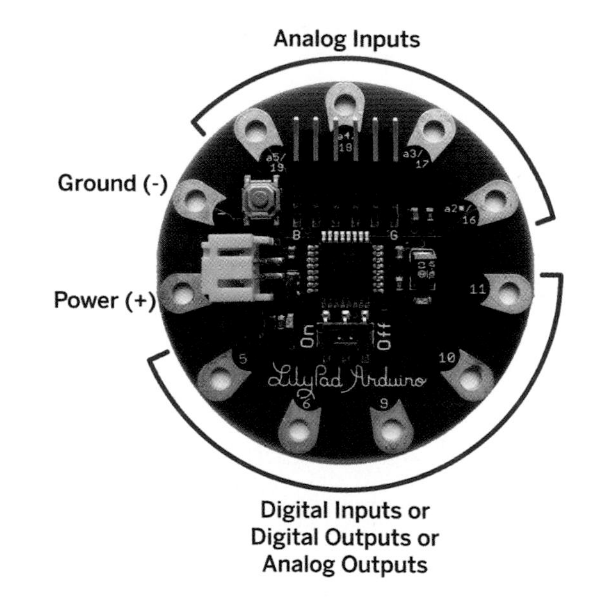

Figure 6-8. *LilyPad Arduino Simple pins*

Table 6-1. LilyPad Arduino Simple pins

Pin Label	Functions
–	Ground
+	Power
5	Digital input/output, PWM
6	Digital input/output, PWM
9	Digital input/output, PWM
10	Digital input/output, PWM
11	Digital input/output, PWM
a2/16	Analog input, digital input/output
a3/17	Analog input, digital input/output
a4/18	Analog input, digital input/output
a5/19	Analog input, digital input/output

When connecting your LilyPad Arduino Simple to your computer, you will need a 5V FTDI board and a USB mini-B cable. It's possible you already own one of these cables. They come with most digital cameras these days. The USB mini-B cable looks like Figure 6-9.

Figure 6-9. *USB mini-B cable*

To prepare for programming, connect your FTDI board to the FTDI headers on the LilyPad Arduino Simple.Then connect the small end of the USB mini-B cable to the FTDI board. It should look like Figure 6-10.

Figure 6-10. *LilyPad Simple ready to program with FTDI board and USB cable*

Some Arduinos don't require FTDI breakout boards. See the LilyPad Arduino USB or the Adafruit Flora for a single-piece solution.

Finally, connect the other end of your USB cable to your computer. Your hardware is ready to be programmed!

Software

Now it's time to get ready to program. First, you need to download the Arduino software. You can find the version that is appropriate for your operating system (*http://arduino.cc/en/Main/Software*).

Next, you need to install the necessary FTDI drivers. You can find FTDI drivers for your operating system (*http://www.ftdichip.com/Drivers/VCP.htm*). Some Arduinos (such as the Arduino Uno) do not require FTDI drivers, but the LilyPad Arduino and LilyPad Arduino Simple do. If you do not install the drivers, you will not be able to program your LilyPad Arduino Simple.

Once your drivers are installed, restart your computer and then open your Arduino program. When you open the Arduino software, you will see the window shown in Figure 6-11.

ProtoSnap LilyPad Development Board

If you want to start by focusing on code and save the circuit construction until later, SparkFun makes a great product called the ProtoSnap LilyPad Development Board (MS MKSF9, SF DEV-11262). ProtoSnap boards include multiple components but create connections between them in the parts of the printed circuit board that would normally be scrap material:

When you first get it, you can use it as is to test your code. Once you want to create your project, you can snap the pieces apart and redo the connections using conductive thread. If you use the ProtoSnap LilyPad Development Board for examples in this chapter, just be sure to change the pin numbers in the code, as some of the connections will be different.

Working with the Flora

Though the examples that follow use the LilyPad Arduino Simple, the Adafruit Flora is an alternative option. Here are notes for how to do this by section:

"Hardware" on page 92
: Note that the Flora does not require the use of an FTDI board and can be connected directly to a computer using a USB mini-B cable.

"Software" on page 95
: The Flora requires a different version of the Arduino software. See Adafruit's Getting Started with Flora (*https://learn.adafruit.com/getting-started-with-flora/overview*) guide for details.

"Hello World" on page 98
: In the Adafruit-Arduino software, select "Adafruit Flora" as the board type and change the LED pin in the code from 13 to 7.

"Digital Output" on page 101
: In both the code and circuit, change from pin 11 to pin 6, 9, 10, or 12.

"Digital Input" on page 104
: In both the code and circuit, change from pin 5 to pin 6, 9, 10, or 12.

"Analog Input" on page 108
: In both the code and circuit, change from pin A2 to pin A7, A9, A10, or A11. Note that these pin numbers are not displayed as such on the Flora board. Use the Flora pinout diagram (*https://learn.adafruit.com/assets/2845*) to determine the correct connections.

"Analog Output" on page 110
: In both the code and circuit, change from pin 11 to pin 6, 9, 10, or 12.

Note: when changing pin numbers in any of these examples, be sure to use the same number in the code as in the circuit.

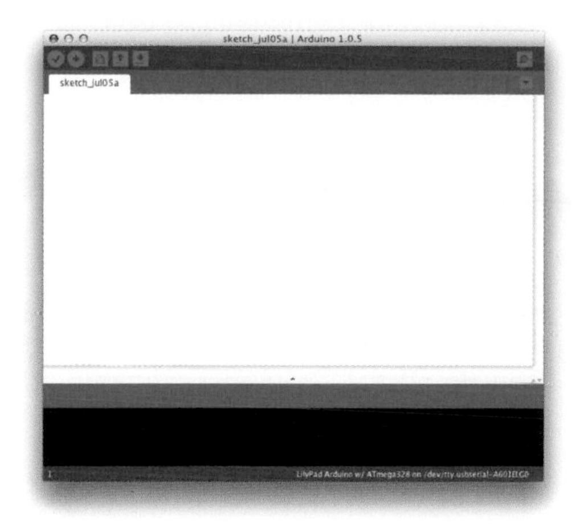

Figure 6-11. *Blank Arduino sketch*

As you mouse over the icons at the top of the window, you will see their various functions. They are as follows:

Verify

Checks the code and indicates if there are any errors in syntax.

Upload

Compiles the code and uploads it to the Arduino board. Once the code is uploaded, it will stay on the Arduino board even when it is unplugged from the computer.

New

Opens a new sketch.

Open

Opens an existing sketch.

Save

Saves your sketch. Note: there is also a "Save As" function available in the "File" menu for when you need to save different versions of your code.

Serial Monitor

Opens the Serial Monitor where you can view data that is being sent and received.

Here are some other things to know as you look at your screen:

- The white area is the text editor in which you will write your code.
- The blue strip below the white area is where you will see status updates when your code is uploading.
- The black area at the bottom is where you will see information about errors.

In Arduino, files are referred to as *sketches*. One of the nice things about working in Arduino is that there are lots of helpful example sketches. To access them, go to File → Examples (see Figure 6-12). This is where you can find a variety of examples to get you started.

Figure 6-12. *The File → Examples menu*

In addition to the example code included with Arduino, you can also find step-by-step explanations of these examples on the Learning page (*http://*

arduino.cc/en/Tutorial/HomePage) on the Arduino website.

To look at the most basic possible Arduino sketch, go to File → Examples → 01.Basics → BareMinimum (Figure 6-13).

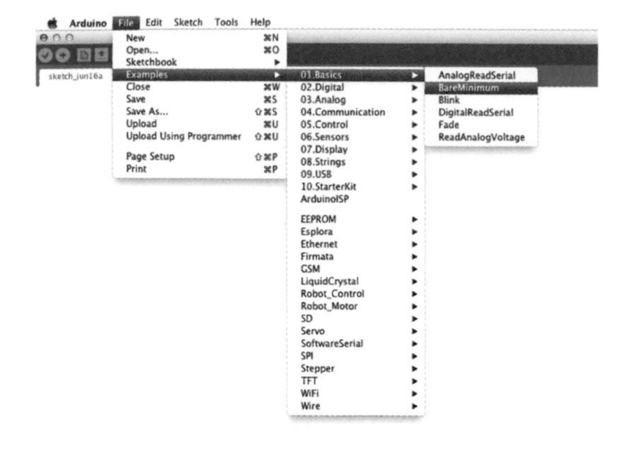

Figure 6-13. *BareMinimum example sketch*

This is the requisite skeleton of any Arduino program. It is a great place to start when you are writing a new sketch. Just remember to save it as a different file name so that you don't overwrite the example.

Things to know:

`setup()`
: This is where you put commands that are to happen only when the program first begins. These happen only once.

`loop()`
: This contains commands that will happen over and over again.

Comments
: Text that is meant to be read by humans but not by the microcontroller goes here. You will see comments both at the start of the program to provide notes, date, and attribution as well as throughout the program to explain what is happening along the way. Comments can also be used to remove lines of code that are not in use. A single line comment is preceded by //. A multiline comment falls between /* and */.

Two great resources for better understanding the Arduino environment and syntax are the "Arduino Development Environment" (*http://arduino.cc/en/Guide/Environment*) page and the "Arduino Reference" (*http://arduino.cc/en/Reference/HomePage*) page. While this book will review some programming strategies specific to wearables, it will not cover the details of Arduino programming. For more on this, you can also check out books like *Getting Started with Arduino* by Massimo Banzi and Michael Shiloh (Make) or *Arduino Cookbook* by Michael Margolis (O'Reilly).

Hello World

"Hello World" is a term used to refer to the simplest possible program that can demonstrate that the system is working. In a typical computer program, this program would write the words "Hello World" to a display device. In Arduino, the equivalent of this is a blinking LED. It's a little something that lets the Arduino say, "Hey world! Here I am!" It also lets you know that your hardware and software are configured properly. Let's get your Arduino to say, "Hello."

For the circuit, you don't need to do anything. There is already an onboard LED that is intended to be used expressly for this purpose.

For the code, there is an example sketch that will fit your needs nicely. Go to File → Examples → 01.Basics → Blink. This will open a sketch called "Blink," as shown in Figure 6-14.

Figure 6-14. *Opening the Arduino "Blink" example*

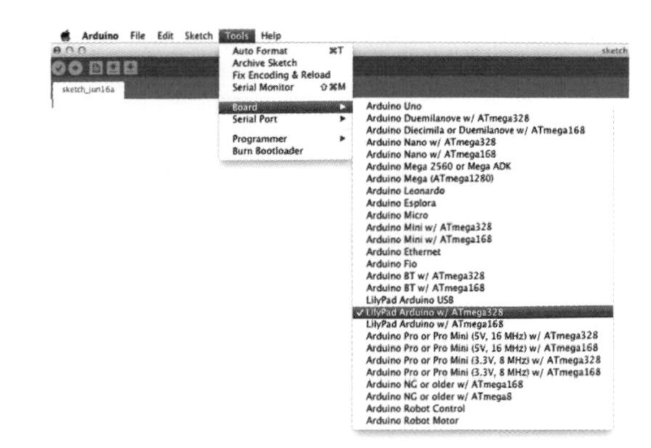

Figure 6-15. *Selecting the board type from the Arduino Tools menu*

Once the hardware is connected and the sketch is open, you need to make sure that everything is set up properly for this code to be uploaded to the Arduino. There are three things you should always check before attempting to upload a program to your Arduino board:

- USB connection
- Board type
- Serial port

I call these "the Magic 3." You should already have your board connected via USB but it's worth checking. It may sound obvious, but in the midst of programming you may have forgotten whether you've plugged your board in or not.

Setting the *board type* enables Arduino to compile the code in such a way that it will work properly on the type of Arduino board that you have. If you do not use the correct board type, your code will not compile properly and you will get an error. To set the board type, go to Tools → Board and from there you will see a list of board options. Depending on the board, this will sometimes include options for processor type or clock speed. For this example, select "LilyPad Arduino w/ ATMEGA 328," as shown in Figure 6-15.

Finally, you need to set your *serial port*. Go to Tools → Serial Port and from there you should see a list of options (Figure 6-16). The LilyPad Arduino Simple will not appear with a pretty name like "Kate's LilyPad Arduino Simple." Rather, it will likely look like "/dev/tty.usbserial-" followed by an assortment of letters and numbers if you are using a Mac, or "COM 3" if you are using a PC. The unique identifier actually corresponds to the FTDI USB-to-serial device rather than the Arduino, so if you swap Arduinos, but use the same FTDI device, that identifier will stay the same. What's important here is you are telling the Arduino program which USB serial port to send the program to. This prevents the Arduino software from attempting to communicate with your mouse, your Bluetooth headset, or your USB-powered mini-fridge where you keep your emergency stash of Fanta. Believe me—that conversation will not go well. If you are having trouble identifying which item on the list is the one you're looking for, you can always use the process of elimination. Look at the list, unplug the device, then look at the list again to see which item has disappeared.

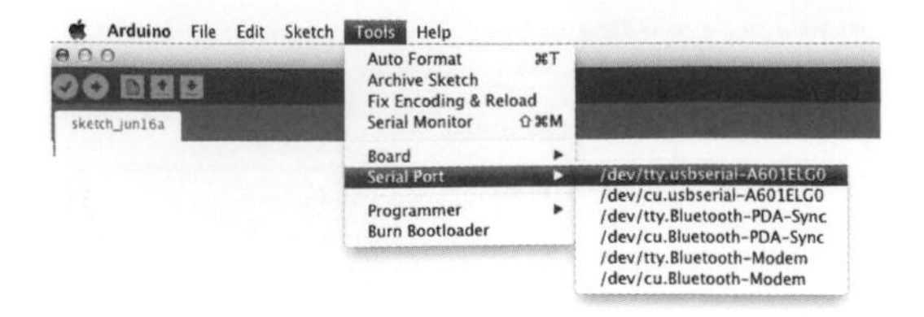

Figure 6-16. *Selecting the serial port from the Arduino Tools menu*

If you do not see anything in the list called "/dev/tty.usbserial-…" or "COM 3" and you are sure your board is plugged in properly, this likely means your FTDI drivers aren't installed. Try installing them again and be sure to restart your computer after doing so.

Once you've checked the Magic 3, you know you're ready to program the Arduino. Let's try to upload the Blink sketch to the Arduino. Here's how:

1. Find the Upload button. It lives in the menu button and is marked with an arrow pointing to the right.
2. Press it!
3. Watch the status update in the blue bar at the bottom of the window. It should change from "Compiling sketch" to "Uploading" to "Done Uploading." If you run into any snags, check out the Arduino Troubleshooting page (*http://arduino.cc/en/Guide/troubleshooting*).
4. Look at the small, green LED that lives on your LilyPad Arduino Simple. It should be blinking (Figure 6-17). Hello Arduino!

Figure 6-17. *LilyPad Arduino Simple with surface-mount LED on pin 13 lit*

Congratulations—you've successfully uploaded your first sketch!

Experiment: Gettin' Blinky

For your "Hello World" exercise, you jumped right in without making any adjustments to the code. You'll take a close look at digital output in the next section, but in the meantime, dip your toes in the water of code by making a few minor tweaks.

Take a look at the Blink example and find the lines that say this:

```
delay(1000);
```

This is what it sounds like—it is a function that delays the program for a set amount of time before it performs its next command. The number in parentheses is the length of the delay in milliseconds (1,000 milliseconds equals a second). So each delay that's used in this code is one second long.

Try changing the length of one of the delays. Upload the code to the Arduino board again and take a look at the behavior of the LED. Has the timing of its blink changed?

Spend some more time playing with the delay values and see what results you get!

Digital Output

Lighting the onboard LED connected to pin 13 is enough to make you say "YAY!", but you'll likely want to be experimenting with different LEDs soon after. Let's look at ways to add an additional LED.

The Circuit

On an Arduino Uno or LilyPad Arduino, you could just connect an external LED to pin 13, but it just so happens that 13 isn't accessible on the LilyPad Arduino Simple, so let's work with pin 11 instead.

First, you can try connecting a LilyPad LED. These are nice because they have everything you need in a compact package. Using alligator clips, make the connections shown in Figures 6-18 and 6-19:

LilyPad Arduino Simple	LilyPad LED
Pin 11	Power (+)
Ground (–)	Ground (–)

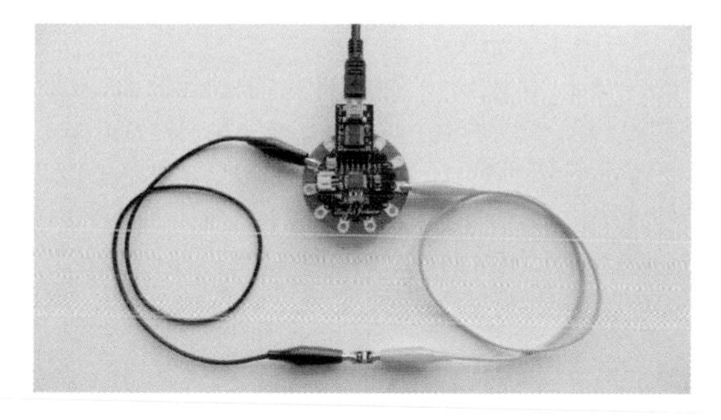

Figure 6-18. *Photo of LilyPad Simple with LilyPad LED on pin 11*

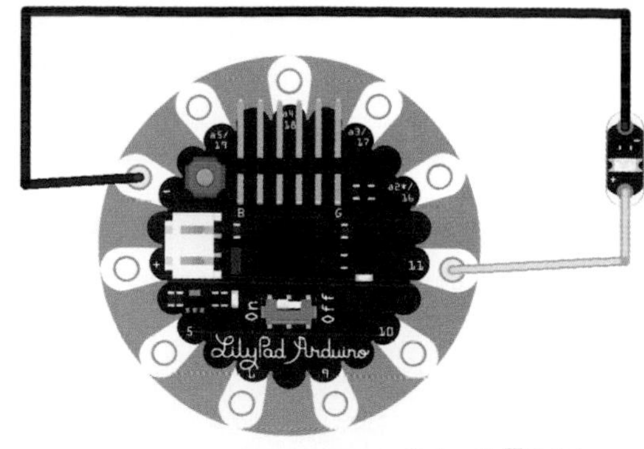

Figure 6-19. *Fritzing diagram of LilyPad Simple with LilyPad LED on pin 11*

Fritzing

Throughout the book, you'll see diagrams created using a circuit design software called Fritzing. While traditional circuit design software usually has a circuit diagram layer and a board design layer, Fritzing adds a third layer called the breadboard layer. This is intended to reflect how the physical circuit will look using prototyping tools like Arduinos, breadboards, and alligator clips. Their parts library includes LilyPad components, and you can also download a Fritzing parts library for Adafruit components including Flora products.

You can also use a through-hole LED, as shown in Figure 6-20.

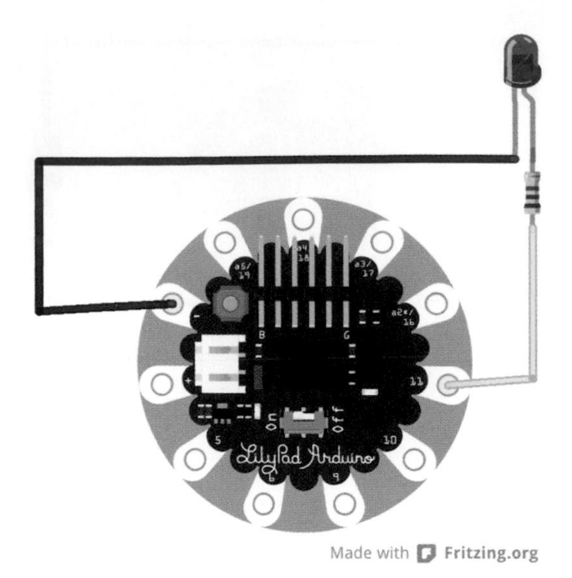

Figure 6-20. *LilyPad Simple with throughhole LED and 220Ω resistor*

Because the output pins on an Arduino can supply up to 40 mA of current, you can also connect two or three LEDs in parallel, depending on the LED (Figure 6-21). When working with LilyPad LEDs, the resistor is included so you don't need to add another.

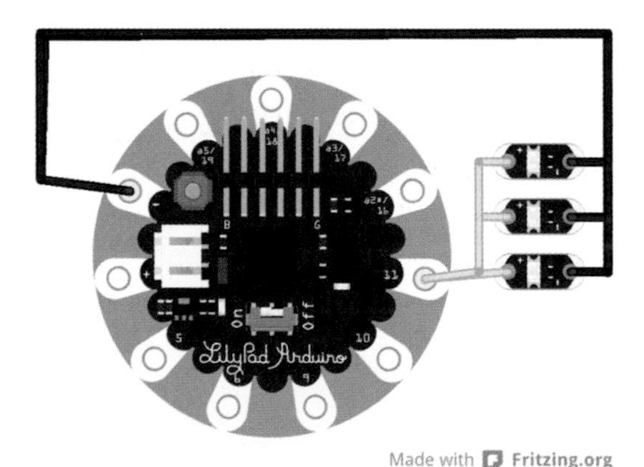

Figure 6-21. *LilyPad Arduino Simple with 3 LEDs in parallel controlled by pin 11 (because these LEDs are controlled by a single pin they will have the same behavior; the necessary resistors are included in the LilyPad LED package)*

The Code

Let's take a closer look at that Blink example. Here are some helpful things to know:

Variables
: These provide a way to name and store values. This could be a changing value, like a reading from a switch or sensor, or a constant value like a particular pin number that you will be using throughout the program. Variables are useful because if you decide to make a change, for example, to which pin you are using for an LED, you only have to make the change in one place in your code (the point at which you define the variable rather than every time you refer to the LED pin number). There are many different variable types that you can read about in the Arduino reference documentation (*http://www.arduino.cc/en/Reference/VariableDeclaration*). At the start of the Blink example, you can see that pin number 13 has been stored in a variable called `led`.

`pinMode(pin, mode)`
: Sets a digital pin as either an input or an output. The two parameters needed are the number of the pin and its mode (i.e., `INPUT` or `OUTPUT`). This command is included in the setup so that the pin's behavior is determined at the start of the program.

`digitalWrite(pin, value)`
: The command used to control a digital output pin. The first parameter is which pin you would like to address. The second is the value which can either be `HIGH` or `LOW`. `HIGH` will turn the pin on, sending out V+ at 40mA. `LOW` will turn the pin off.

Now that you have a better understanding of what's going on in the Blink example, go ahead and change the `led` variable to specify pin 11:

```
int led = 11;
```

Upload your new code and voilà! The newly connected LED should light accordingly.

Power

When you first upload and run these examples, your board will be receiving power from your computer via USB (Figure 6-22). But what if you unplug it?

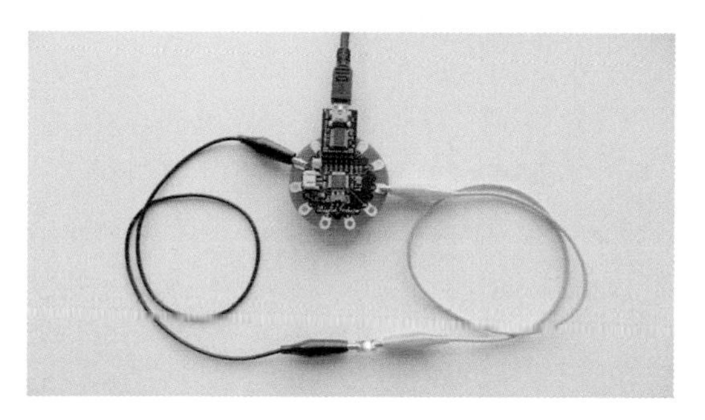

Figure 6-22. *LilyPad Arduino Simple circuit powered via USB*

One of the nice attributes of the LilyPad Simple is that it include a JST connector for battery connections. This board will accept an input voltage range of 2.7–5.5V. Any 3.7V rechargeable battery is a suitable power source, as shown in Figure 6-23.

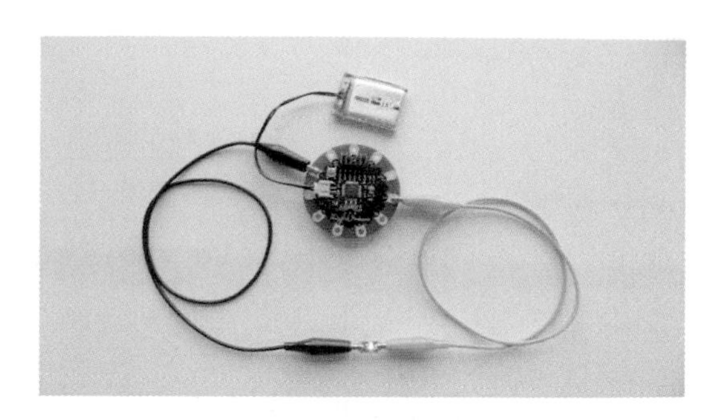

Figure 6-23. *LilyPad Arduino Simple circuit powered with a 3.7V rechargable battery*

You can also use alligator clips to connect a 2x or 3x AA or AAA battery pack to the + and – pins. More on power options and considerations in Appendix B.

When connecting an alternative power source, be sure that the red wire goes to + and the black to –. If you reverse the connections, you may fry the microcontroller and render the board unusable.

Experiment: Morse Code Messages

Even a simple blinking LED can take on great meaning in the right circumstances (Figure 6-24).

Morse code is method of transmitting messages with short and long pulses of sound or light. A dash (long pulse) is usually three times the length of a dot (short pulse).

Figure 6-24. *Dinner suggestion shirt (illustration by Jen Liu)*

Using Table 6-2, write a program that sends a message by way of the blinks of the LED. Think about what it would be like if you mounted this LED on a piece of your clothing. What would you want it to say?

Table 6-2. Morse Code translation guide

Character	Code	Character	Code	Character	Code
A	._	J	.___	S	...
B	_...	K	_._	T	_
C	_._.	L	._..	U	.._
D	_..	M	__	V	..._
E	.	N	_.	W	.__
F	.._.	O	___	X	_.._
G	__.	P	.__.	Y	_.__
H	....	Q	__._	Z	__..
I	..	R	._.		

Digital Input

Thus far, you've been working exclusively with outputs. But in microcontrollerland it is important to understand the difference between outputs and inputs.

Outputs are pins where information is delivered *from* the microcontroller in the form of varying voltage. Various types of actuators (e.g., LEDs, motors, and speakers) can be connected to output pins and will use the voltage to perform different actions. LEDs will light up, motors will spin, and buzzers will beep.

Inputs are pins where you can connect devices that supply information *to* the microcontroller. Such devices include switches and various types of sensors. Information is fed to the microcontroller in the form of varying voltage.

Now that you've gotten some experience with digital outputs, let's give digital inputs a try. Based on what you learned in Chapter 3, you have a good idea of what a switch is, how it works, and what types are available to you. But how do you connect them to a microcontroller?

The Circuit

When connecting a switch to a microcontroller, you can connect it from any digital input pin to either power (+) or ground (–), depending on what kind of logic structure you want to create. If the switch is connected to power, the pin will read "HIGH" when the switch is closed and "LOW" when it's open. If the switch is connected to ground, the logic will be reversed ("LOW" when closed, "HIGH" when open).

Connecting the switch is not enough to complete your circuit. When the switch is closed, you will have a solid connection to power or ground, depending on how you've wired it. But when the switch is open, the input pin will *float*. A floating pin has no reliable reference and thus can produce erratic values that will likely interfere with the reliability of your program. The way to prevent floating pins is with a *pull-up* or *pull-down* resistor. This resistor is of a large enough value that when the switch is closed, current will follow the path of the switch, but when it is open it will act as a spring that gently pulls the input back to its resting state.

If the switch is connected to power, you can use a pull-down resistor connected from the digital input pin to ground. If the switch is connected to ground, use a pull-up resistor connected to power. Within this context, something in the range of a 10KΩ resistor will usually do the trick. Figures 6-25 and 6-26 show what these digital input circuits look like with the LilyPad Arduino Simple.

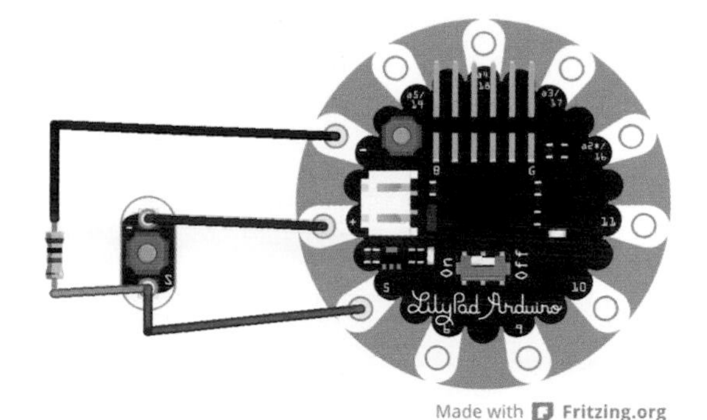

Figure 6-25. *LilyPad Arduino Simple with switch and pull-down resistor*

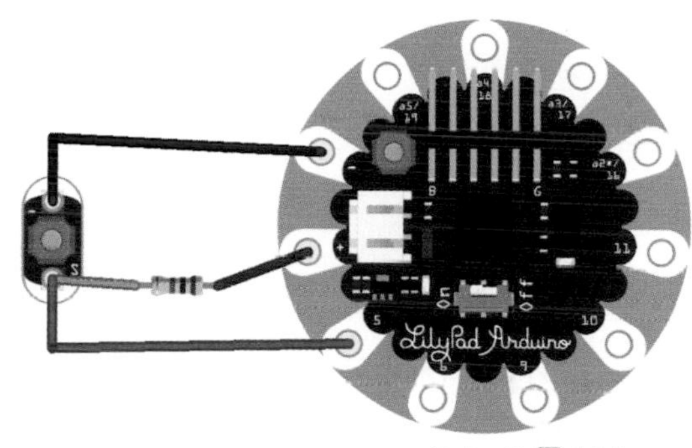

Figure 6-26. *LilyPad Arduino Simple with switch and pull-up resistor*

If you want to reduce the amount of wiring you have to do, the LilyPad's ATmega chip actually has an internal pull-up resistor on the digital pins that you can activate with the command `pinMode(pin Number, INPUT_PULLUP)`. The circuit for this is shown in Figure 6-27.

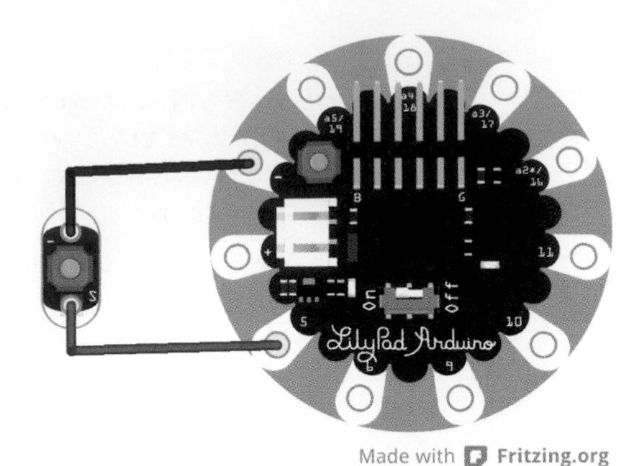

Figure 6-27. *LilyPad Arduino Simple with switch wired for use with internal pull-up resistor; you can also use two alligator clips without the button and simply connect and disconnect the exposed clips at the loose ends*

Whichever method you choose, these will all allow the value of the switch to be read by a digital input pin on the microcontroller.

For this example, let's use the wiring for use with the internal pull-up resistor illustrated in Figure 6-27.

The Code

Now that you have a switch connected, how can you write a program that can tell what the switch is doing? Here are a few more Arduino commands that will help you to read the value of a digital input:

`pinMode(pin, mode)`
: This is something that you encountered earlier with digital output. Generally speaking, with digital input you would set the mode to `INPUT`. This will work with circuit examples that use external pull-down or pull-up resistors. However, if you would like to use the *internal* pull-up resistor, then set the mode to `INPUT_PULLUP`.

`digitalRead(pin)`
: This is the opposite of the `digitalWrite()` command. Rather than controlling a pin by sending voltage out, this allows you to read the voltage coming into a pin. The only parameter you need to provide is the pin number. However, you do need a place to store the information that is read, so this command is usually used in combination with a variable—for example, `buttonState = digitalRead(buttonPin);`.

In order to read values that are coming into the microcontroller, you need to print it to some sort of display. Because the Arduino has no built-in visual display, you can use USB-serial communication and the serial monitor in the Arduino software to view what sort of values you're getting in. Here are the new commands you need to know to accomplish this:

`Serial.begin(speed)`
: By including this in the `setup()` function, this initializes the serial connection and sets the speed of communication. A standard rate that you'll often find in examples is 9600 baud. This will work well when communicating between your Arduino and computer.

`Serial.println(val)`
: This transmits a value followed by a carriage return (a character sent when you press Enter or Return). In your case, the value will be the switch value.

Now that you understand what's going on, go ahead and run this code:

```
/*
Make: Wearable Electronics
Digital Input example
 */
//variable for the digital input pin
int buttonPin = 5;
//variable for the reading from the button
int buttonValue = 0;

void setup() {
  // initialize serial communication
  // at 9600 bps
  Serial.begin(9600);
  // set pin as input
```

```
  // use internal pull-up resistor
  pinMode(buttonPin, INPUT_PULLUP);
}

void loop() {
  // read input pin:
  buttonValue = digitalRead(buttonPin);
  // print button value:
  Serial.println(buttonValue);
  delay(100);
}
```

Upload the sketch to your Arduino board. Then, while the board is still connected via USB, open the Serial Monitor and you will see the switch values change as you press and release the button (Figure 6-28).

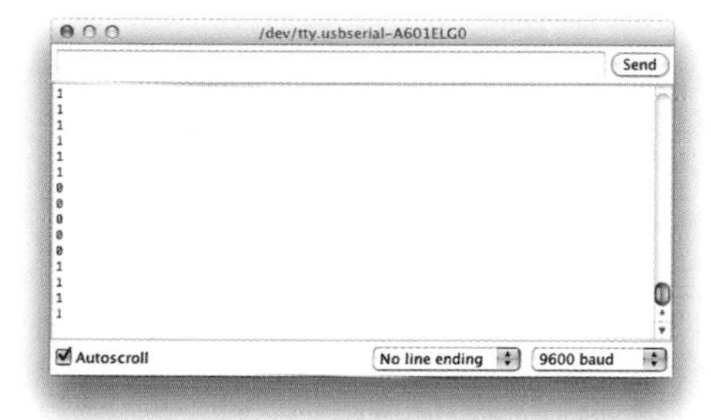

Figure 6-28. *Button values as seen in the Arduino Serial Monitor*

When you press the button, the value should change to "0." Otherwise it will be "1."

See also:

- Arduino Digital Read Serial example (*http://arduino.cc/en/Tutorial/DigitalReadSerial*)

You've now conquered digital input *and* serial communication in one fell swoop. With that under your belt, let's try a new experiment.

Experiment: Button as Controller

Now that you know how to work with both inputs and outputs, you can create a relationship between them. Control structures in the Arduino syntax allow you to establish such relationships.

First off, let's put together a circuit that includes both an input and an output (see Figure 6-29).

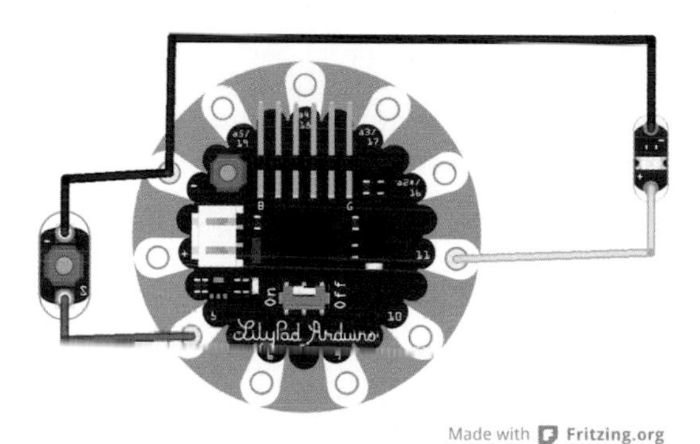

Figure 6-29. *LilyPad Simple with digital input and digital output*

Now you just need to connect them in the code. The simplest structure to work with is the `if` statement. It goes something like this:

```
//if button reads high
if (buttonValue == HIGH)
{
// turn LED on
 digitalWrite (LEDpin, HIGH);
}
```

Within this `if` statement, `HIGH` means "1" and `LOW` means "0." Keep in mind that with a pull-down circuit, the switch will read `HIGH` when it's closed. With a pull-up circuit, it will read `HIGH` when it's *open*.

The only problem with this is that once a program starts, if the switch ever reads as `HIGH`, the LED would stay on forever. There are no instructions if the switch is `LOW`. In most cases, you will need to use an `else` clause, which provides instructions for what to do if the initial condition isn't met:

```
// if switch reads high
if (buttonValue == HIGH)
{
// turn LED on
  digitalWrite (LEDpin, HIGH);
// otherwise
} else {
// turn LED off
  digitalWrite (LEDpin, LOW);
}
```

Using this information and the code examples provided in the digital input and output sections, create a program that allows the button to control the lighting of the LED. Once that's up and running, try switching the logic so the relationship between the button press and the LED light is reversed.

Once that's working, try creating a program that causes the LED to blink while the button is pressed.

Analog Input

Another key concept when getting to know your microcontroller pins is the difference between digital and analog.

Digital refers to a binary state. On or off. High or low. Voltage flowing or not flowing. 1 or 0. There are only two possible states. There are both digital inputs (such as a switch) and outputs (which could turn an LED on or off).

Analog refers to pins that can accommodate a range of values. With analog inputs, you can connect sensors such as a light sensor that can tell you if it's light, dark, or somewhere in between. With an analog output, you can accomplish more varied effects, such as an LED that can fade from on to off.

When trying to understand the difference between a digital and analog input, you can think about the traditional interface devices for home lighting (Figure 6-30). A regular light switch that you would find on a wall is similar to a digital input. It can only turn the lights on or off. But a dimmer is similar to an analog input. It provides enough information so that you can tweak the lights levels in order to create a specific mood.

Figure 6-30. *The on/off switches on the left would be considered a digital input; the dimmer on the right would be considered an analog input*

The Circuit

A common sensor to get started with for analog input is a light sensor such as a phototransistor. It just so happens that there is a LilyPad light sensor available (DEV-08464, shown in Figure 6-31). This sensor will output between 0 and V+ depending on the light level it senses, with 0V indicating the darkest and V+ indicating the brightest.

Figure 6-31. *LilyPad Light Sensor*

If you look closely at the light sensor, you will see that the pins are marked with +, –, and S. This gives you some hints about how to connect your light sensor to your LilyPad Arduino. Go ahead and make the connections shown in Figure 6-32 using alligator clips.

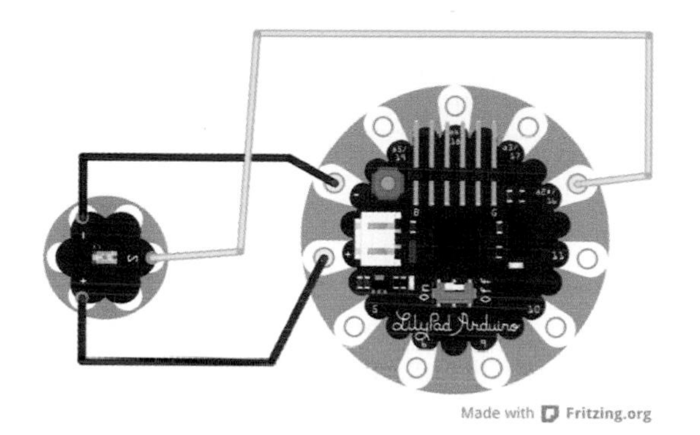

Figure 6-32. *LilyPad Arduino Simple with light sensor on pin A2*

On the LilyPad light sensor, S stands for the signal that is being produced based off of the light levels: between 0 and V+. On the LilyPad Arduino Simple (or any Arduino), when the pin number is proceeded by an A, it indicates that the pin is an analog input pin. For this example, you could also use pin A3, A4, or A5, but you would need to adjust the code accordingly.

Once your circuit is complete, connect your FTDI board to the LilyPad and to your computer, then get yourself ready to program.

The Code

The code for reading an analog input is quite similar to that for a digital input, with the exception of this command:

`analogRead(pin)`
: This reads the value of a specified analog pin. The pin can either be referred to as just the number ("2") or with the "a" preceding it ("a2").

`pinMode()` does not need to be set for an analog input pin.

Here's the code:

```
/*
Make: Wearable Electronics
Analog Input example
 */

// initialize variable for light sensor reading
int lightSensorValue = 0;

// initialize variable for light sensor pin
int lightSensorPin = A2;

void setup() {
  // initialize serial communication at 9600 bps
  Serial.begin(9600);
}

void loop() {
  // read pin and store value in a variable:
  lightSensorValue = analogRead(lightSensorPin);
  // print the light sensor value:
  Serial.println(lightSensorValue);
  // delay between readings:
  delay(100);
}
```

Check your board type and serial port, and upload the code. Open your Serial Monitor, make sure your baud rate is set to 9600, and you should see sensor values on the screen! (See Figure 6-33.)

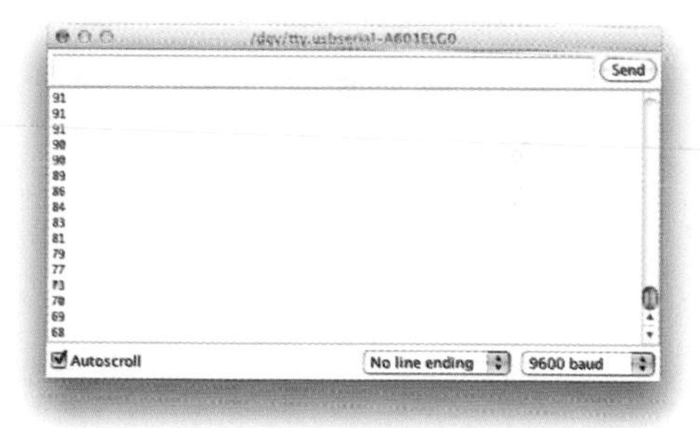

Figure 6-33. *Light sensor values as seen in the serial monitor*

Notice how as you cover and uncover the sensor, the values on screen change. Try moving the circuit toward a very bright light and try covering it completely. As the light gets brighter, the values should go up, and as it gets darker, they should go down. Find out what the broadest range of values you can observe is.

This shows you in a very basic form how to read sensor values with the Arduino. In Chapter 7, I go more in-depth into what sensors are and how to work with them.

See also:

- Arduino Analog Read Serial tutorial (*http://arduino.cc/en/Tutorial/AnalogReadSerial*)
- Arduino Analog Read Voltage tutorial (*http://arduino.cc/en/Tutorial/ReadAnalogVoltage*)
- Arduino Analog Input tutorial (*http://arduino.cc/en/Tutorial/AnalogInput*)

Experiment: Sensor as a Switch

Sensors can act as switches, too. This is this snippet of code that you used back in "Experiment: Button as Controller" on page 107 to allow a switch to control an LED:

```
if (buttonValue == HIGH) // if switch reads high
{
  digitalWrite (LEDpin, HIGH); // turn LED on
} else { // if switch reads low
  digitalWrite (LEDpin, LOW); // turn LED off
}
```

You can modify this for use with a sensor. For example:

```
if (lightSensorValue > 500)
// if light sensor reads greater than 500
{
  digitalWrite (LEDpin, HIGH); // turn LED on
} else { // otherwise
  digitalWrite (LEDpin, LOW); // turn LED off
}
```

Modify the Analog Input example so that changes in light levels will control the LED. Use the values you see in the Serial Monitor to determine appropriate value to use in your code.

Analog Output

On the output end, analog allows you to provide a range of values rather than simply turning something on or off. This means that you can brighten or dim an LED with subtlety or even control the speed of a motor.

But whereas an analog input pin reads a set range of voltages, despite what you might think, an analog output pin does not produce a range of voltages. Instead, it *simulates* a change in voltage to create an analog effect by pulsing 5 volts in differing duty cycles. This effect is called pulse width modulation (PWM). If the pin is quickly switched back and forth between 0V and 5V, it creates the effect as if it were outputting 2.5V, and so on.

Arduinos have limited pins that are able to perform pulse-width modulation. On an Arduino Uno, they are marked with a tilde (~). On the LilyPad Arduino Simple board, they are unmarked but it just so happens that all of the digital input/output pins (5, 6, 9, 10, 11) can also perform PWM so you have lots of options to choose from.

If you are ever unsure which pins on an Arduino can perform PWM, just check the product page.

The Circuit

Because some digital input/output pins can also function as PWM pins, you'll be using the same circuit you used in the Digital Output example. Go ahead and re-create the circuit shown in Figure 6-34.

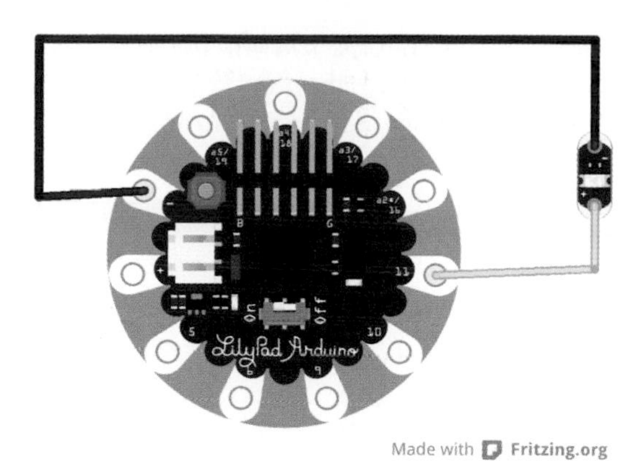

Figure 6-34. *LilyPad Simple with LilyPad LED on pin 11*

The Code

The command you use to control analog output is this:

`analogWrite(pin, value)`
: The pin is the number of the pin you'd like to control. The value can be between 0 and 255 with 0 being 0V and 255 being V+. If you would like to do something like brighten and dim an LED, you can incrementally move it through different values.

Go ahead and upload this code to see the LED turn on at a variety of brightnesses:

```
/*
Make: Wearable Electronics
Analog Output example
 */

int LEDpin = 11; // LED is connected to pin 11

void setup()  {
  pinMode(LEDpin, OUTPUT); // sets pin as output
}

void loop()  { // LED completely off
  analogWrite(LEDpin, 0);
  delay(100);
  analogWrite(LEDpin, 50);
  delay(100);
  analogWrite(LEDpin, 100);
  delay(100);
  analogWrite(LEDpin, 150);
  delay(100);
  analogWrite(LEDpin, 200);
  delay(100); // LED at full brightness
  analogWrite(LEDpin, 255);
  delay(100);
}
```

This is a very simple way to start out working with analog output. By employing more complex programming methods, you can achieve more sophisticated behaviors.

See also:

- Arduino Fading tutorial (*http://arduino.cc/en/Tutorial/Fading*)

Experiment: Sensitive System

Many basic interactive projects create a relationship between the values of an analog input and the values of an analog output. Figure 6-35 shows a circuit that includes both a light sensor and an LED.

Using `if` statements and the `analogWrite()` command, use the brightness of an LED to reflect the changes in the values of a light sensor.

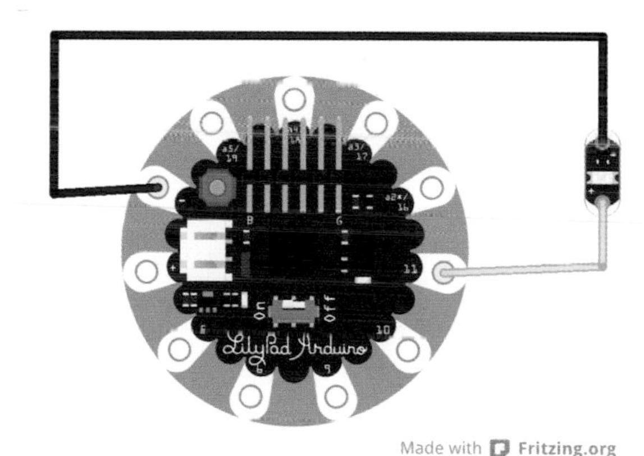

Figure 6-35. *LilyPad Arduino Simple with Light Sensor and LilyPad LED*

What's Next

This is the most basic of introductions to working with microcontrollers by using the LilyPad Arduino Simple. As you begin to develop projects, keep in mind that there is a broad range of Arduino and Arduino-compatible products out there that might better meet your needs. You will encounter some of them in the coming chapters. You'll also further explore the plethora of sensors and actuators available for use in combination with microcontrollers for your wildly imaginative wearable electronics projects.

Sensors 7

Simply stated, a sensor is an electronic component that measures some aspect of the physical world and converts that measurement into varying electrical characteristics, namely voltage or resistance. Sensors can sense things like light, movement, temperature, and touch. They are exciting because they make the physical world perceivable by computers—even tiny computers like microcontrollers.

As you get deeper into the realm of interactivity, it's worth considering the wide range of sensors available to you. In this chapter, you'll take a moment to consider how you can best listen to what's happening in, on, and around the body through technological means. You'll encounter an assortment of body-centric sensors and look at some simple ways to work with the data they produce.

Working with Sensors

There are both conceptual and technical factors to consider when working with sensors.

Getting to Know Your Sensor

Looking at a sensor that is new to you can be intimidating, exciting, or overwhelming. It is easy to look at the name of a newly released sensor, think, "That's exactly what I need!", order it, get it home, and realize that it is incompatible with your project.

Just like in any good relationship, it's worth sniffing out your prospective sensor before making the big commitment. You can always, at the very least, have a virtual introduction to it through datasheets, product descriptions, reviews, forums, and tutorials. And if you're part of a hacker, maker, or educational community, you may very well know someone who has the sensor you're considering. Ask if you can borrow it and take it out for a spin before taking the plunge and getting one of your own. While electronic sensors are significantly cheaper than they used to be, it's still an investment, so it's worth doing your research.

When encountering a sensor for the first time, whether it be online, in a catalog, or in real life, here is the type of information you should be looking for:

Connector type
: Depending on the manufacturer, breakout board, and intended use, sensors will have different connection types (Figure 7-1). If there are headers, are they male or female? And do they use breadboard spacing or something else? Is it a standardized plug and socket set such as JST? Or a proprietary connector that is

specific to the manufacturer? Does it have sew tabs because it is intended for use with conductive thread? These factors will likely determine what other materials you will need to build your circuit.

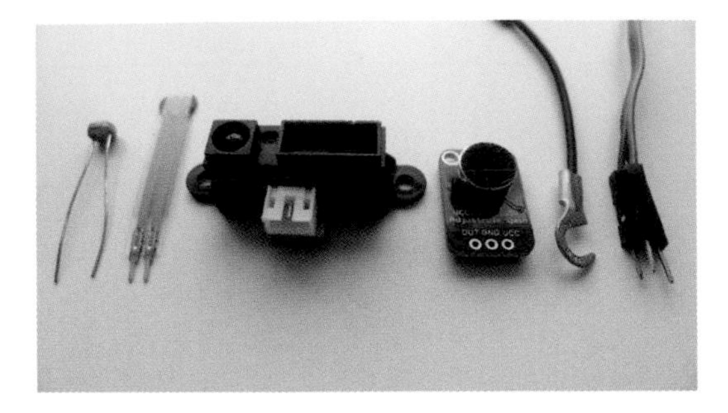

Figure 7-1. *Sensors with different connectors; left to right: legs, terminals, JST connector, pins, hook, male headers*

Sensitivity

What is the range of values your sensor senses? For instance, smaller force-sensing resistors (FSRs) can sense as little as 2 grams of force (Figure 7-2), whereas larger ones have a sensing range of 100 grams to 10 kilograms (Figure 7-3).

Figure 7-2. *This smaller force-sensing resistor is sensitive to even the lightest touch of a finger*

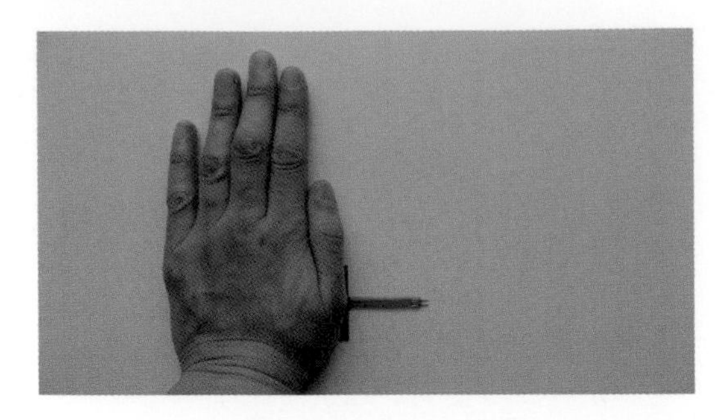

Figure 7-3. *This larger force-sensing resistor is sensitive to a higher range of pressure, making it appropriate to be pressed firmly, leaned on, or stepped on*

Accuracy

How accurate is your sensor? FSRs are very sensitive, but not always accurate. Depending on their range of sensitivity, they provide good answers to questions like, "Is this being pressed or not?" or, "Is this being stood on or not?" but they do not provide the level of accuracy needed if you were building a precise food scale or a body scale.

Shape, size, and weight

What are the physical characteristics of your sensor (see Figure 7-4)? These greatly affect how a component can be worn on the body. Be sure to look at dimensions and technical drawings to determine if a sensor will fit in with your design.

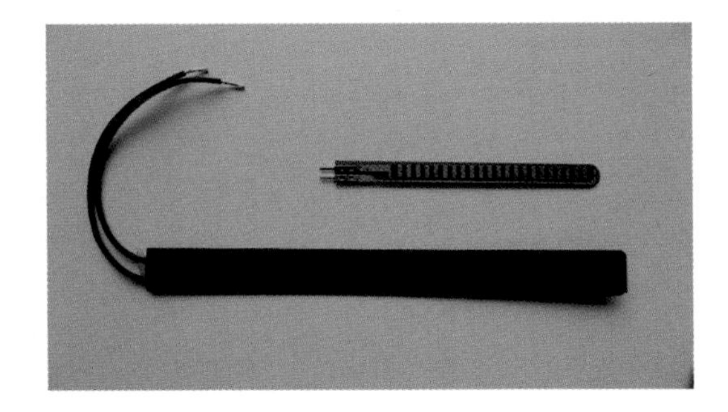

Figure 7-4. *Flex sensors come in different lengths; a shorter flex sensor might be more appropriate for a project with limited surface area*

Sensor output

What kind of information does your sensor provide and how? The sensor's output is what gets read by the microcontroller (Figure 7-5). You've learned so far that a varying voltage output can be read by the analog input pins on the Arduino. Later in this chapter, you will also learn how to read varying resistance and how a sensor transmits data via serial communication.

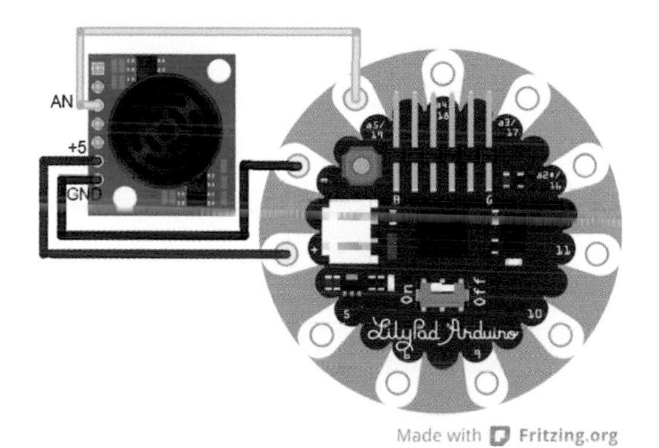

Figure 7-5. *Maxbotix Ultrasonic sensors feature multiple outputs, including analog voltage, serial communication, and pulse width modulation (digital pulses)*

Voltage Divider Circuit

Some of the sensors you work with are *variable resistors*. A variable resistor is a component that changes resistance in response to a changing condition. Variable resistors that you will encounter later in this chapter include flex sensors, stretch sensors, and light sensors.

The problem with trying to read a variable resistor with a microcontroller is that a microcontroller's analog input reads varying *voltage*, not varying resistance. Luckily, there is a simple circuit that enables a variable resistor to produce varying voltage: a *voltage divider* circuit (shown in Figure 7-6).

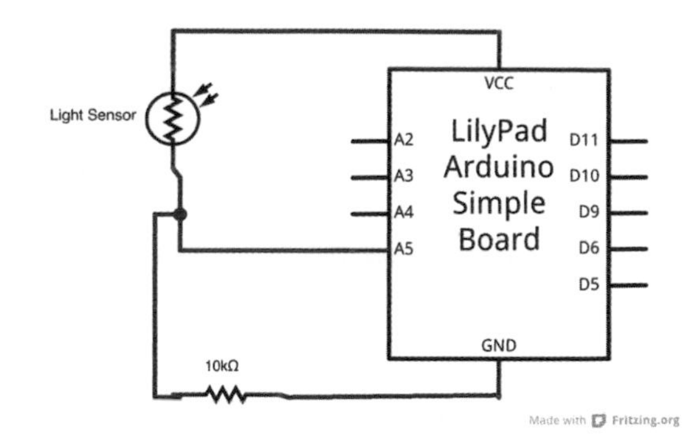

Figure 7-6. *Voltage divider circuit diagram with variable resistor and fixed 10KΩ resistor*

This circuit pairs a variable resistor connected from power to the analog input with a fixed resistor connected from ground to the same analog input. The fluctuating ratio of their resistances creates a varying voltage between them. For your purposes, the fixed resistor just needs to be in the same order of magnitude as the variable resistor. Many of the variable resistors you encounter in this chapter will work just fine with a fixed 10KΩ resistor.

You've actually already worked with a variable resistor in Chapter 6: the LilyPad Light Sensor. If you look closely in Figure 7-7, you can see the surface mount fixed resistor and the traces going to the pads.

Figure 7-7. *Closeup of LilyPad light sensor; note the resistor on the right as well as how the circuit board traces create a voltage divider circuit*

Communicating with I2C

All of the LilyPad sensors provide a varying voltage which can be read by an analog input pin on the Arduino. But Flora sensors communicate sensor values to the microcontroller through different means: I2C. While this book won't provide a comprehensive orientation to I2C, it's helpful for you to know at least some of the basic details.

Short for Inter-Integrated Circuit and referred to as "I-squared-C" or "I-two-C", I2C is a two-wire serial communication protocol.

The following connections are made between any I2C sensor and the microcontroller (shown in Figure 7-8):

SCL	Serial clock pin—pulses on this pin provide the timing for the communication
SDA	Serial data pin—the wire on which the actual data is sent and received
gnd	Share a common ground with the microcontroller
3V	Flora sensors require 3V power, but check the datasheet for other I2C sensors

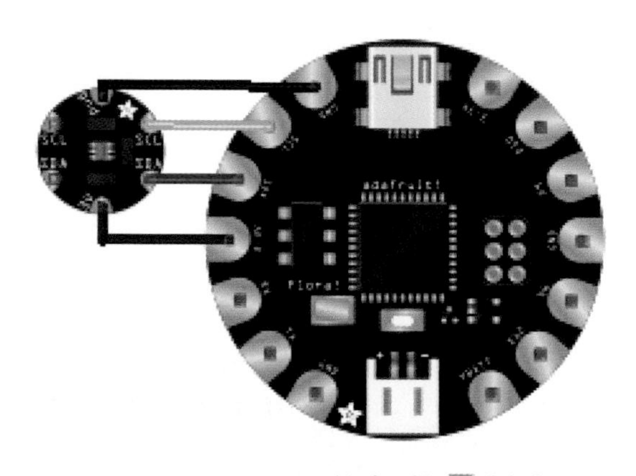

Figure 7-8. *Flora sensor connections*

On the Flora, SCL and SDA are clearly marked. On the Arduino Uno and the LilyPad Arduino, SDA is A4 and SCL is A5. For other boards, check their datasheets.

A neat aspect of working with I2C devices is that they can be chained (Figure 7-9) so that they don't take up lots of pins on the Flora. This also greatly reduces the amount of wiring. I2C sensors (and other devices) usually have a predetermined address. The address is used by the microcontroller to speak to a particular device in the chain of I2C devices. This information can be found in the device's datasheet.

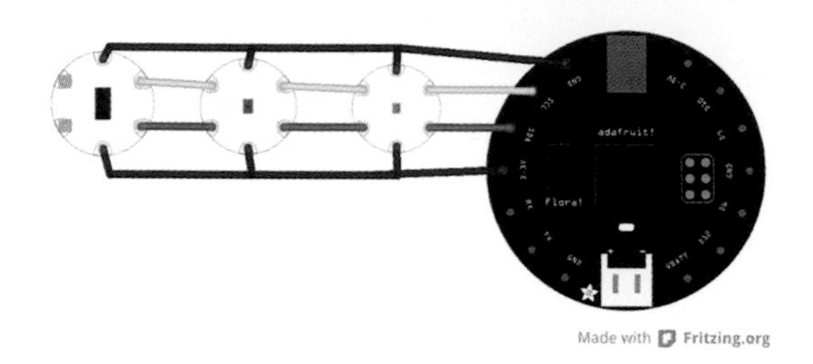

Figure 7-9. *Circuit layout of Flora sensor chain*

On the software side, there is an Arduino library called "Wire" that enables Arduinos to communicate with I2C devices and handles the nitty-gritty details of this protocol so you don't have to. *Libraries* are additional packages of code that can be added in to support different functionalities or tasks. Some are included with your Arduino download ("standard libraries") and some you need to download and install yourself ("contributed libraries"). Wire is a standard library. In addition, Adafruit provides very useful and sophisticated libraries for many of its sensors, which are contributed and typically user-installed.

While I2C is a significantly more advanced serial communication protocol, Adafruit provides excellent documentation for working with the Flora sensors, so it is fairly straightforward to get up and running quickly without having a comprehensive understanding of how the details of the I2C protocol work.

Working with Sensor Data

Once you're able to get access to the sensor data, you then need to figure out what to do with it. In this section, I explore some concepts for making sense of sensor data, including thresholds, mapping, calibration, constraining, and smoothing.

Let's use the light sensor circuit from the Sensitive System experiment in the previous chapter (Figure 6-32) for the following examples. Here are the parts you will need:

- LilyPad Arduino Simple (SF DEV-10274)
- LilyPad LED (SF DEV-10081)
- LilyPad Light Sensor (SF DEV-08464)
- FTDI Board (AF 284, SF DEV-10275)
- USB mini-B cable (AF 899, DK WM5163-ND, RS 55010682, SF CAB-11301)
- Alligator clip test leads (AF 1008, RS 278-1156, SF PRT-11037)

Thresholds

A *threshold* can be used to set a boundary between one condition and another. You can think of it like a border between two countries or a fence between two yards. Setting the boundary makes it easier to distinguish one thing from the other.

When working with a range of sensor values, sometimes it's helpful to indicate what different ranges within those values mean. Thresholds are a good way to get started.

Say you're working with a LilyPad light sensor. It's great to have a bunch of numbers flying by on the Serial Monitor, but how can you use them to make something happen? For instance, if you want an LED to turn on when it's dark and off when it's light, you need to define what "dark" is.

By looking at the values in the serial monitor and exposing the sensor to varying conditions (in this instance, turning the lights on and off), you can select a *threshold* value. This is a number above which you would consider the condition to be "light" and below which to be "dark" (see Figure 7-10).

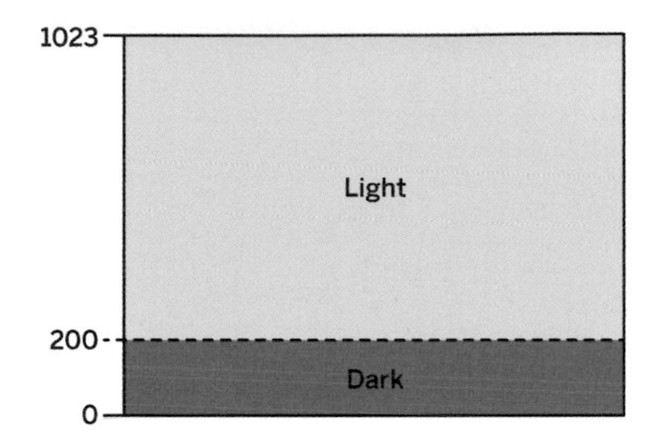

Figure 7-10. *With a threshold set at 200, all values above are considered "light" and all below "dark"*

You can implement a threshold in your code through the use of an `if` statement:

```
// if it is "dark"
if(lightSensorValue<200){
  //Turn LED on
  digitalWrite(LEDpin, HIGH);
}
// if it is "light"
else{
  //Turn LED off
  digitalWrite(LEDpin, LOW);
}
```

This example will turn an LED connected to pin 11 on when it is dark and off when it is light. Here is the complete sketch:

```
/*
Make: Wearable Electronics
 Single Threshold example
 */

//initialize variables
int lightSensorValue = 0;
int lightSensorPin = A2;
int LEDpin = 11;
```

```
void setup() {
  //initialize serial communication:
  Serial.begin(9600);
}
void loop() {
  // read the light sensor pin and
  // store value in a variable:
  lightSensorValue = analogRead(A2);

  // if it is "dark"
  if(lightSensorValue<200){
    //Turn LED on
    digitalWrite(LEDpin, HIGH);
  }
  // if it is "light"
  else{
    //Turn LED off
    digitalWrite(LEDpin, LOW);
  }

  // delay between readings:
  delay(100);
}
```

You can also have multiple thresholds (Figure 7-11) and use the `else if` conditional statement.

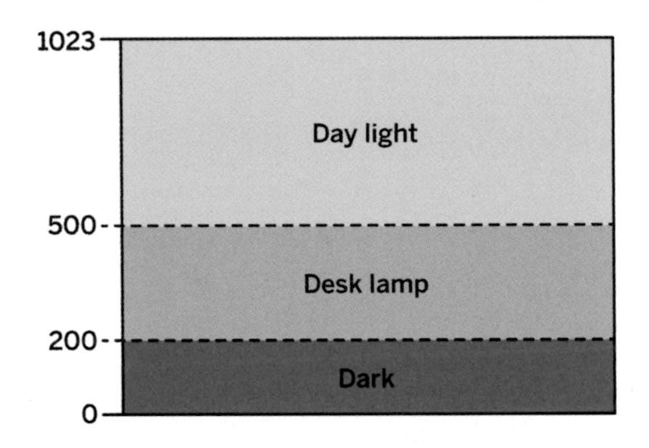

Figure 7-11. *Multiple thresholds set at 200 and 500*

The following sketch uses two thresholds and prints out a description of the light level as well as the raw sensor value in the Serial Monitor:

```
/*
Make: Wearable Electronics
Multiple Threshold example
*/

//initialize variables
int lightSensorValue = 0;
int lightSensorPin = A2;
int LEDpin = 11;
int threshold1 = 500;
int threshold2 = 200;

void setup() {
  //initialize serial communication:
  Serial.begin(9600);
}
void loop() {
  // read the light sensor pin and
  // store value in a variable:
  lightSensorValue =
  analogRead(lightSensorPin);

  // print the light sensor value
  Serial.print("Light Sensor Value: ");
  Serial.print(lightSensorValue);

  // get ready to print light level
  Serial.print(", Light Level: ");

  //if the value is greater than
  // threshold #1
  if(lightSensorValue>threshold1){
    Serial.println("daylight");
  }
  //if the value is less or equal to
  // threshold #1 and greater than
  // threshold #2
  else if(lightSensorValue>threshold2){
    Serial.println("desklamp");
  }
  //if the value is equal to or less than
  // threshold #2
  else{
    Serial.println("dark");
  }

  // delay between readings:
  delay(100);
}
```

Give this code a try and modify the values so that they better match your environment.

Figure 7-12. *"Capacity Indicator Bag" by Sally Chan uses an FSR and a 3-LED display to provide a visual indication of the weight of the bag's contents*

Mapping

Mapping is a way to translate a value from one range of numbers to another. It can be used to create a direct relationship between an input and an output. For instance, the value provide by a light sensor could control the *brightness* of an LED (as opposed to turning the LED on and off as you did earlier).

To accomplish this, there is a very useful Arduino function called `map()`. It looks like this:

```
map(value, fromLow, fromHigh,
    toLow, toHigh)
```

`value` is the value that you would like to map. `fromLow` and `fromHigh` is the low and high end of the original data set. `toLow` and `toHigh` are the low and high values of the mapped data set. If you were to map the full range of analog input (0 to 1023) to the full range of analog output (0 to 255), you'd use the following line of code (illustrated in Figure 7-13):

```
map(lightSensorValue, 0, 1023, 0, 255)
```

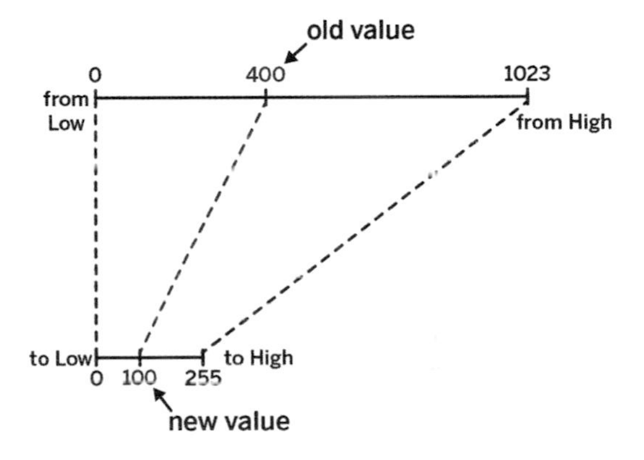

Figure 7-13. *Mapping a value from 0-1023 to 0-255*

A complete sketch looks something like this:

```
/*
Make: Wearable Electronics
 Mapping example
 */

//initialize variables
int lightSensorValue = 0;
int lightSensorPin = A2;
int LEDpin = 11;
int mappedLightSensorValue = 0;

void setup() {
  //initialize serial communication:
  Serial.begin(9600);
}
void loop() {
  // read light sensor pin and
  // store value in a variable:
  lightSensorValue =
```

```
  analogRead(lightSensorPin);
  //map sensor value
  mappedLightSensorValue =
    map(lightSensorValue, 0, 1023, 0, 255);
  //set analog output accordingly
  analogWrite(LEDpin, mappedLightSensorValue);

  // print the sensor and mapped sensor values:
  Serial.print("Light Sensor Value: ");
  Serial.print(lightSensorValue);
  Serial.print(", Mapped Light Sensor Value: ");
  Serial.println(mappedLightSensorValue);

  // delay between readings:
  delay(100);
}
```

Many sensors don't have values that fully occupy the 0 to 1023 range. If you have a light sensor whose lowest value is 25 and highest is 940, you can change the *from* values accordingly (see Figure 7-14):

```
map(lightSensorValue, 25, 940, 0, 255)
```

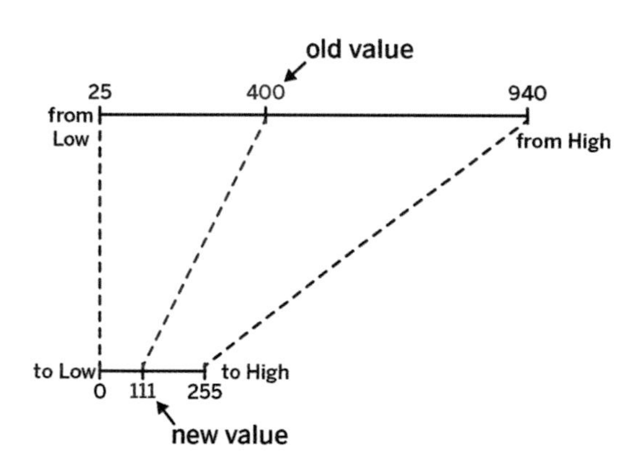

Figure 7-14. *Mapping a value from 25–940 to 0–255*

If you are working with a LilyPad light sensor as your input and using the mapped value to control an LED on the analog output pin, you would have an LED that brightens and dims in a way that mimics the conditions of the room. If you wanted to inverse the relationship so that the LED gets brighter as the room gets darker, you can simply flip the *to* values, as illustrated in Figure 7-15:

```
map(lightSensorValue, 25, 940, 255, 0)
```

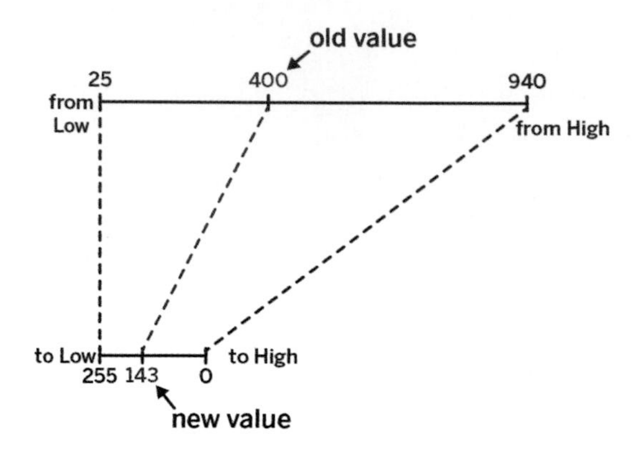

Figure 7-15. *Mapping a value from 25–940 to 255–0; this will invert the relationship*

See also:

- Arduino `map()` reference page (*http://arduino.cc/en/Reference/Map*)
- Arduino In, Out Serial example (*http://arduino.cc/en/Tutorial/AnalogInOutSerial*)

Calibration

Calibration is a way to fine-tune your code so that it is responsive to a specific set of conditions. The range of what a sensor senses will differ based on its environment and context (Figure 7-16). The amount of light available in your bedroom or studio will differ greatly from that on the street or in a park. A force-sensing resistor will read different values when stepped on by a 5-year-old than a 50-year-old. If you know you'll be using your project in varying contexts, it's worth including a calibration routine in your code. You can determine the highest and lowest possible values and configure the rest of your program accordingly.

This is great to include in your setup, but you can also create a calibration routine that is triggered by a button, should you need to recalibrate without restarting the Arduino entirely.

To see an example of this, in Arduino go to File → Examples → 03.Analog → Calibration. In the code, change the analog input pin to A2 and the digital output pin to pin 11. Then upload the code to your Arduino board.

This example looks for the highest and lowest values that occur during the first five seconds that the program is running. Once the Arduino is programmed, in order to recalibrate the sensor values, press the reset button and expose the sensor to the highest and lowest light conditions during the following five seconds.

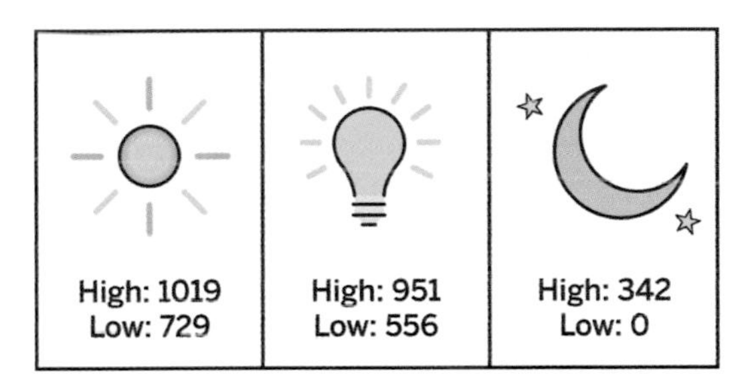

Figure 7-16. *The highs and lows of a light sensor value can differ according to the current conditions; calibration can help with this*

See also:

- Arduino calibration example (*http://arduino.cc/en/Tutorial/Calibration*)

Constraining

Sometimes your sensor will provide readings that fall outside of your desired range (Figure 7-17). For these cases, Arduino provides a function called *constrain*. The three parameters needed are the data that is being constrained, the lowest value you would like to keep, and the highest value you would like to keep. If there are any values that are below or above the specified range, the `constrain` function will convert them to the lowest or highest desired values, respectively.

In practice, it might look something like this:

```
/*
Make: Wearable Electronics
 Constrain example
  */
//initialize variables
int lightSensorPin = A2;
int lightSensorValue = 0;
int constrainedLightSensorValue = 0;

void setup() {
  //initialize serial communication:
  Serial.begin(9600);
}

void loop() {
  //read light sensor pin and store
  // value in a variable:
  lightSensorValue = analogRead(lightSensorPin);
  //constrain the light sensor values
  // to 300 to 650
  constrainedLightSensorValue =
    constrain(lightSensorValue, 300, 650);

  //print the results:
  Serial.print("Light Sensor Value ");
  Serial.print(lightSensorValue);
  Serial.print
  (", Constrained Light Sensor Value: ");
  Serial.println(constrainedLightSensorValue);

  // delay between readings:
  delay(100);
}
```

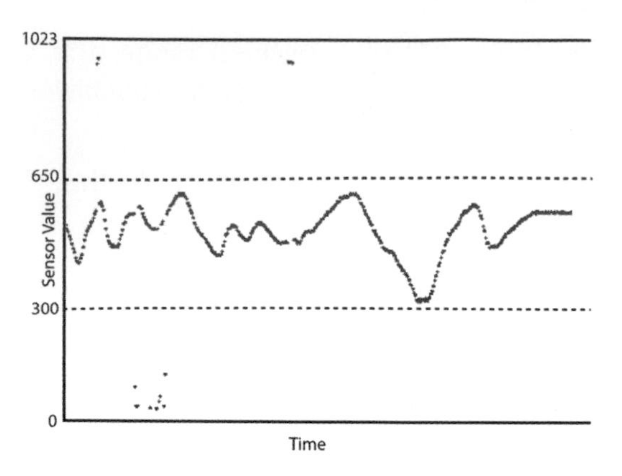

Figure 7-17. *The constrain() function allows sensor readings to be constrained within a set range*

See also:

- Arduino `constrain()` reference page (*http://arduino.cc/en/Reference/Constrain*)
- Arduino calibration example (*http://arduino.cc/en/Tutorial/Calibration*)

Smoothing

While some sensors produce data that is smooth and predictable, others offer a dataset that's rougher around the edges (Figure 7-18). Smoothing can help turn an erratic datastream into something cleaner and easier to work with.

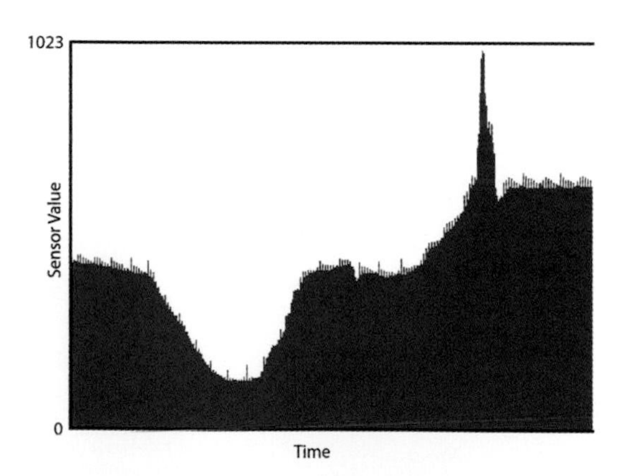

Figure 7-18. *Some data requires smoothing*

To give this a try in Arduino, go to File → Examples → 03.Analog → Smoothing. Once the sketch is open, change the analog input pin to A2 and upload the sketch to your Arduino. Open the serial monitor and see what the new, smoothed sensor data looks like!

See also:

- Arduino smoothing example (*http://arduino.cc/en/Tutorial/Smoothing*)
- Arduino's `runningAverage` class (*http://playground.arduino.cc/Main/RunningAverage*)

Graphing

Looking at data in the serial monitor can be a good place to start but sometimes it's helpful to see a visual representation of how the data changes over time. Check out the Arduino Graph example (http://arduino.cc/en/Tutorial/Graph) for more information.

Experiment: Wooo! Shirt

Using the light sensor circuit that you've been working with in this section, incorporate the circuit into a shirt with the light sensor positioned in the armpit of the shirt.

Here's some starter code for you to modify. Upload it to your Arduino, and then put the shirt on:

```
/*
Make: Wearable Electronics
 Wooo! Shirt Experiment
 */

//initialize variables
int lightSensorValue = 0;
int lightSensorPin = A2;
int LEDpin = 11;
int wooThreshold = 120;
```

```
void setup()
{
  //initialize serial communication:
  Serial.begin(9600);
  pinMode(ledPin, OUTPUT);
}

void loop(){
  // read the value from the sensor
  lightSensorValue = analogRead(lightSensorPin);

  //if the arm is up
  if(lightSensorValue>wooThreshold){
    //print Wooo!
    Serial.print("Wooo!");
    //Turn LED on
    digitalWrite(LEDpin, HIGH);
  }
  // if the arm is down
  else{
    // print boo
    Serial.println("boo ");
    //Turn LED off
    digitalWrite(LEDpin, LOW);
  }

  Serial.print(" Sensor Value: ");
  Serial.println(lightSensorValue);

  delay(100);    // delay for 1/10 of a second
}
```

This code prints a "Wooo!" when it detects light and a "boo" when it does not. Based on what you've learned about the concepts of thresholds, mapping, calibration, smoothing, and constraining, create a program that reliably prints "Wooo!" when your arm is raised.

Figure 7-19. *Wooo! shirt (illustration by Jen Liu)*

Keep in mind that using the Serial Monitor as a feedback mechanism is great for prototyping but it will keep you tethered to the computer. Later on, you can build a more creative response into your design using LEDs or other possible outputs that you'll learn about later in Chapter 8 so that you can "Wooo!" more effectively in the wild.

What to Sense

It's easy to hear about a cool sensor and decide to do a project with it.

"Oh, there's a really neat X sensor that just came out. I should obviously do an X project!"

But this leads to an interaction that's designed around the technology rather than *technology that's designed around a particular interaction*.

When working with sensors, a good place to start is to think about *what you're trying to sense*. What is the motion, action, or condition? What is the context and environment? What are the important aspects to consider? Then you can ask questions like these:

- *"What different sensor (or sensors) could I use?"*
- *"What do I want to measure?" (sound, light, pressure, presence, etc.)*
- *"Where should the sensors live?"*
- *"What should I be looking for in the data I am gathering from them?"*

In the following section, you'll look at some possibilities for what to sense and a starting selection of sensors that will fit the bill. But keep in mind that this is just the tip of the iceberg. Once you have a project idea in mind, you should go out and research what's available to best help your idea come to life.

Flex

Bodies are bendy and it just so happens that *flex sensors* sense a flex or a bend (Figure 7-20). They're very good for areas of the body that bend in a broad, round arc. They work well on elbows, knees, fingers, and wrists. They are variable resistors and need to be used in combination with a voltage divider circuit (see Figure 7-21) in order to be read by a microcontroller.

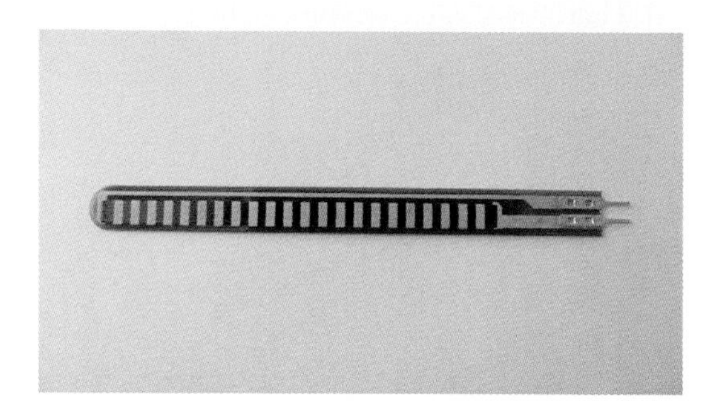

Figure 7-20. *Flex sensor*

Here are some factors to consider when choosing a flex sensor:

Length
: Flex sensors come in different lengths, usually 2.2 inches (AF 1070, SF SEN-10264) or 4.5 inches (AF 182, SF SEN-08606). Use whichever best fits your application. For instance, the longer ones work well with fingers, but the shorter ones might be more appropriate for toes.

Directionality
: Flex sensors can be single or bi-directional. Bidirectional flex sensors (RO[1] RB-Ima-11) sense flex in both directions, whereas the single-direction can only sense flex in one direction. Single direction is fine for many human joints like fingers, elbows, and knees. But the bidirectional are useful for joints like the wrist,

1. RobotShop (for a complete list of supplier abbreviations, see "About Part Numbers" on page xiv).

where the bend can take place in both directions.

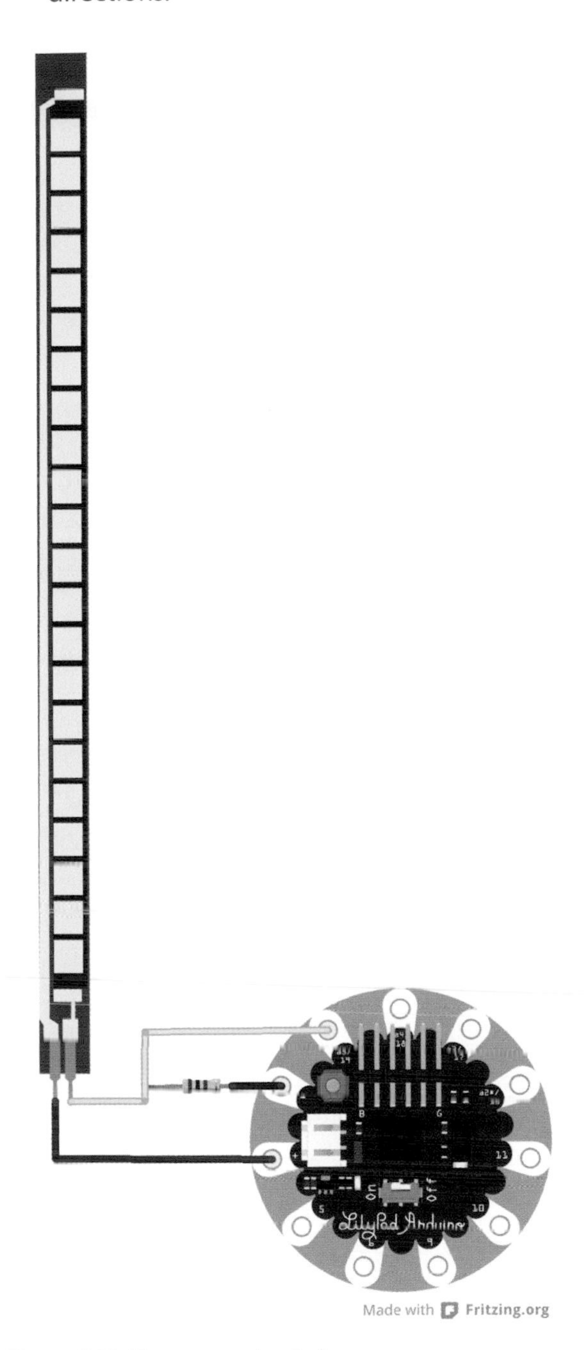

Figure 7-21. *Flex sensor circuit diagram*

Resistance range

Some flex sensors are also available in different resistance ranges (RO RB-Ima-24, RO RB-Ima-25). For your purposes, there are no real advantages or disadvantages that come with the options in this category. Just be sure that you're using a resistor of the appropriate size in your voltage divider circuit.

The biggest challenges in working with flex sensors are positioning and protection. In order to get an accurate reading of the flex of your elbow, the sensor needs to be positioned in the same place on your elbow every time. Creating a secure pocket for the sensor can help with this, as shown in Figures 7-22 and 7-23.

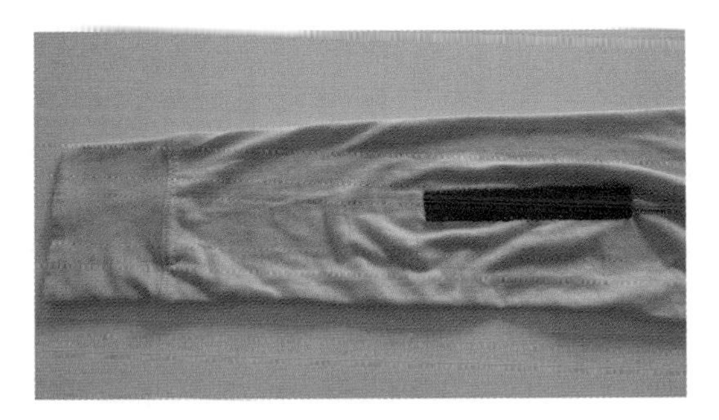

Figure 7-22. *Sleeve to hold flex sensor in place*

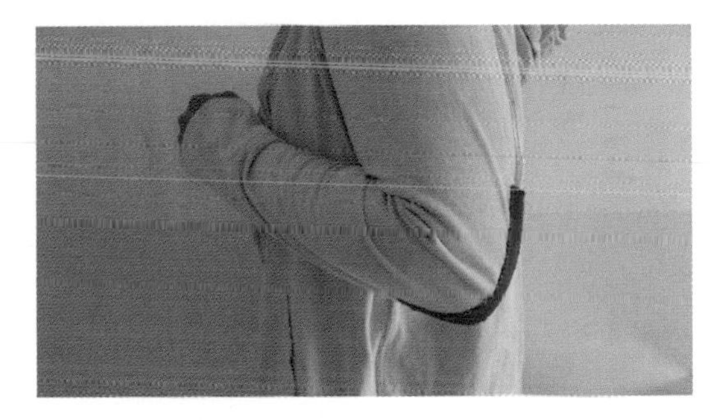

Figure 7-23. *Flex sensor on a bent elbow*

The other thing to consider is that while flexing is a pretty rigorous and strenuous activity, many flex systems are fairly delicate, particularly at their connection terminals. Be sure to protect your connections. Reinforce with heat shrink, and protect them with some sort of material.

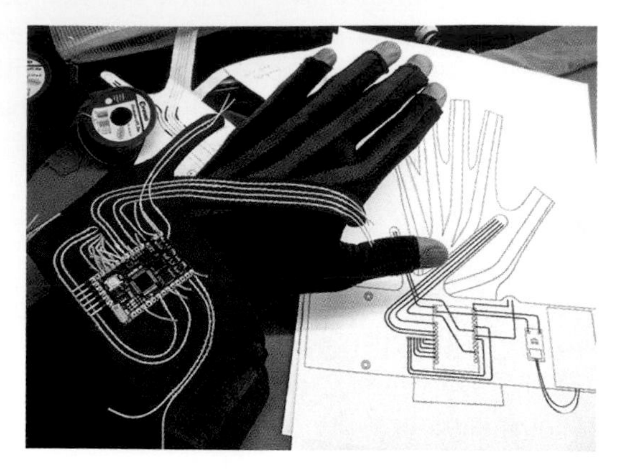

Figure 7-24. *"The Gloves Project" uses flex sensors to create experimental gestural music; the project is developed by Rachel Freire, Imogen Heap, Seb Madgwick, Tom Mitchell, Hannah Perner Wilson, Kelly Snook, and Adam Stark (photograph by Hannah Perner Wilson)*

Force

Bodies often touch and get touched. One way to sense touch is through the use of *force-sensing resistors*, or *FSRs* (Figure 7-25). FSRs have a makeup that's similar to flex sensors but are configured to be sensitive to pressure rather than bending. They are also variable resistors and have delicate connections similar to flex sensors (see Figure 7-26).

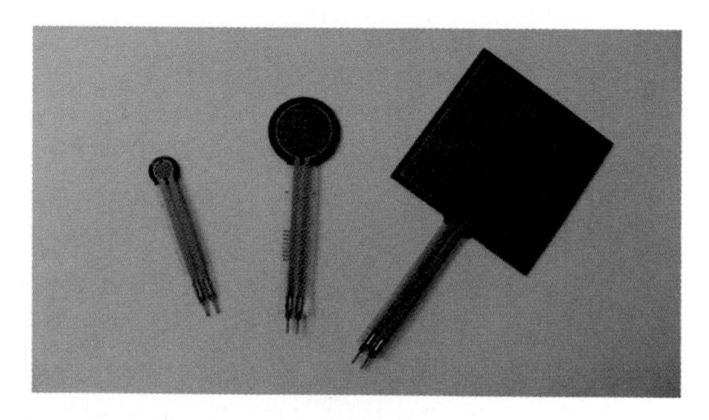

Figure 7-25. *Force-sensing resistors*

They come in different shapes and sizes. Different types are suited for different applications. See Table 7-1 for details.

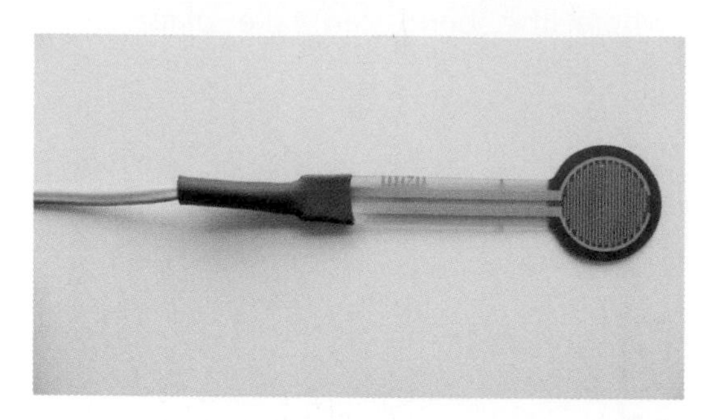

Figure 7-26. *Heatshrink tubing is used to protect the delicate solder connections between the sensor and wires*

Table 7-1. FSR comparison

Type	Sensing area	Part number	Notes
Small (round)	0.16″ diameter	AF SEN-09673	Very versatile; best for sensing touch at highly specific locations
Medium (round)	0.5″ diameter	AF 166, SF SEN-09375	Excellent for sensing the pressure of a fingertip
Large (square)	1.75x1.5″	AF 1075, SF SEN-09376	Sits well on the top of a hand, shoulder, ball of the foot
Long	0.25x24″	AF 1071, SF SEN-09674	Great for sensing pressure along the length of an arm or leg

Figure 7-27 shows a circuit diagram, and Figure 7-28 shows how you can keep an FSR secured inside a pocket.

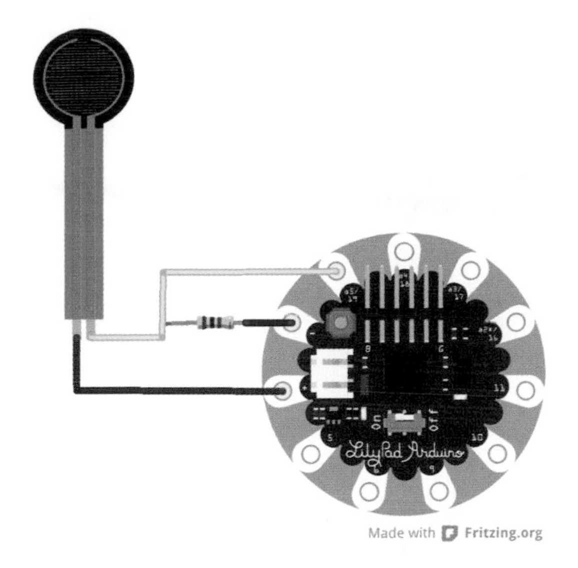

Figure 7-27. *FSR circuit diagram*

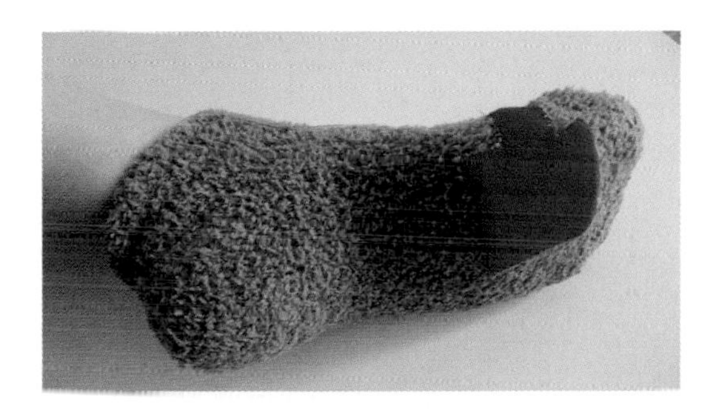

Figure 7-28. *A pocket sewn onto the sock helps keep the FSR securely in place*

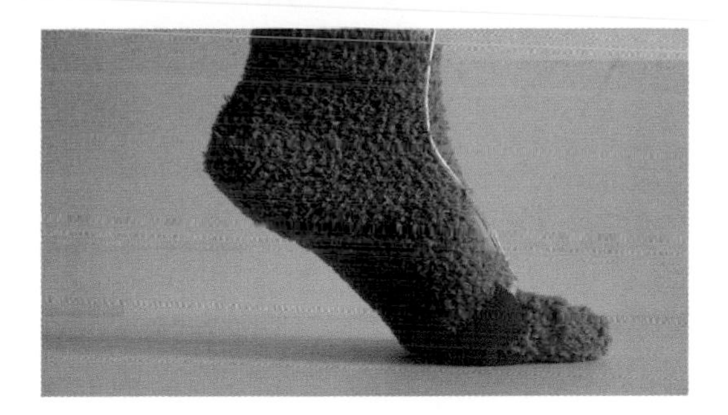

Figure 7-29. *When pressure is applies to the ball of the foot the change in sensor data can be read by the microcontroller*

Figure 7-30. *Rachael Kess's "Snowman" mask blushes when the wearer touches her cheek*

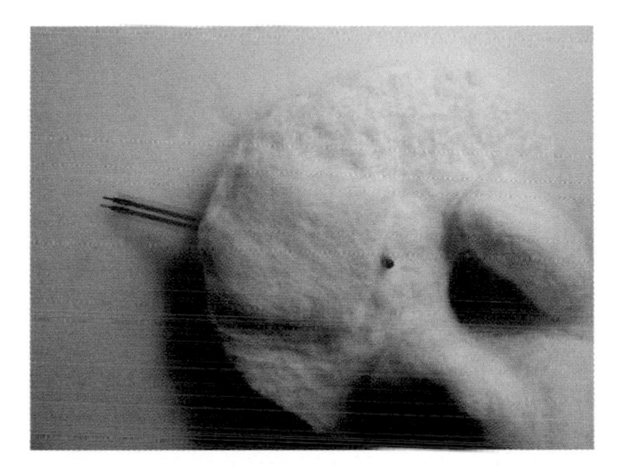

Figure 7-31. *Work in progress image from Rachael Kess's "Snowman"; FSR incorporated into a hand-felted mask*

Stretch

From the bend of a knee to the expansion and contraction of a rib cage with each breath, properly positioned stretch sensors (as shown in Figure 7-32) can capture the fluctuating nuances and curves of the human form. A stretch sensor is simply a conductive rubber cord whose resistance decreases the more it gets stretched. This is yet another example of a variable resistor.

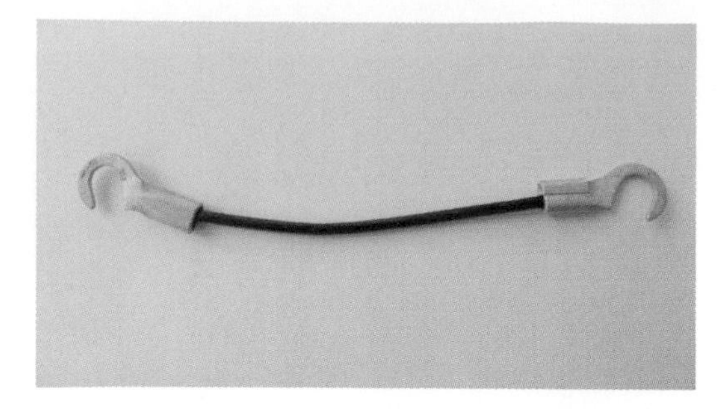

Figure 7-32. *Stretch sensors with hooks attached*

Stretch sensors come precut at different lengths with hooks crimped to either end for easy connection (RO RB-Ima-12, RB-Ima-13, RB-Ima-14, RB-Ima-15, RB-Ima-16, RB-Ima-17, RB-Ima-18), or you can buy it by the meter and cut it to whatever length you need (AF 519). (See Figure 7-33.)

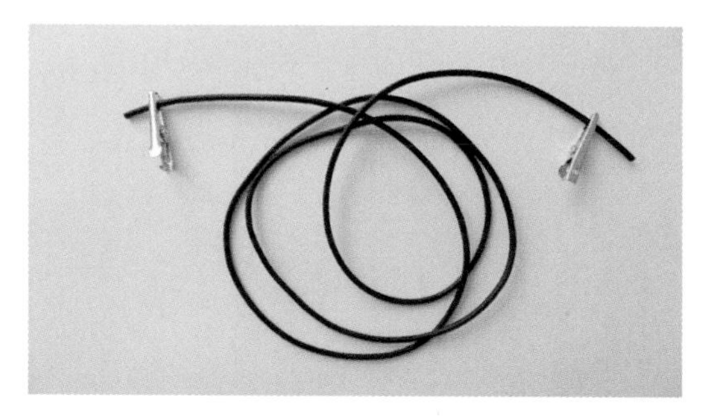

Figure 7-33. *Stretch sensor with hardware for customization*

Stretch sensors are a fun material to work with. They can also be elegantly incorporated into textiles through knitting or weaving.

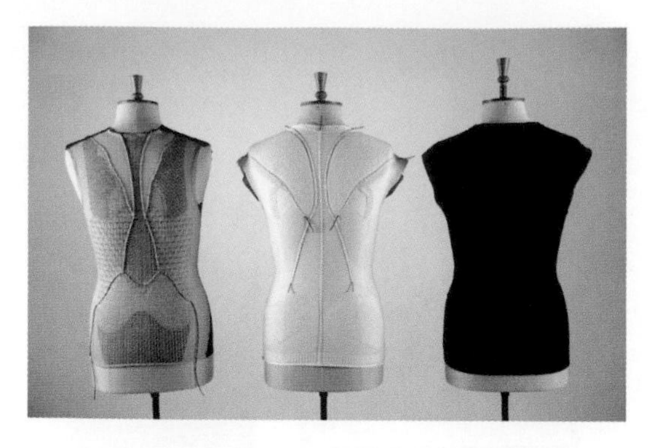

Figure 7-34. *"Aeolia" by Sarah Kettley, with Tina Downes, Martha Glazzard, Nigel Marshall, and Karen Harrigan, explores the process of incorporating stretch sensors into garments through weaving, knitting, and embroidery techniques (photograph by Tina Downes and Catherine Northall)*

Movement, Orientation, and Location

People are active and mobile creatures. They reach for things they want, turn toward loud noises, and crouch down to coax the cat from under the bed. When creating wearables that react to events such as these, it is helpful to be able to sense movement.

A cheap and easy way to sense movement is through the use of tilt switches (shown in Figure 7-35; see "Tilt Switches" on page 55).

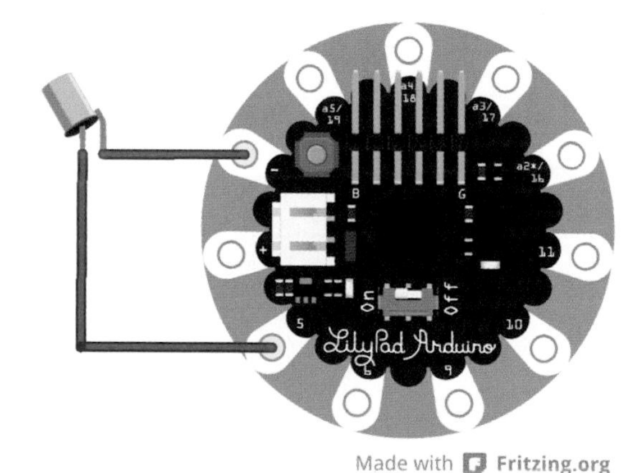

Figure 7-35. *A basic tilt switch can be read by a digital input pin*

But there are also far more sophisticated sensors that you can use. *Accelerometers* measure acceleration or changes in speed of movement. They can also provide a good measurement of tilt due to the changing relationship to gravity.

Accelerometers have a set number of axes—directions in which they can measure. The ones shown in Figure 7-36 are *three-axis accelerometers*, meaning they can measure acceleration on the x, y, and z plane.

Figure 7-36. *Accelerometers: LilyPad Accelerometer, ADXL 335, Flora Accelerometer*

The ADXL335 triple-axis accelerometer is available in a variety of form factors—both on a standard breakout board (AF 163, SF SEN-09269) as well as on a LilyPad board (SF DEV-09267). These boards contain the same chip but are intended for different uses (conductive thread circuit versus breadboard circuit). This is an analog accelerometer, meaning that they output varying voltage for each axis reading. These three outputs can be connected to three different analog inputs on the Arduino. Connections are shown in Figures 7-37 and 7-38.

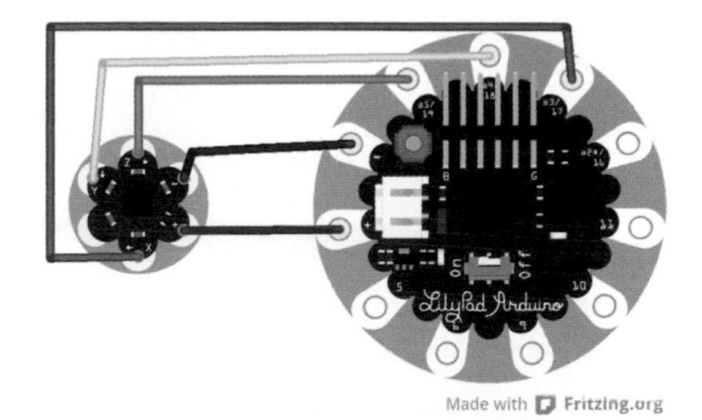

Figure 7-37. *LilyPad Accelerometer circuit layout*

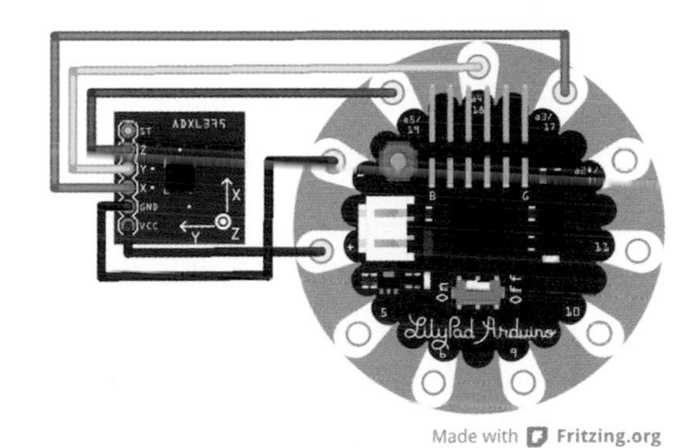

Figure 7-38. *ADXL335 breakout circuit layout*

When working with analog accelerometers, in order to get actual acceleration readings, the raw data from the analog input needs to be interpreted. If need be, see the sensor datasheet for further information on how to do this. If you're just working with relative tilt, observing the changes in sensor data in the serial monitor is often good enough to get you started.

Figure 7-39. *The accelerometer shirt by Leah Buechley uses accelerometer data to control the color of an RGB LED*

There are also digital accelerometers that communicate their data over a serial interface. These require one fewer connection between the sensor and the circuit board; and although they are a little more complex on the code front, they often provide more functionality and slightly less noisy data.

An example of a digital accelerometer is the Flora accelerometer (AF 1247). This module actually includes both an accelerometer and compass. Because it is digital, is it able to provide a lot of data with only four connections. This, like all Flora sensors, uses I2C as the serial communication method. The magnetometer on board senses magnetic north or the direction of whatever the strongest magnetic field is. This can be extremely useful when you want to determine which way a person is facing. Figure 7-40 shows a circuit diagram for this component.

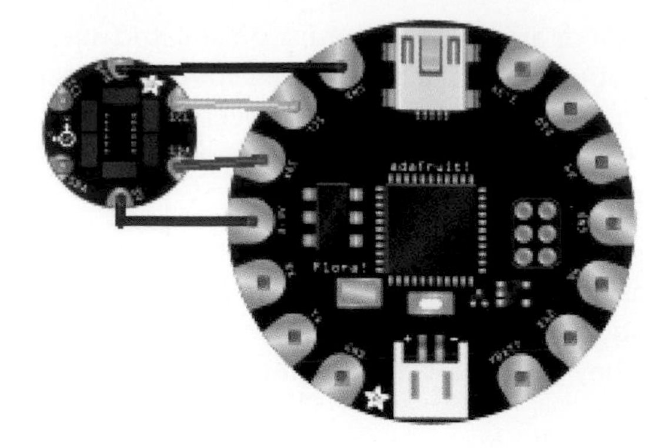

Made with Fritzing.org

Figure 7-40. *Flora accelerometer/compass circuit diagram*

If tilt, motion, and orientation aren't enough, and you want your wearable to know where you are on the planet, GPS is the way to go. Just like your car, bike, or phone, your jacket or disco pants can have GPS, too. There are a number of Arduino-compatible GPS units available, but the Flora GPS (Figure 7-41) is a compact and sewable option.

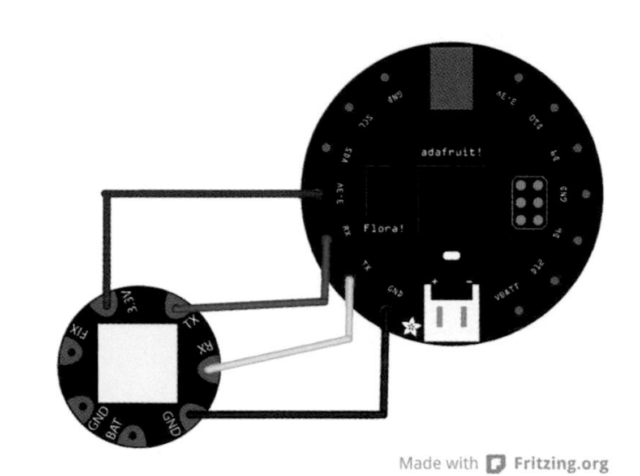

Made with Fritzing.org

Figure 7-41. *Circuit layout for Flora GPS (AF 1059)*

Figure 7-42. *Flora GPS Jacket by Adafruit, Becky Stern, and Tyler Cooper (photographed by Collin Cunningham for Adafruit)*

See also:

- SparkFun accelerometer buying guide (*https://www.sparkfun.com/pages/accel_gyro_guide*)
- LilyPad Accelerometer example (*http://lilypadarduino.org/?p=384*)
- LilyPad Accelerometer Shirt (*http://bit.ly/UcA1pS*)
- Flora Accelerometer tutorial (*http://bit.ly/1zMP3Up*)
- Flora GPS tutorial (*http://learn.adafruit.com/flora-wearable-gps*)
- Flora GPS Jacket (*http://learn.adafruit.com/flora-gps-jacket*)

Heart Rate and Beyond

Your heart beats faster when you're excited, and your skin gets clammy when you're nervous. Besides sensing your environment and your movements, you can also use sensors to learn more about what is happening within someone's body. A great place to start sensing these biometrics is pulse or heart rate.

Optical heart rate sensors (Figure 7-43), such as the Pulse Sensor Amped (AF 1093, SF SEN-11574), are a small, lower-cost solution for measuring pulse. This type of sensor measures the mechanical flow of blood, usually in a finger or earlobe. It contains an LED that shines light into the capillary tissue and a light sensor that reads what is reflected back. It produces varying analog voltage that can be ready by the analog input on any Arduino (Figure 7-44).

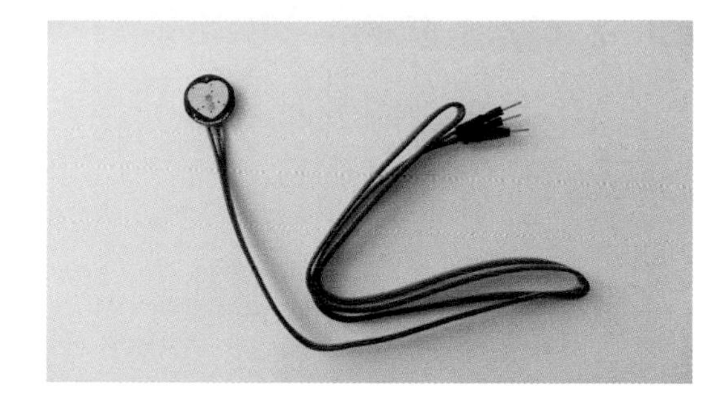

Figure 7-43. *Pulse sensor*

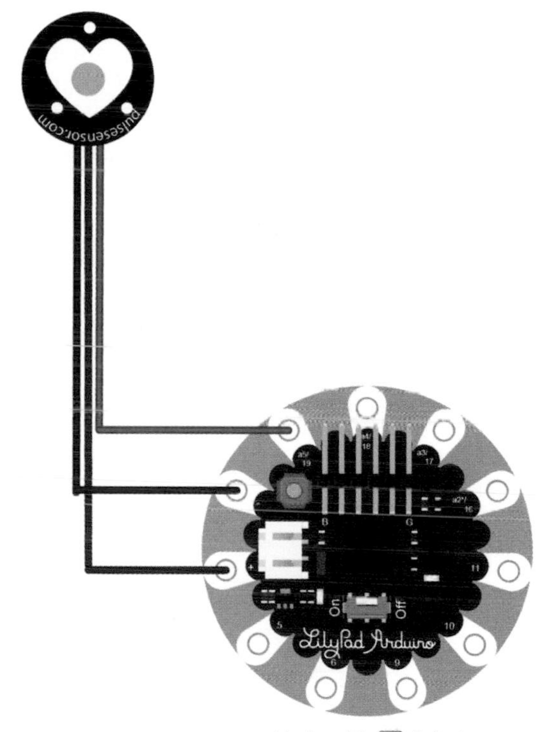

Figure 7-44. *Pulse sensor circuit diagram*

Figure 7-45. *"Heart Strings" by Jackson McConnell and Imman Pirani uses the Pulse Sensor Amped to add another layer of exchange to Skype; each wearer feels the heartbeat of the person on the other end of the video call*

Chest strap heart monitors are a more expensive but more accurate solution for measuring heart rate. They measure the actual electrical frequency of the heart through two conductive electrodes (oftentimes made of conductive fabric) that must be pressed firmly against the skin. Polar produces heart rate monitors that wirelessly transmit a signal with every heartbeat. There are multiple options for receiving this wireless signal. The Polar Heart Rate Monitor Interface (SF SEN-08661) receives the wireless signal from the heart rate monitor band and shares it with the Arduino via I2C (see Figures 7-46 and 7-47). The Heart Rate Educational Starter Pack (AF 1077) includes a simpler setup with a receiver whose output pin pulses high when a heartbeat is detected.

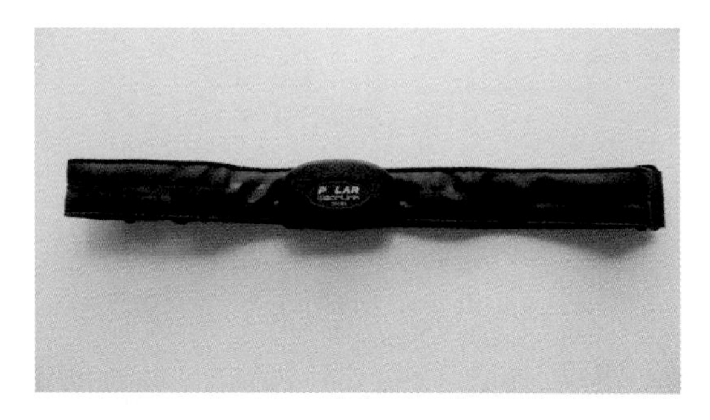

Figure 7-46. *Polar heart rate monitor band*

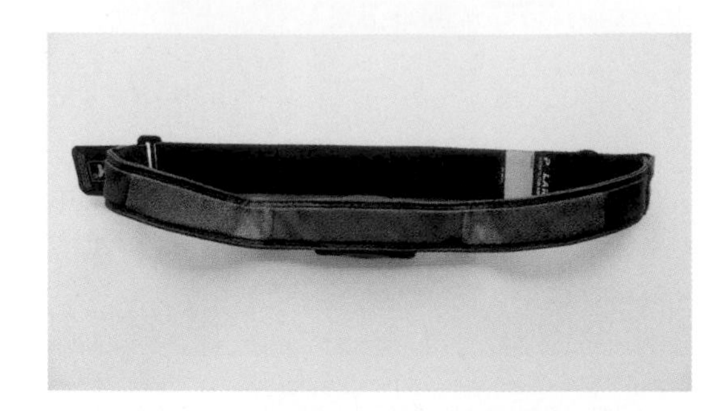

Figure 7-47. *Polar heart rate monitor band (inside)*

Figure 7-48. *"Heart Spark" by Eric Boyd is a custom-designed printed circuit board necklace that receives a signal from a Polar heart rate monitor band and blinks in unison with the wearer's heartbeat*

Beyond heart rate, there are many other biological signals you might want to measure. Here are a few:

Galvanic skin response (GSR)
: A method of measuring the conductivity of the skin. Changes in this conductivity can indicate a response to physical or psychological stimulus. GSR sensors are used in classic lie detectors. A GSR sensor can be built with some basic inexpensive electronic components.

Electromyography (EMG)
: A method of measuring of muscle activity by detecting its electrical potential. The Muscle V3 Sensor Kit (SF SEN-11776) provides varying analog voltage so you can easily read muscle activity with an Arduino analog input pin.

Electroencephalography (EEG)
: A method of measuring electrical activity in the scalp. EEG headsets are often used in thought-controlled computing applications.

To learn more about how to work with these types of sensors, check out Make Volume 26 for the "Biosensing" article by Sean Montgomery (Figure 7-49) and Ira Laefsky.

Figure 7-49. *Sean Montgomery creates a variety of biometric-data-driven wearables; he is pictured here wearing his "Thinking Cap," which responds to fluctuations in EEG signals*

Figure 7-50. *"PyroKinesis" by Seth Hardy uses EEG readings to enable wearers to control a flame effect with their brainwaves*

See also:

- "Polar Heart Rate Monitor Interface + Arduino" (*http://bit.ly/UcAfxh*)
- "Pulse Sensor Getting Started Guide" (*http://bit.ly/1oMP41P*)

Proximity

Sometimes you will want to know how close or far away something is from the body. Proximity sensors are useful for detecting nearby objects, walls, or even other people (Figure 7-51). When selecting a proximity sensor, it is worth considering what your desired range or detecting distance is, as well as what sort of beam width you need to monitor.

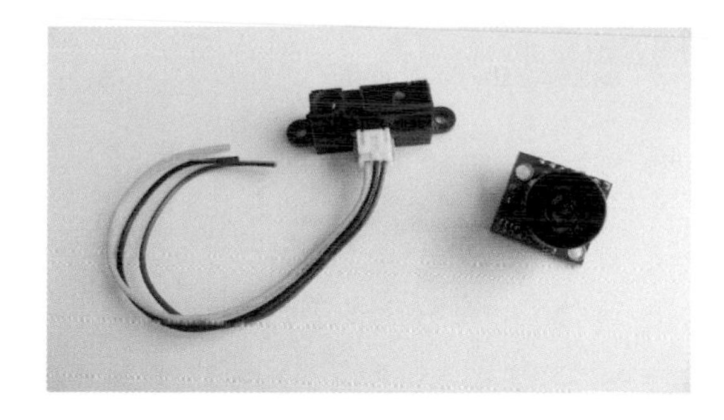

Figure 7-51. *Proximity sensors*

There are two types of proximity sensors that are fairly easy to get up and running: infrared or ultrasonic.

Infrared or IR sensors (Figure 7-52) use light to measure proximity. The sensor sends out a beam of infrared light (invisible to the human eye) that bounces off the object in front of it and is read by the sensor. IR sensors are the less expensive option for proximity sensors but are more easily tricked by heat and sunlight. They tend to have shorter, more focused distance ranges, such as 3 to 30cm (SF SEN-08959), 10 to 80cm (AF 164, SF SEN-00242), or 20 to 150cm (AF 1031, SF SEN-08958).

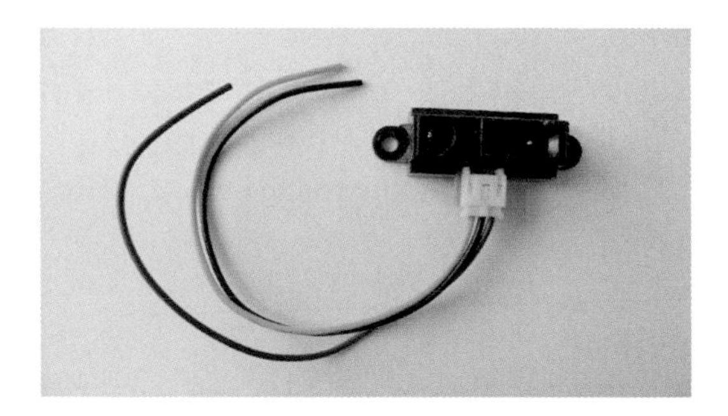

Figure 7-52. *Infrared proximity sensor*

These sensors can be directly connected to the analog input on any Arduino board, as shown in Figure 7-53.

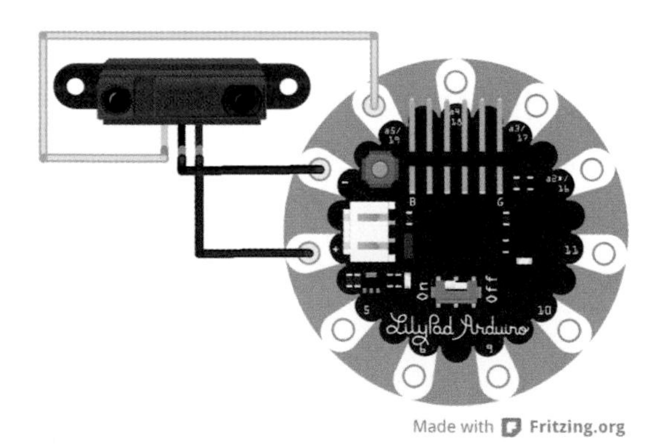

Figure 7-53. *IR circuit diagram*

Ultrasonic sensors (Figure 7-54) work similarly except that instead of sending out light, they send out ultrasonic sound (which can't be heard by humans). The sound bounces off whatever is proximate and returns to the sensor. The proximity is determined by the length of time it takes for the sound to return. Maxbotix manufacturers a sophisticated line of ultrasonic sensors that meet a range of needs from the most basic to highly sensitive and rugged. They come in a variety of beam widths, have long sensing ranges (0 to 150 inches on their most basic model: AF 979, SF SEN-08502), and even are available with waterproof outdoor housings. Ultrasonic sensors are more expensive and bulkier than IR sensors, but they are more precise and harder to trick. A sample circuit is shown in Figure 7-55.

Figure 7-54. *Ultrasonic proximity sensor*

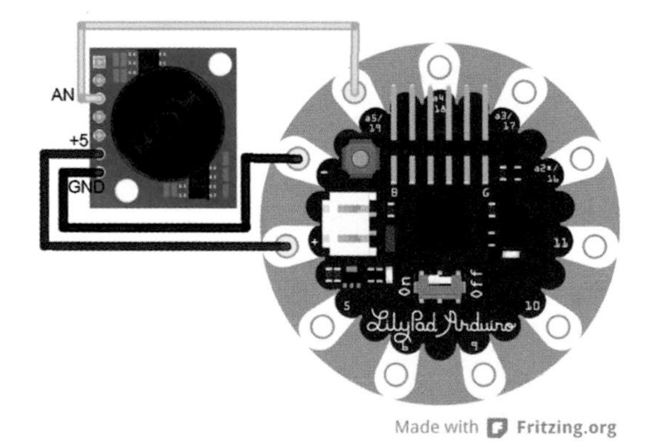

Figure 7-55. *Ultrasonic circuit diagram*

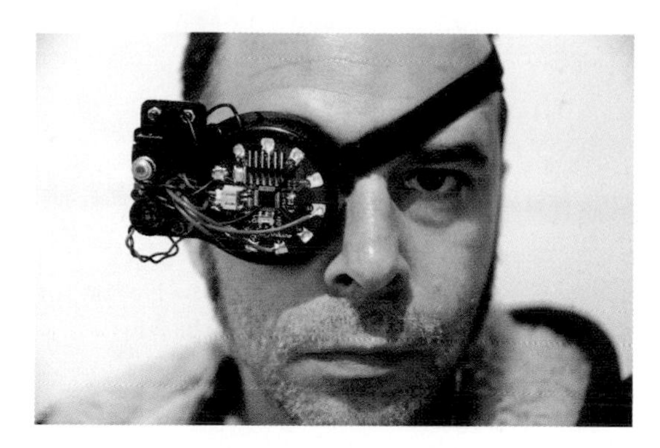

Figure 7-56. *"Augmented Vision" by Greg McRoberts is a wearable seeing aid device that uses flashing RGB LED to represent fluctuating data gathered by an infrared heat sensor and ultrasonic distance sensor*

Light

Remember your old friend the light sensor? You used a LilyPad Light Sensor (SF DEV-08464) in your first analog input example, but light sensors come in many other forms (Figure 7-57).

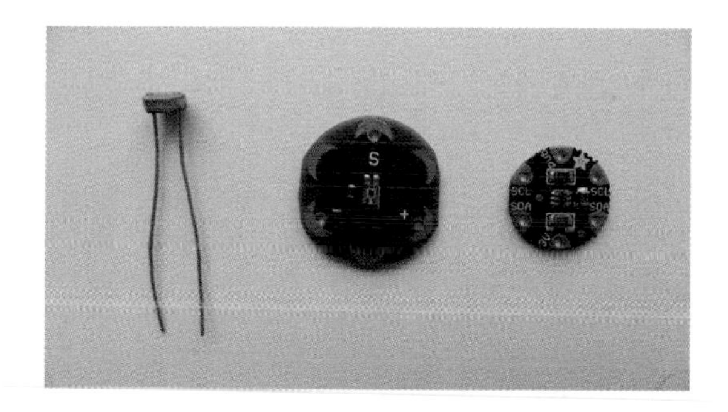

Figure 7-57. *Photocell, LilyPad Light Sensor, Flora Light Sensor*

The most basic type of light sensor is the photocell (AF 161, SF SEN-09088). Its resistance varies based on the level of light it senses. Some have resistance that increases as the light level increases, but some have the reverse relationship. This can be quickly determined by viewing the sensor values in the serial monitor (see the example in "Analog Input" on page 108). Figure 7-58 shows the circuit design for using this sensor.

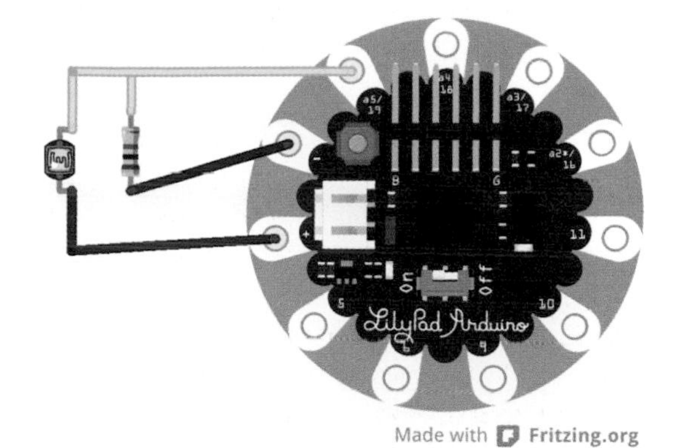

Figure 7-58. *Photocell with voltage divider circuit connected to LilyPad Arduino Simple*

The photocell is a great sensor to work with because it is small, easy to manipulate, and incredibly inexpensive. It can be used to sense ambient light levels, but it can also be used for less intuitive purposes like determining whether a jacket is open or closed or if the heel of a shoe is on the ground or in the air.

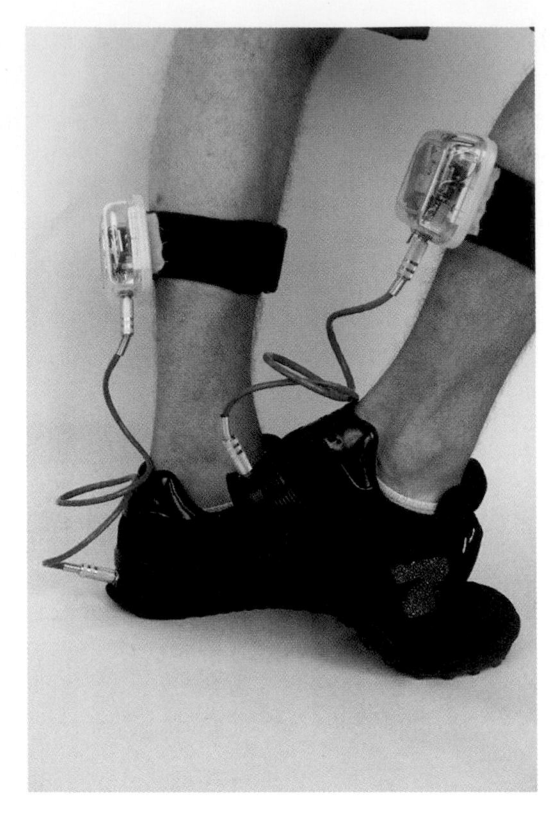

Figure 7-59. *"Perform-o-shoes" by Andrew Schneider are music-controlling footwear that have a photocell embedded in the bottom of the heel; the higher the shoe is off the ground, the faster the music track will play*

The Flora Lux Sensor (AF 1246) is a more sophisticated light sensor (see Figure 7-60). It measures infrared, full-spectrum, and human-visible light, which means which means that your wearable can know the difference between daylight, artificial light, or even light that's invisible to humans. This sensor has an I2C interface.

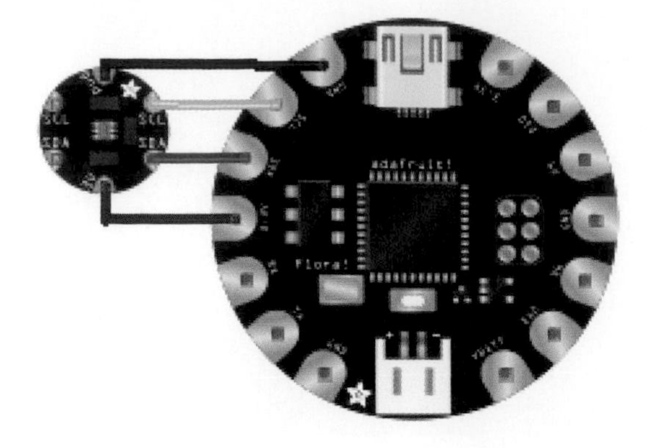

Figure 7-60. *Flora Lux Sensor circuit diagram*

See also:

- Adafruit's Flora Lux Sensor tutorial (*https://learn.adafruit.com/flora-lux-sensor*)

Color

Color is hugely important in the worlds of design and fashion. Whether it be a chameleon effect or a dynamic effort to stand out from the crowd, the ability to sense color enables a garment to be able to perceive and react to its stylistic context.

There are many color sensors out there, but the Flora Color Sensor (AF 1356) has the added bonus of being sewable (see Figure 7-61) and having an onboard LED that helps to illuminate the object whose color you are trying to sense.

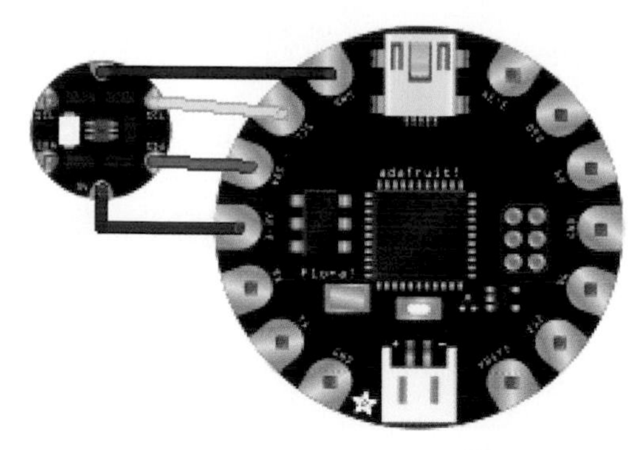

Figure 7-61. *Flora Color Sensor circuit diagram*

Figure 7-62. *Intended to get fourth- and fifth-grade girls interested in wearable computing and programming, Glowbowz by Jaymes Dec are the world's first programmable hair bow; RGB LEDs sewn into Glowbowz can be programmed to change colors based on any sensor data; this version of Glowbowz uses a color sensor to match the bow's color to whatever outfit the wearer wants*

See also:

- Adafruit's Chameleon Scarf tutorial (*https://learn.adafruit.com/chameleon-scarf*)

Sound

Sounds can provide significant clues about what is going on around you. By detecting sound level, you can create wearables that are more sensitive to their environment, like a scarf that purrs when it is whispered to or a collar that pops up in response to loud noises.

A simple microphone can act as a great sensor for audio-reactive projects. For getting started with reading an audio signal in Arduino, a small electret microphone will do the trick. These little guys can't be plugged directly into an analog input—their fluctuating signal is measured in microvolts, which is far too subtle for the ears of your microcontroller. But they are available on breakout boards (AF 1064, SF BOB-09964) that feature an amplifier chip and other components (see Figures 7-63 and 7-64) that allow it to be directly connected to an Arduino. The Adafruit model features a *trimpot* (a knob that can be adjusted with a screwdriver) on the back that allows you to make adjustments to the gain on the fly. Figure 7-65 shows a circuit you can use, and you can also use conductive thread, as shown in Figure 7-66.

Figure 7-63. *Electret Microphone Amplifier—MAX4466 with Adjustable Gain (front)*

Figure 7-64. *Electret Microphone Amplifier—MAX4466 with Adjustable Gain (back)*

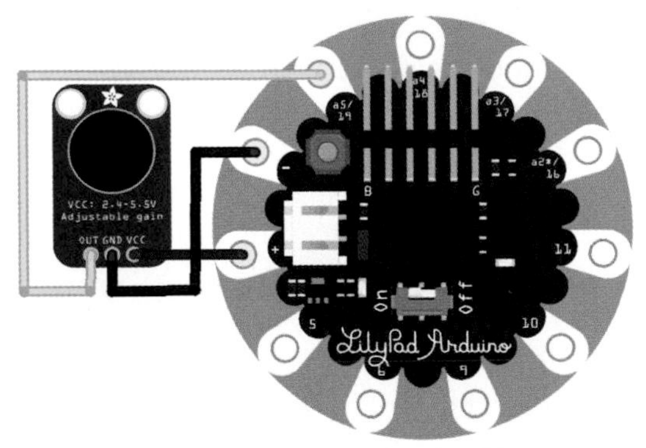

Figure 7-65. *Electret Microphone Amp circuit layout*

Figure 7-66. *Holes meant for headers can also be used for conductive thread connections*

The microphone signal can be read by an analog input pin, but you need to do further calculations to determine the amplitude (volume) that the mic is detecting. Here's some code to get you started:

```
/*
Make: Wearable Electronics
 Mic Example
 Based on "Example Sound Level Sketch for the
 Adafruit Microphone Amplifier"
 http://bit.ly/1qlN7hk
 */

int micPin = A2;

// Sample window width in mS (50 mS = 20Hz)
int sampleWindow = 50;

void setup(){
  Serial.begin(9600);
}

void loop() {
    // Start of sample window
  unsigned long startMillis = millis();
  int amplitude;
  int micReading;
  int maxReading = 0;
  int minReading = 1024;

  // collect mic readings and find the
  // max and min
  while (millis() - startMillis < sampleWindow){
    micReading = analogRead(micPin);

    if (micReading > maxReading){
      maxReading = micReading;
      //save the maximum reading
    }
    else if (micReading < minReading){
      minReading = micReading;
      // save the minimum reading
    }
  }

  //find the amplitude
  amplitude = (maxReading - minReading);
  Serial.println(amplitude);
}
```

See these other examples for details:

- *Arduino Cookbook*, Recipe 6.7
- Adafruit Microphone Amplifier Breakout tutorial (*http://bit.ly/UcBcpq*)

Temperature

Clothing provides warmth and protection. It makes sense that responsive clothing might want to react to temperature. Temperature sensors can be used to sense both environmental conditions as well as the warmth of the body. There are many temperature sensors available from analog to digital, to high temperature to waterproof, to noncontact. They are also sometimes combined with sensors for barometric pressure, humidity, and altitude.

Each of these units work in its own way and requires a bit of research and testing. An easy place to start is with a thermistor. A thermistor is a variable resistor and can be connected to the Arduino with a simple voltage divider circuit (see Figure 7-67).

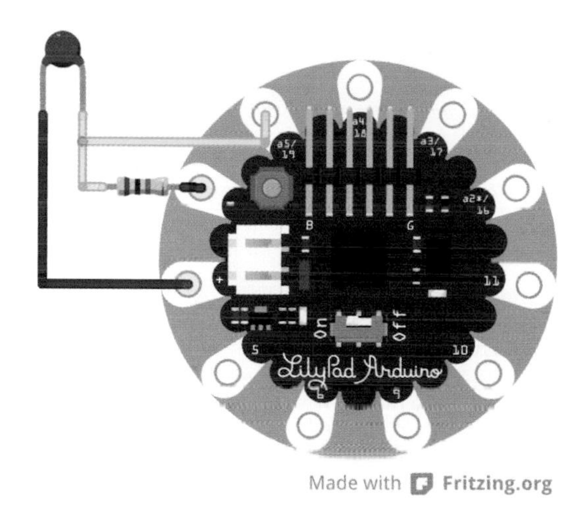

Figure 7-67. *Thermistor circuit layout*

Thermistors are not the most precise temperature sensors, but they are excellent for rough temperature comparison. For example, you can easily use a threshold to create a distinction between what is considered hot and what is considered cold.

For more precise reading, try working with a LilyPad Temperature Sensor (SF DEV-08777, shown in Figure 7-68), or a TMP36, a simple analog temperature sensor (Figure 7-69).

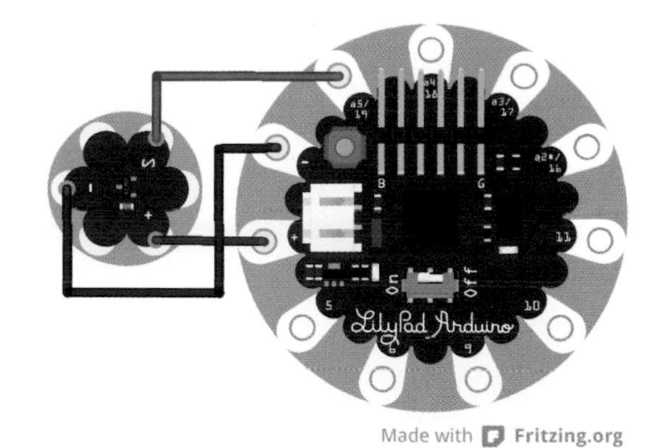

Figure 7-68. *LilyPad Temperature Sensor circuit layout*

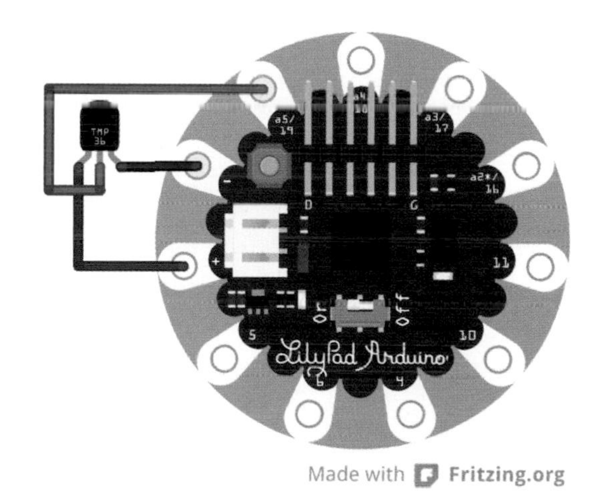

Figure 7-69. *TMP36 circuit layout*

The following code will work with either the LilyPad Temperature Sensor or the TMP 36. Be sure to change the `supplyVoltage` variable to whatever voltage you are working with in your circuit (see Figures 7-70, 7-71, 7-72, and 7-73).

```
/*
Make: Wearable Electronics
 Temperature Sensor example
 */

// This is a reference voltage for your power
// supply. Measure it with a multimeter when
// running and change to the correct voltage.
float supplyVoltage = 3.7;
```

```
int tempSensorPin = A2;
int tempSensorValue;
float tempSensorVoltage;


// the setup routine runs once when you press
// reset:
void setup() {
  // initialize serial communication at 9600
  // bits per second:
  Serial.begin(9600);
}


void loop() {
  // read the temperature sensor value
  tempSensorValue = analogRead(tempSensorPin);
  ;

  // convert the reading to voltage based off
  // the reference voltage
  float tempSensorVoltage =
  (tempSensorValue * supplyVoltage)/1024.0;

  // convert the reading to Celsius
  // converting from 10 mv per degree with
  // 500 mV offset
  float temperatureC =
  (tempSensorVoltage - 0.5) * 100 ;
  // to degrees ((tempSensorVoltage - 500mV)
  // times 100)

  // print in Celsius
  Serial.print("Degrees C: ");
  Serial.print(temperatureC);

  // convert to Fahrenheit
  float temperatureF =
  (temperatureC * 9.0 / 5.0) + 32.0;

  // print in Fahrenheit
  Serial.print(", Degrees F: ");
  Serial.println(temperatureF);

  delay(100);
}
```

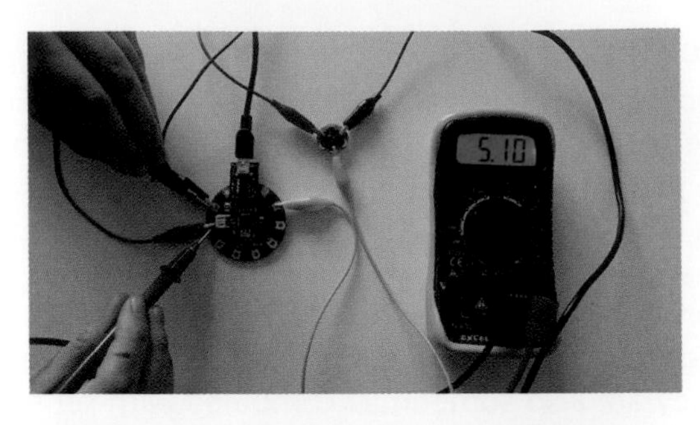

Figure 7-70. *Measuring the voltage from a 5V FTDI board*

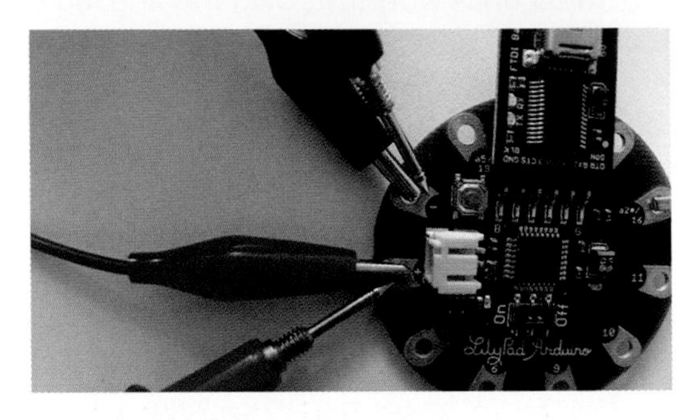

Figure 7-71. *Measuring the voltage from a 5V FTDI board (detail)*

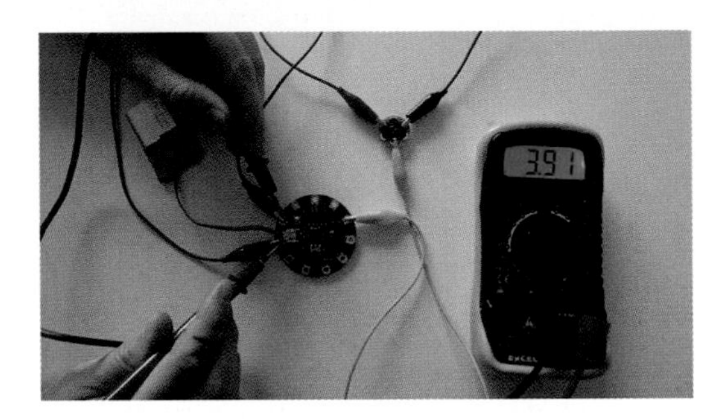

Figure 7-72. *Measuring the voltage from a 3.7V lithium polymer battery*

Figure 7-73. *Measuring the voltage from a 3.3V FTDI board*

See also:

- Adafruit's TMP 36 Temperature Sensor Overview (*http://bit.ly/UcBvAp*)
- SparkFun Inventor's Kit, example 7
- *Arduino Cookbook*, Recipes 6.8 and 13.5

DIY Sensors

In addition to manufactured sensors, you can also create your own. As you saw earlier, a variable resistor is simply something that changes resistance in response to a changing condition. Think about this from a material perspective, and you can end up with some pretty interesting results.

Figure 7-74. *"Felt Stretch Sensor" by Lara Grant*

There are many DIY techniques for sandwiching a semi-resistive material between two pieces of conductive material (Figure 7-75). The semi-resistive material can be a plastic (like Velostat) or a fabric (like some made by Eeonyx), and the conductive material can be conductive fabric, thread, yarn, wire, mesh, or anything else you can dream up. Use a connection to each conductor as the two sides of a variable resistor, and you can monitor the change in values as you apply pressure to—or flex—the sensing sandwich that you've created.

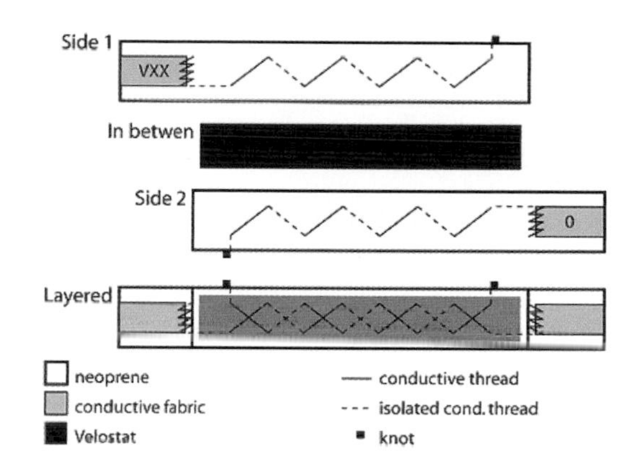

Figure 7-75. *Flex sensor assembly diagram by Hannah Perner-Wilson*

You can also develop a material that is itself a variable resistor. Many artists and makers have experimented with felting together sheep's wool and conductive fibers (such as steel or copper wool). The addition of the nonconductive sheep's wool to the mix creates electrical resistance. The more you compress the mixed wool, the closer the conductive fibers become to each other—thus lowering the resistance and increasing conductivity.

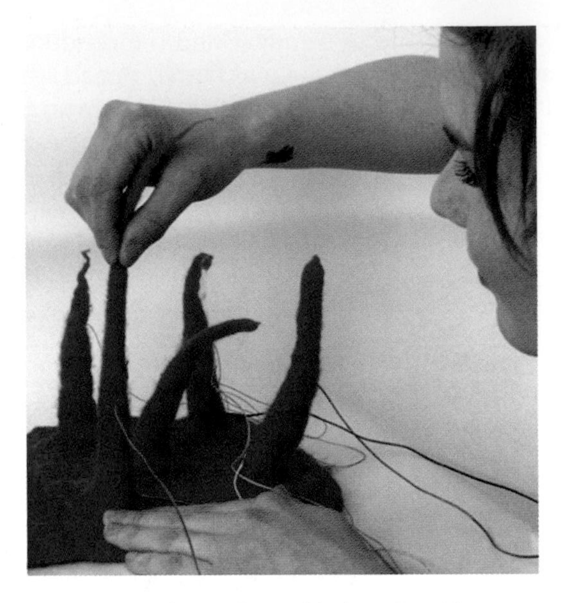

Figure 7-76. *"Felt Stroke Sensor" by Lara Grant*

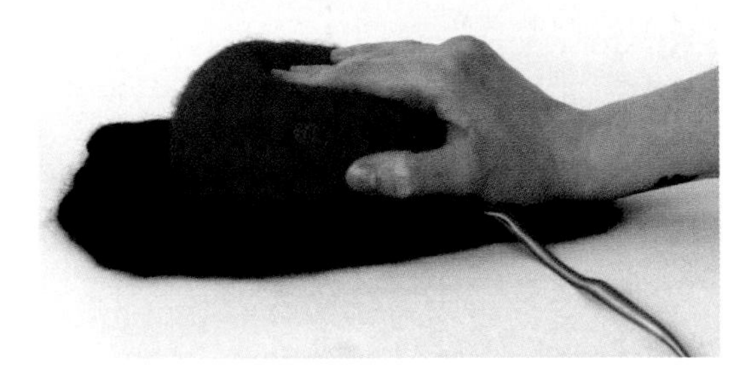

Figure 7-77. *"Felt Pressure-Sensitive Button" by Lara Grant*

A similar effect can be achieved with knitting or crocheting somewhat conductive yarns (Figure 7-78). The more the knit is stretched or pressed, the more highly conductive it becomes.

Figure 7-78. *Knitted Pressure Sensors from How To Get What You Want*

Hannah Perner-Wilson and Mika Satomi maintain a website called How To Get What You Want (*http://www.kobakant.at/DIY*), which is home to a vast repository of DIY Wearable Technology documentation. Check out their "Sensor" section for a helpful collection of tutorials on how to make your own sensors.

Experiment: Body Listening

The interfaces you use tend to target specific areas of the body, such as hands, fingers, and feet. But what are other parts or areas of the body that aren't properly considered? For this experiment, create an interface for a part of the body that you think is not listened to enough.

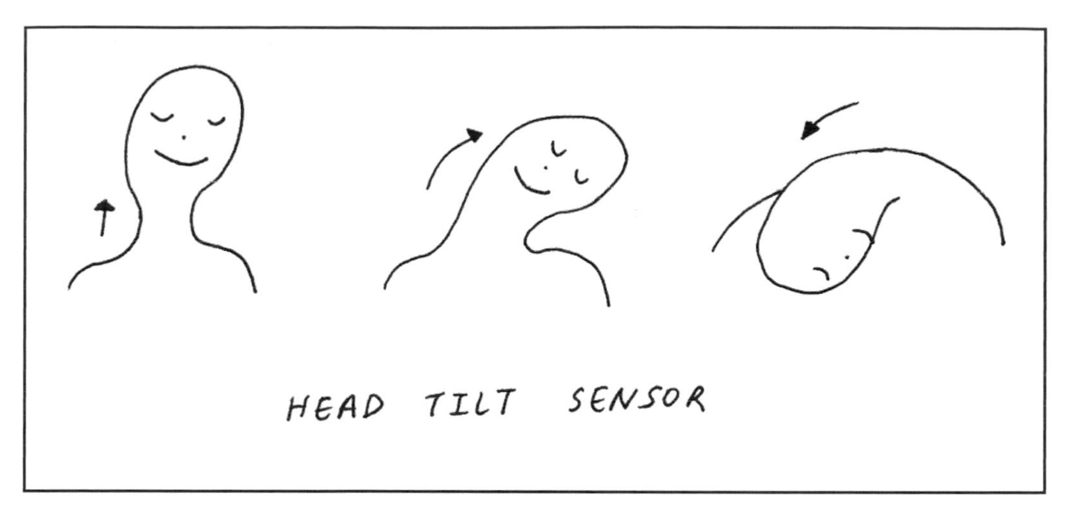

Figure 7-79. *Head Tilt Sensor (illustration by Jen Liu)*

Here's a process to follow:

1. Decide on a body part or area of focus.
2. Make a list of five ways you can sense or listen to that area.
3. Pick one approach that you can easily prototype.
4. Prototype it.
5. Try out your invention.
6. Make adjustments to code and hardware as needed.
7. Repeat until you think it listens well.

Figure 7-80. *"Kegel Organ" by Erin Lewis allows the user to play a musical instrument through contractions of the pelvis floor muscles*

Other Sensors

This introduction to sensors is really meant as a springboard to launch you into the deep and beautiful pool of sensor possibilities. Remember to start from your concept and work out from there. "Is there a sensor that senses X?" is a great question to bring to a search engine, an online forum, or your neighborhood nerd friend. From there, let the datasheet be your guide and you'll be on your way to producing smartly sensitive wearable systems.

Figure 7-81. *"Concussion Helmet" by Michael Vaughan provides a visual indication when hockey players have been hit too hard in the head to return to the game*

Actuators

8

Actuators are the things that go boom, blink, and bzzzzt. They are the things that make things happen. In this chapter, I cover actuators that produce a range of outcomes, including light, sound, movement, and heat. Through the use of these components, you'll be able to produce garments that can glow, shake, and sing.

Figure 8-1. *"Freestyle SoundKits" by Jessica Thompson are wearable sound pieces that generate and broadcast electronic beats as users move through the urban environment*

Light

Whether you're a cyclist or a fashionista, there are times when being seen can make all the difference. Here I review a variety of ways to wear light.

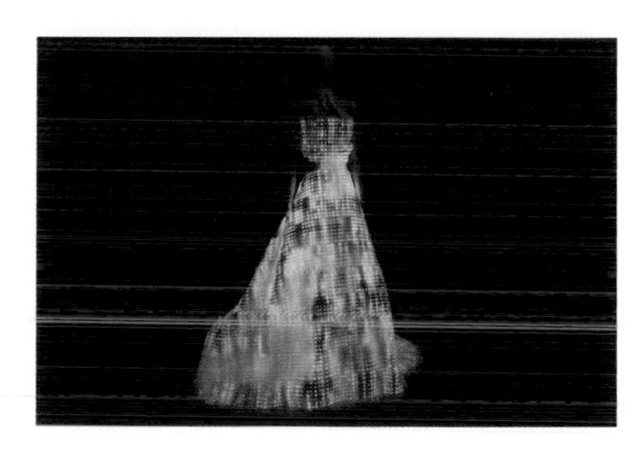

Figure 8-2. *"The Galaxy Dress" designed by CuteCircuit (photograph by JB Spector, Museum of Science and Industry of Chicago)*

Basic LEDs

You first encountered LEDs in Chapter 1 and you've been using them as a basic output ever since. Let's take a moment to get to know LEDs a little bit better.

First of all, LEDs, like most electronic components, come in different types of packages (see

Figure 8-3). Through-hole LEDs are easy to handle and prototype with, but surface mount LEDs tend to integrate more delicately with the design of garments.

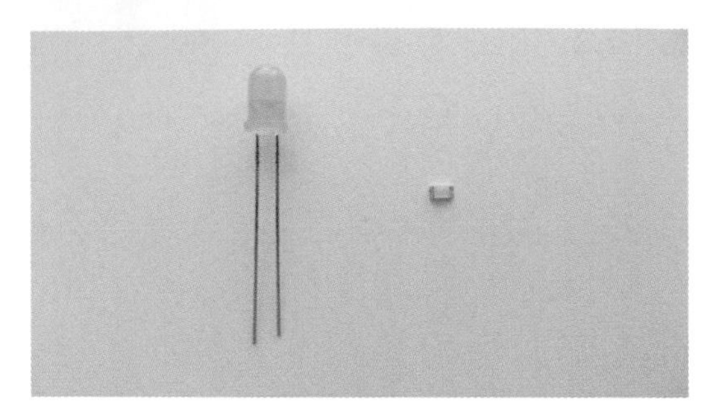

Figure 8-3. *LED packaging types: through-hole (left) and surface mount (right)*

Each type of package comes in many different sizes and sometimes even in different shapes (see Figures 8-4 and 8-5).

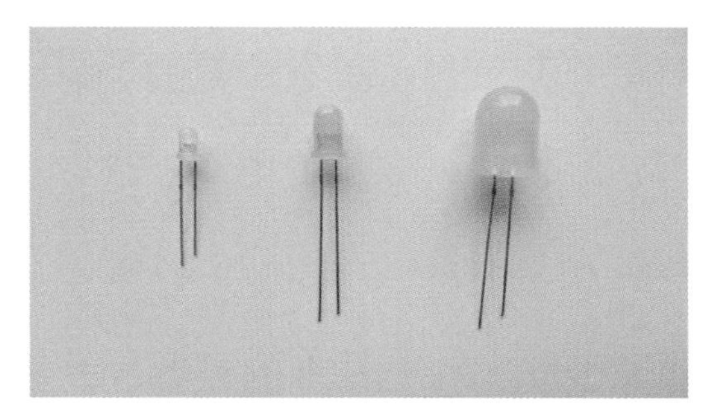

Figure 8-4. *Through-hole LED sizes: 3mm, 5mm, and 10mm*

Figure 8-5. *Surface mount LED sizes: 7805 and 1206 packaging*

LEDs also differ by color, brightness, and viewing angle. Be sure to consult the product description and datasheet of the LEDs you are working with to get the details of how they'll look and what they need to get glowing.

There are many options for controlling LEDs. As you know from the examples in Chapter 6, you can use a single pin of a LilyPad Arduino to control three LilyPad LEDs in parallel (Figure 8-6). These three LEDs will behave in the same way.

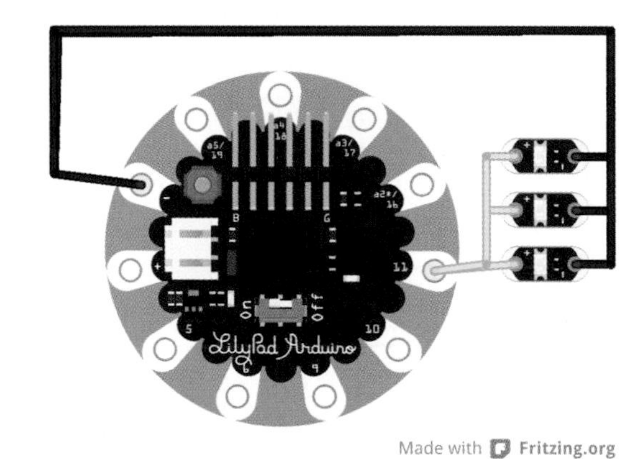

Figure 8-6. *LilyPad Arduino Simple with three LilyPad LEDs in parallel controlled by pin 11*

If you would like these LEDs to have different behaviors, you would have to use three different digital output pins, as shown in Figure 8-7.

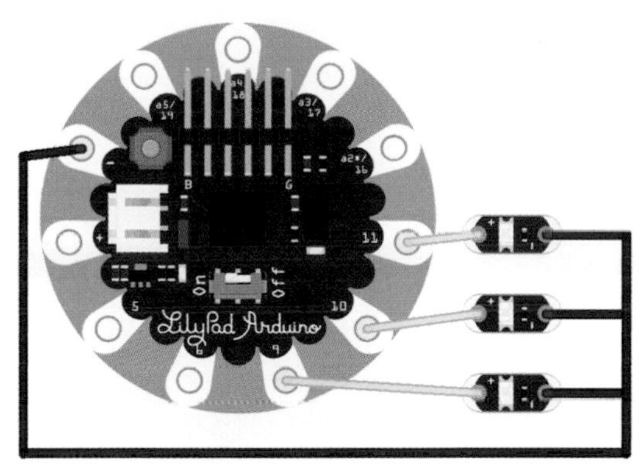

Figure 8-7. *LilyPad Arduino Simple with LilyPad LEDs on pins 9, 10, 11—individually controllable*

Each output pin can power up to three LilyPad LEDs, so you can connect up to 27 LEDs (see Figure 8-8) to a LilyPad Arduino Simple to be controlled by its nine digital output pins. Just be sure you are working with a battery that can supply the necessary current for the LilyPad Arduino and LEDs (approximately 400mA).

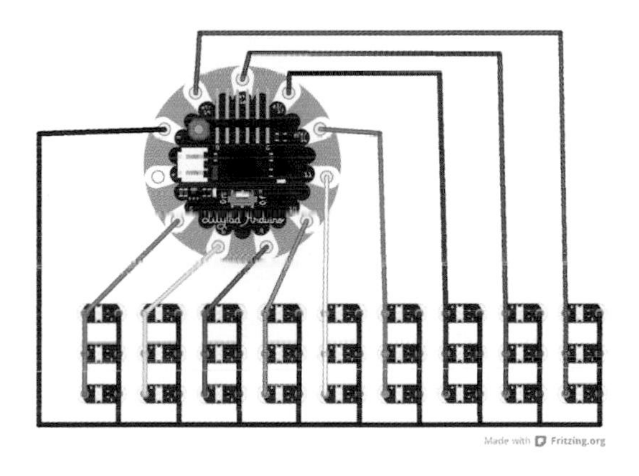

Figure 8-8. *Controlling a large number of LEDs with the LilyPad Simple*

If you would like to control a large number of LEDs with a single pin, there are some low-voltage LED string lights available (see Figures 8-9 and 8-10). These also can be powered by a 3V coin cell battery like the CR2032. Figure 8-11 shows a circuit you can use with an LED string, and you can see it lit in Figure 8-12.

Figure 8-9. *LED string light (SF COM-11751)*

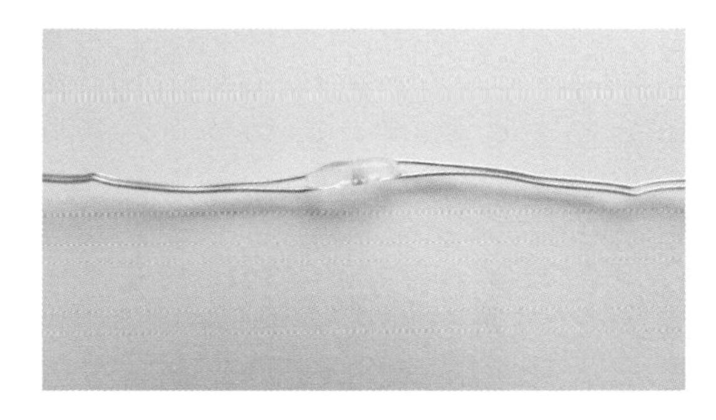

Figure 8-10. *LED string light, detail*

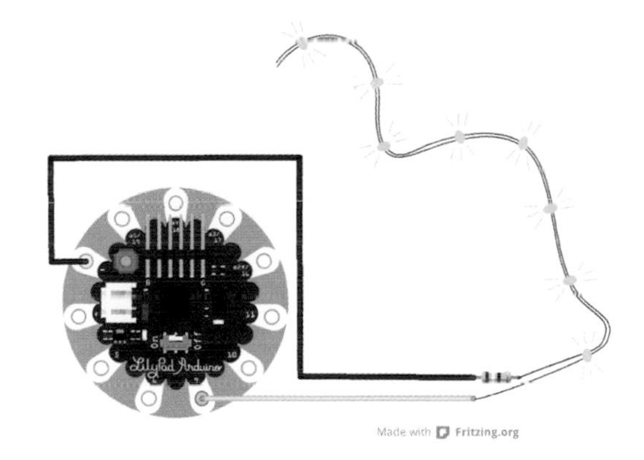

Figure 8-11. *LED string lights on pin 9*

Figure 8-12. *LED string light, lit*

Finally, if you need to control a large number of basic LEDS individually, this can be accomplished through techniques such as *multiplexing*, *Charlieplexing*, or the use of components such as shift registers and PWM extender chips. For more information on these options, check out the Visual Output chapter in the *Arduino Cookbook* by Michael Margolis.

There are also LEDs that can light in multiple colors like an RGB LED (see Figure 8-13). RGB stands for "red, green, blue." These LEDs have four pins—three that correspond to each color and a fourth that is either a common anode (meant to connect to power) or common cathode (meant to connect to ground, shown in Figure 8-14). The color that the LED displays depends on the intensity of the PWM signal of each color pin.

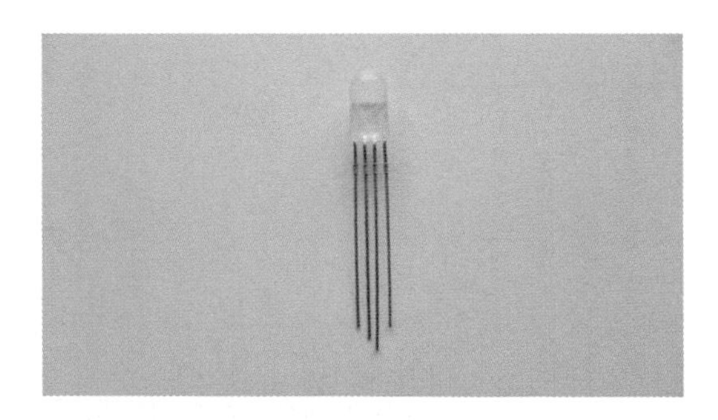

Figure 8-13. *RGB through-hole LED*

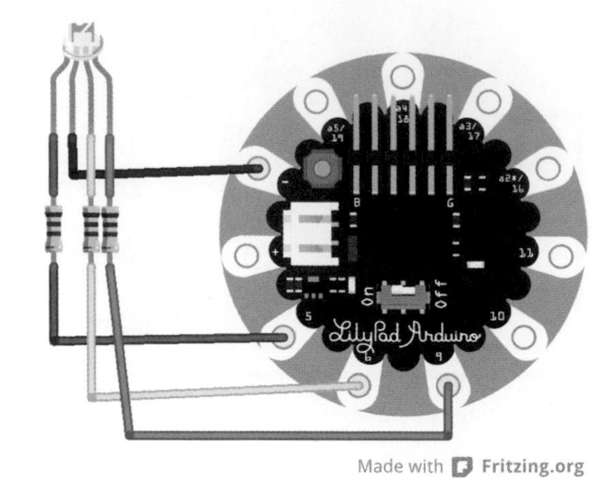

Figure 8-14. *Circuit layout for an RGB through-hole LED with a common cathode*

The LilyPad TriColor LED (SF DEV-08467) is shown in Figure 8-15, and you can see a circuit diagram in Figure 8-16. For instructions on how to use this LED, check out the tutorial on the LilyPad Arduino website (*http://lilypadarduino.org/?page_id=548*).

Figure 8-15. *LilyPad TriColor LED*

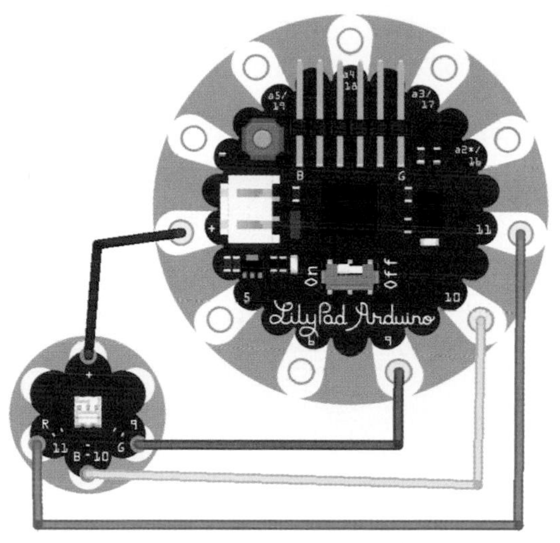

Figure 8-16. *Circuit layout for LilyPad Tricolor LED; note that this module has a common anode*

There are a variety of ways to wear LEDs, whether it be for safety, style, or making a statement. Figures 8-17 through 8-19 show some examples.

Figure 8-17. *"The Sessile Handbag" by Grace Kim merges technology with natural forms: hand-felted "barnacles" are combined with embroidered LEDs (photographed by Jeannie Choe)*

Figure 8-18. *LED eyelashes by Soomi Park apply LEDs directly to the body, intended to create the illusion of larger eyes*

Figure 8-19. *Jacket Antics by Barbara Layne from Studio SubTela feature LED matrixes that work in tandem to create a multibody display (photographed by Hesam Khoshneviss)*

Basic LEDs are just one way to get started with illuminated clothing. In the following sections, I review additional tools that can be used to create wearable light.

Addressable LEDs

When working with LEDs, you sometimes want to create a visual effect that is bright, bold, and extremely dynamic. The Flora RGB Smart NeoPixel (Figure 8-20) is one of the most versatile modules in the Flora toolkit. It consists of wearable, sewable, easily wired, individually addressable, ultra-bright, multicolored LEDs. What more could you want out of a light-emitting diode?

Figure 8-20. *Flora Neopixels, V2*

The NeoPixels are meant to be used in combination with the Flora main board. They require three connections—power, ground, and a connection to either a digital output pin (for the first NeoPixel) or the NeoPixel in the chain before it (for the NeoPixels that follow). See the circuit layout diagrams in the examples that follow to see how these connections are made.

For the software, you will need to work with Adafruit's special version of the Arduino IDE as well as an additional library for the NeoPixels. Follow the Getting Started with Flora (*http://bit.ly/UcBWKY*) tutorial in the Adafruit Learning System for the most up-to-date instructions.

Once your Flora is up and running, you'll be ready to get going with the Neopixel. Here are some examples.

One NeoPixel example

To get started, let's light up a single NeoPixel. Once you understand the basics, then you can let the fanciness explode.

Parts:

- (1) Flora (AF 659)
- (1) Flora RGB Smart NeoPixel version 2 (AF 1260)
- Alligator clip test leads (AF 1008, RS 278-1156, SF PRT-11037)
- USB mini-B cable (AF 899, DK WM5163-ND, RS 55010682, SF CAB-11301)
- 3.7V lithium-ion polymer rechargeable battery (AF 258, SF PRT-00339)

The circuit layout is shown in Figure 8-21.

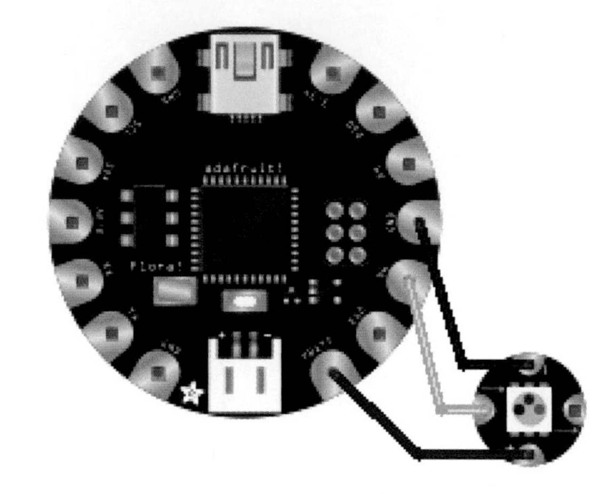

Figure 8-21. *Flora with one NeoPixel*

Once your circuit is assembled, program your Flora with the following code:

```
/*
Make: Wearable Electronics
 Flora NeoPixel example with 1 pixel
 */

#include <Adafruit_NeoPixel.h>

// The digital pin used to control the
// pixel strip
int pinNumber = 6;

// The number of pixels in the strip
int numberOfPixels = 1;

Adafruit_NeoPixel strip =
  Adafruit_NeoPixel(numberOfPixels,
  pinNumber, NEO_GRB + NEO_KHZ400);

void setup() {
  // initialize pixel strip
  strip.begin();
  // set pixels to off to begin
  strip.show();
```

```
}

void loop() {
  // set pixel 0 to red
  strip.setPixelColor(0, 255, 0, 0);
  strip.show();
  delay(500);

  // set pixel 0 to green
  strip.setPixelColor(0, 0, 255, 0);
  strip.show();
  delay(500);

  // set pixel 0 to blue
  strip.setPixelColor(0, 0, 0, 255);
  strip.show();
  delay(500);

  // turn pixel 0 off
  strip.setPixelColor(0, 0, 0, 0);
  strip.show();
  delay(1000);
}
```

There are a few commands in this code that are worth explaining:

```
Adafruit_NeoPixel(numberOfPixels, pinNumber,
NEO_GRB + NEO_KHZ400);
```

This command has three parameters: the number of pixels, the pin number, and the pixel type flag (don't change that one). Be sure to adjust this if you change pins or the number of pixels you are using:

```
strip.setPixelColor(0, 255, 0, 0);
```

This is used to set the pixel color—big surprise! You need to use this command to set each pixel individually—the first parameter is the pixel number (starting with 0) and the second, third, and fourth are the red, green, and blue values:

```
strip.show();
```

Once all of your pixels have been set, this command lights the entire strip with the predetermined colors. Color changes will not appear until the `strip.show()` command.

If your pixel does *not* light up, double-check the connections in your circuit and make sure that you have the library properly installed (see the Flora RGB Smart NeoPixels (*http://bit.ly/UcCvoh*) tutorial for details).

Now that you know how to light up a single pixel, let's try three!

Multiple pixel example

Parts:

- (1) Flora (AF 659)
- (3) Flora RGB Smart NeoPixel version 2 (AF 1260)
- Alligator clip test leads (AF 1008, RS 278-1156, SF PRT-11037)
- USB mini-B cable (AF 899, DK WM5163-ND, RS 55010682, SF CAB-11301)
- 3.7V lithium-ion polymer rechargeable battery (AF 258, SF PRT-00339)

A nice part of working with NeoPixels is that they chain very easily, as shown in Figure 8-22. Be sure to pay attention to the direction of the arrows on the NeoPixels when assembling this circuit. They should all face away from the Flora board. Check out the alligator clip version of the circuit in Figure 8-23 and the sewn version in Figure 8-24.

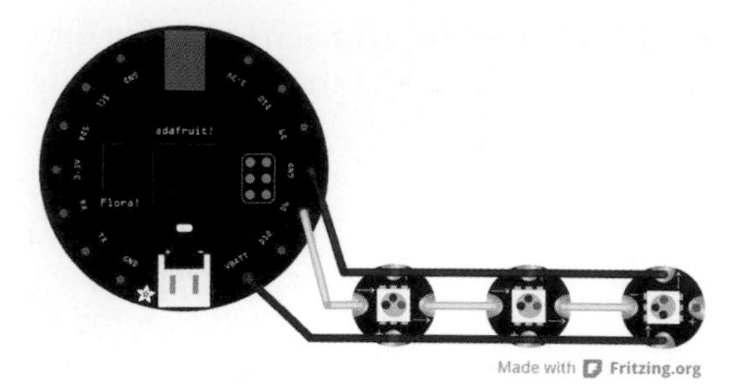

Figure 8-22. *Flora NeoPixel circuit diagram*

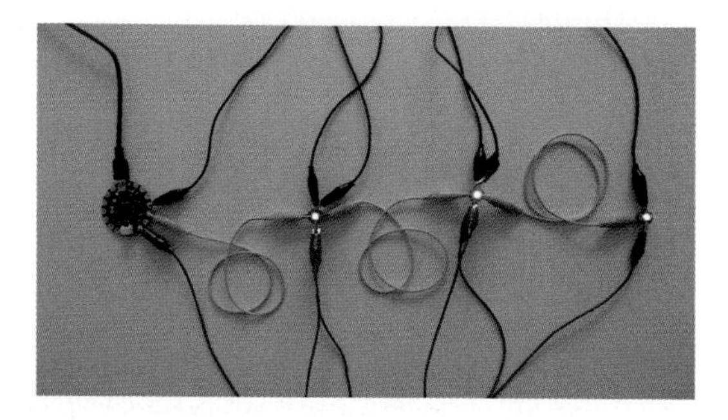

Figure 8-23. *Flora NeoPixels connected with alligator clips*

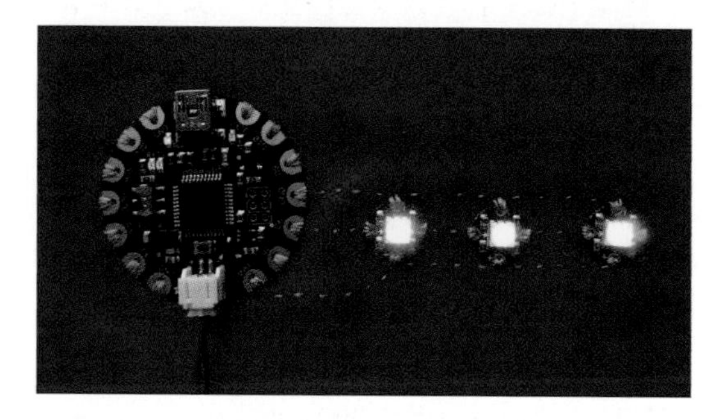

Figure 8-24. *Flora with three NeoPixels sewn with conductive thread*

You won't see much change in the code except that now you are setting the colors of multiple pixels before showing the new configuration of the strip. Here's the code:

```
/*
Make: Wearable Electronics
 Flora NeoPixel example with 3 pixels
 */

#include <Adafruit_NeoPixel.h>

// The digital pin used to control the
// pixel strip
int pinNumber = 6;

// The number of pixels in the strip
int numberOfPixels = 3;

Adafruit_NeoPixel strip =
  Adafruit_NeoPixel(numberOfPixels, pinNumber,
  NEO_GRB + NEO_KHZ400);

void setup() {
  // initialize pixel strip
  strip.begin();
  // set pixels to off to begin
  strip.show();
}

void loop() {

  // set pixel 0 to yellow
  strip.setPixelColor(0, 255, 255, 0);
  // set pixel 1 to pink
  strip.setPixelColor(1, 255, 51, 153);
  // set pixel 2 to yellow
  strip.setPixelColor(2, 255, 255, 0);
  strip.show();
  delay(1000);

  // set pixel 0 to pink
  strip.setPixelColor(0, 255, 51, 153);
  // set pixel 1 to yellow
  strip.setPixelColor(1, 255, 255, 0);
  // set pixel 2 to pink
  strip.setPixelColor(2, 255, 51, 153);
  strip.show();
  delay(1000);

  // turn pixel 0 off
  strip.setPixelColor(0, 0, 0, 0);
  // turn pixel 1 off
  strip.setPixelColor(1, 0, 0, 0);
  // turn pixel 2 off
  strip.setPixelColor(2, 0, 0, 0);
  strip.show();
  delay(1000);

}
```

For more complex behaviors, check out Adafruit's Flora RGB Smart NeoPixels (http://bit.ly/UcCvoh) tutorial.

Because these Pixels are individually addressable and because it is so easy to quickly add more, the possibilities of what you can do with these Pixels are endless. Just use your imagination to explore what lighting effects you would like to create!

Figure 8-25. *Adafruit's LED Ampli-Tie by Adafruit, Becky Stern, Limor Fried, and Phillip Burgess is available as a tutorial (http://learn.adafruit.com/led-ampli-tie) (photographed by Adafruit and John de Cristofaro)*

Fiber Optics

In addition to components that generate light, there are also materials that can transmit light. For a different approach to lighting, let's take a look at fiber optics (Figure 8-26).

Figure 8-26. *Fiber-optic strands*

Fiber optics or optical fibers are flexible, transparent fibers that can transmit light. They are used for applications that range from sophisticated high-speed communication systems to magical light-up wands that you can get at your local summer carnival. Fiber optics come in either *end-glow* or *side-glow*. Apply light to one end of the strands, and you'll see light at the other ends or along the sides.

What's neat about fiber optics is that LEDs are often used as their light source. This is great for you because it makes use of your existing knowledge of LEDs. To look at an example, let's check out my super awesome fiber-optic headband that I got from an electronics surplus site (Figure 8-27).

Figure 8-27. *Fiber-optic headband*

It includes two LEDs as light sources to illuminate two bundles of fiber optics. If you take a closer look (Figures 8-28 and 8-29), you can see how this is assembled.

Figure 8-28. *The LED housing positions the LED so its light is pointed directly into the ends of the fiber-optic strands*

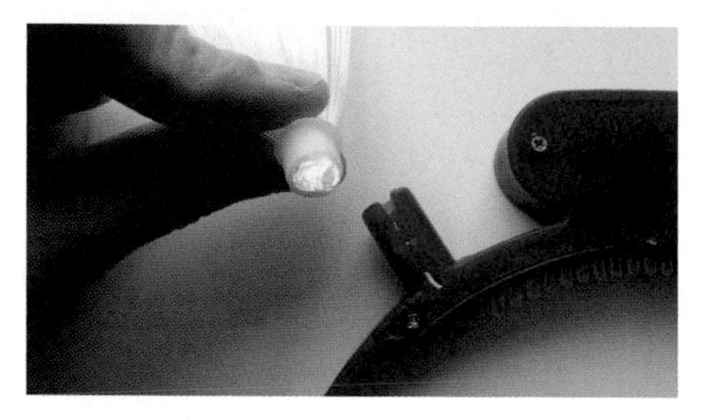

Figure 8-29. *The plastic ring around the fibers holds them together in a tight bunch*

Because of their amazing flexibility and light-transmitting properties, many artists and designers have been incorporating fiber optics into their designs, particularly through the practice of weaving. For example, see Figures 8-30 through 8-32.

LilyPad Pixel Board

The LilyPad Pixel Board (SF DEV-11891) also makes use of the Adafruit NeoPixel library:

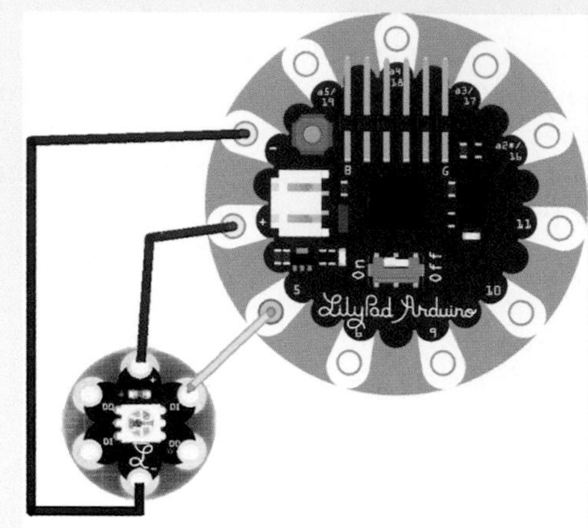

They can be used with a standard Arduino installation, but you will need to download and install the necessary library (*http://bit.ly/UcCLUa*).

Here's the pin layout:

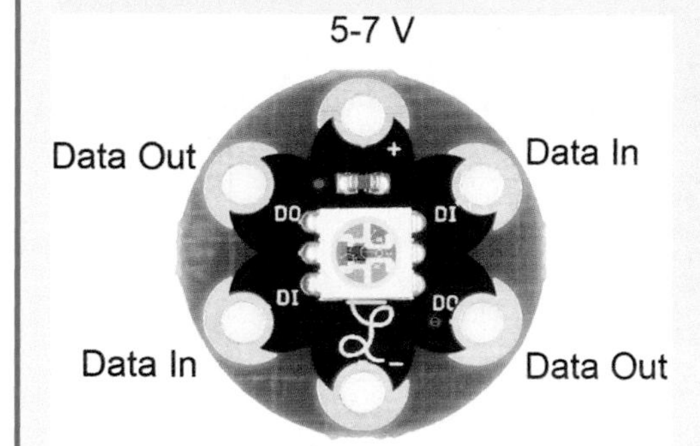

And here's a circuit diagram:

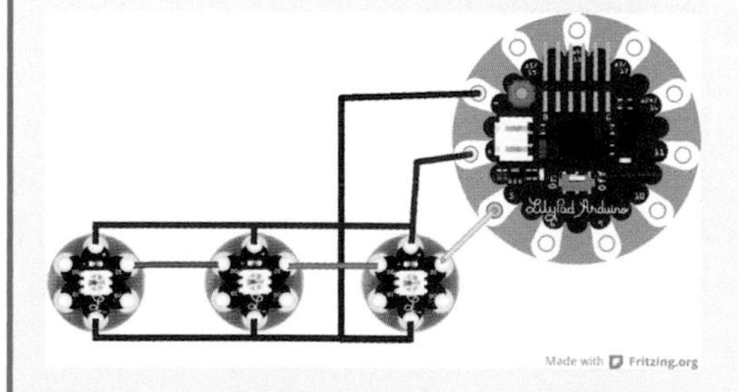

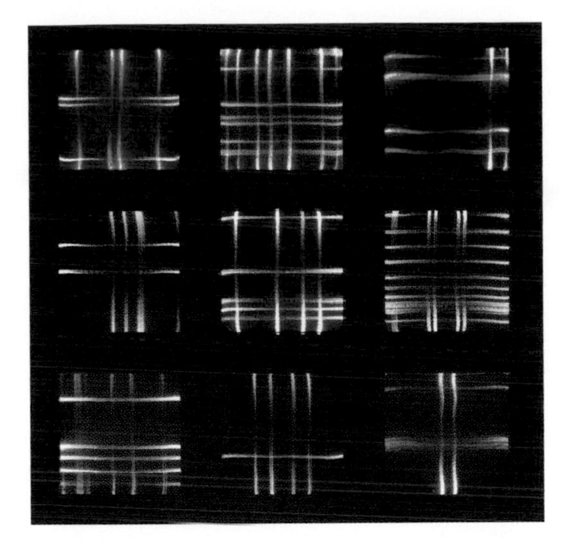

Figure 8-30. *"50 Different Minds" by LigoranoReese is a handwoven, fiber-optic tapestry that changes colors and patterns in response to Internet activity; this sequence, "Comings and Goings," interprets arrivals and departures from nine of the busiest airports in the U.S. Data sponsored by Flightstats Inc. (custom software by Luke Loeffler)*

Figure 8-31. *"50 Different Minds," detail*

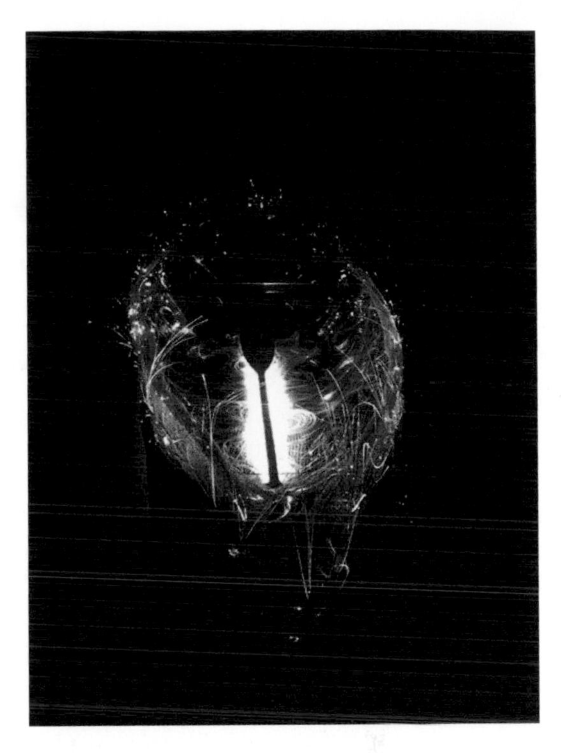

Figure 8-32. *"Vessel" by Erin Lewis is a woven, fiber-optic canoe that displays wind-gust data from Lake Ontario through changing light*

For those who are not well-versed in the practice of weaving, there are manufactured fiber-optic textiles (see Figure 8-33) that are becoming more widely available.

Figure 8-33. *Fiber-optic fabric*

To light up this 40 × 75 cm textile, all you need is a single LED (preferably a super bright). Let's look at how to assemble this setup.

Parts:

- Fiber-optic fabric (SF COM-11594)
- (1) super bright LED (AF 754, SF COM-00531)
- Heat shrink tubing
- Heat gun

When handling fiber optics, be careful to not crease them (see Figure 8-34), as this will permanently affect how they transmit light. Be gentle with the material and be sure to roll, not fold the textile when it is not in use. Bubble wrap (Figures 8-35 and 8-36) is a big help, too.

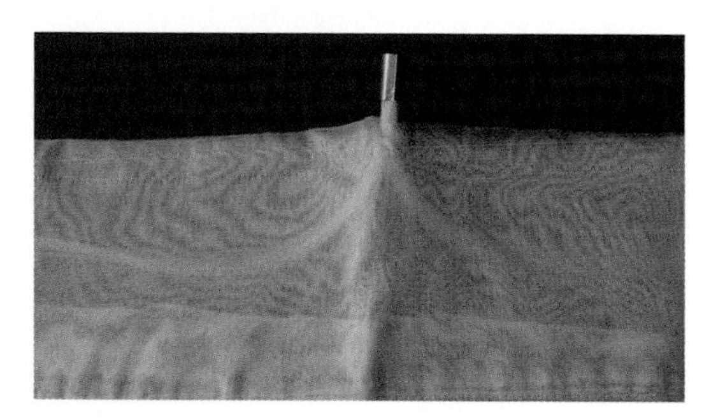

Figure 8-34. *At the end of the textile, you can see a pocket where the fiber optics are gently gathered for the bundling at the endpoint*

Figure 8-35. *When storing the textile, you can use a bit of bubble wrap to support it in a roll*

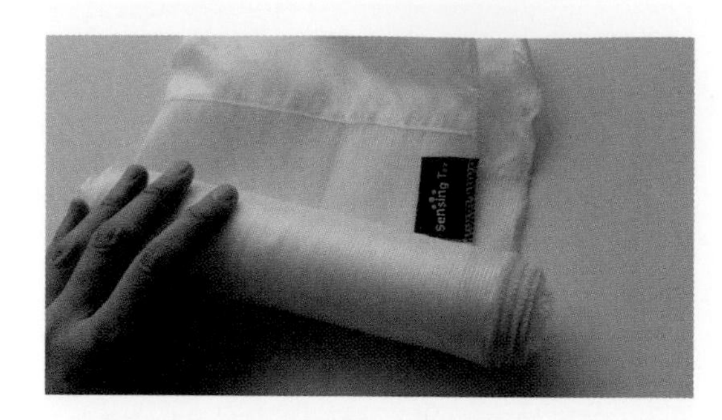

Figure 8-36. *Ready for storage*

The easiest way to attach an LED to a fiber-optic bundle is to use a bit of heat shrink tubing. Check out the stages of the process in Figures 8-37 through 8-42.

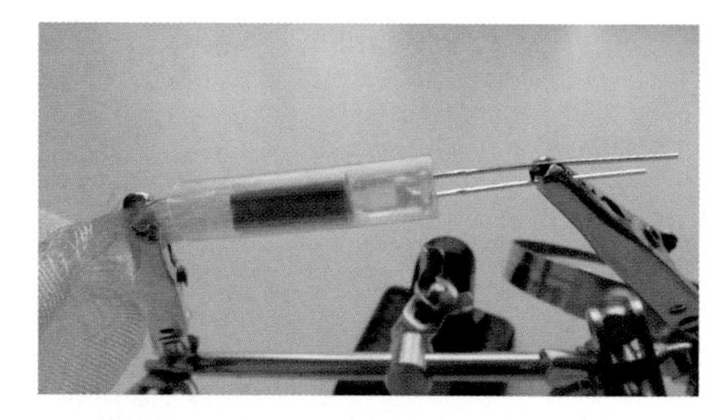

Figure 8-37. *Cover the LED and fiber-optic bundle with an appropriately sized piece of heat shrink tubing; secure the setup in place using helping hands*

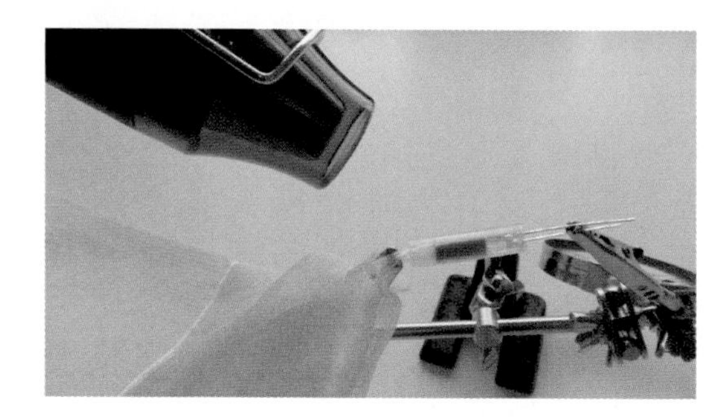

Figure 8-38. *Use a heat gun to shrink the tubing*

Be sure to aim the heat gun toward the heat shrink tubing but away from the length of the fiber-optic filament. Plastic fiber optics can melt when exposed to intense heat.

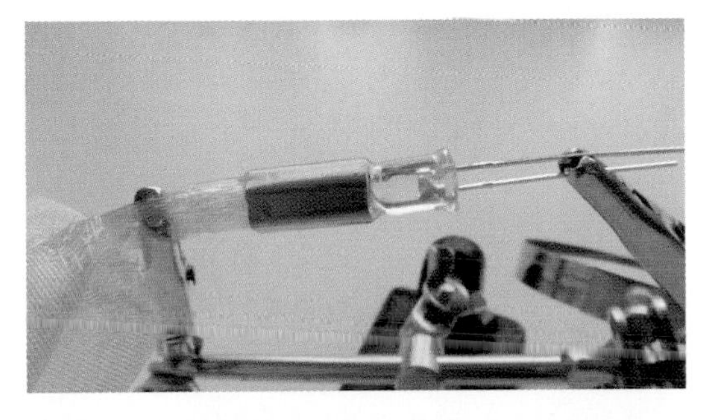

Figure 8-39. *The tubing should look snug around both the LED and the fiber optic bundle*

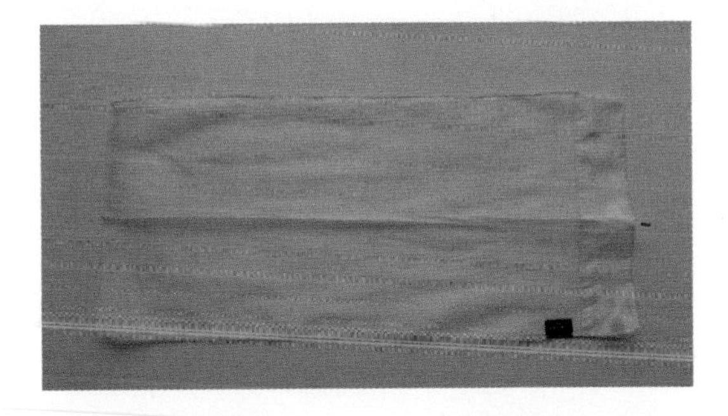

Figure 8-40. *Your fiber-optic textile is ready to be lit!*

Figure 8-41. *Fiber-optic textile, lit*

Figure 8-42. *Fiber-optic textile, lit*

Once your LED is secured to the fiber-optic textile, it can be powered or controlled using the standard means you would use for any LED, such as with a digital output pin on a LilyPad Arduino. From there, you can figure out how you might feature this material in a design of your making.

Figure 8-43. *Fiber-Optic Dress by Moon Berlin (photographed by Patrick Jendrusch)*

Electroluminescent Materials

Electroluminescent (or *EL*) materials, as shown in Figure 8-44, emit light when current is applied (Figure 8-45). These materials usually consist of a conductor (such as copper) coated with phosphor and come in the form of a wire, tape, or panel.

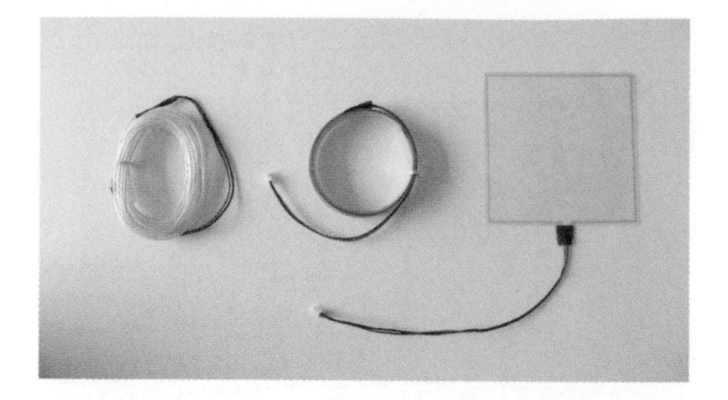

Figure 8-44. *Three type of EL materials: wire, tape, and panel*

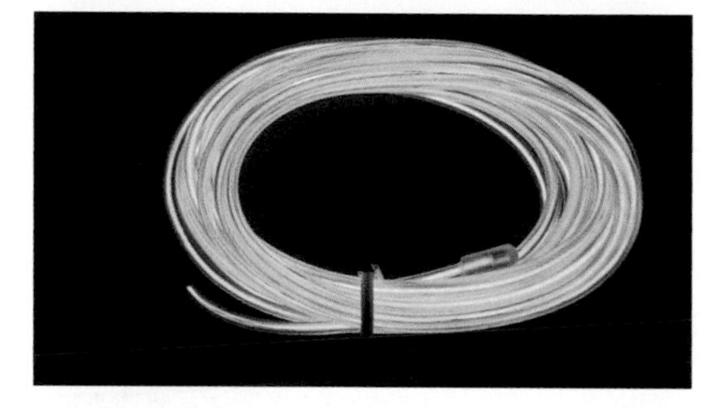

Figure 8-45. *EL wire, lit*

EL wire works well for creating complex patterns or for fitting into small spaces. Standard EL wire can be handstitched (see Figure 8-46) using regular or transparent thread to maintain a particular shape.

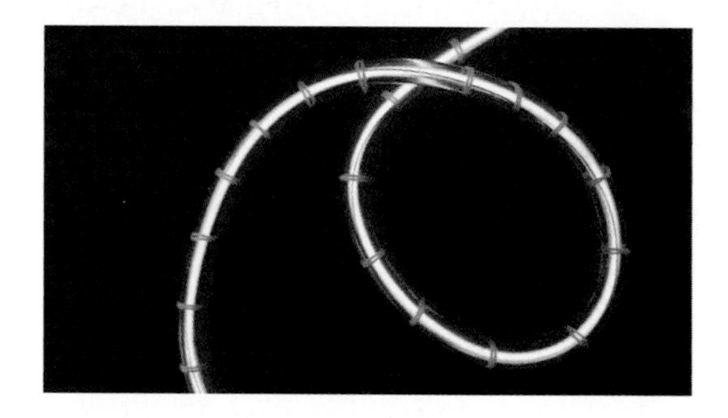

Figure 8-46. *Handstitched EL wire*

Some EL wire is also available as welted piping, meaning there is additional material that you can sew directly through in order to hold it in place (see

Figures 8-47 and 8-48). This makes it extremely easy to elegantly add it into any seam (Figure 8-49).

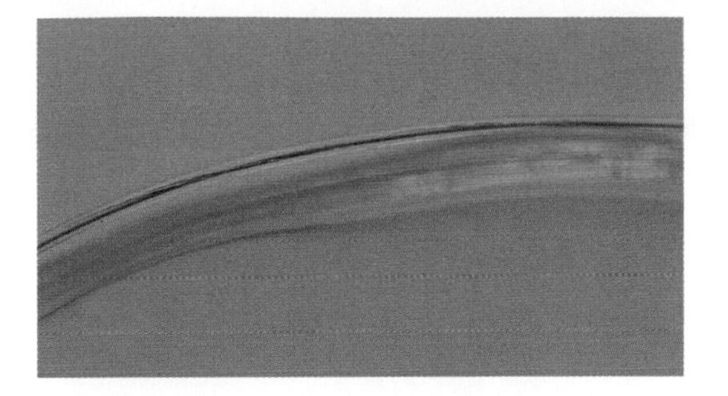

Figure 8-47. *Sewable Electroluminscent (EL) Wire Welted Piping (AF 675)*

Figure 8-48. *Sewing down EL wire with a sewing machine*

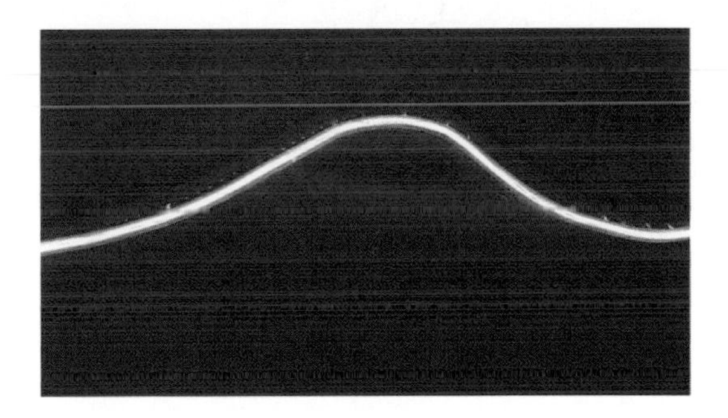

Figure 8-49. *EL wire incorporated into a seam*

EL tape and panels are great for creating bold visual statements (see, for example, Figures 8-50 and 8-51).

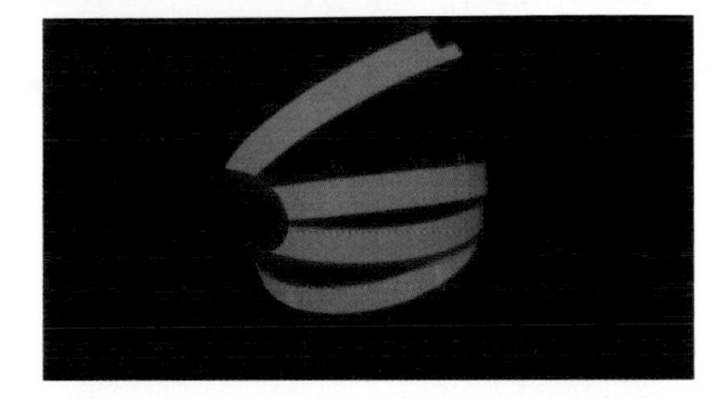

Figure 8-50. *EL tape, lit*

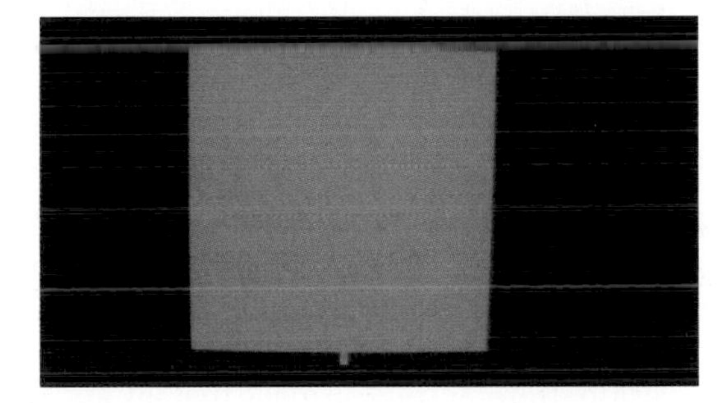

Figure 8-51. *EL panel, lit*

They can also be dramatically transformed through the use of stencils or cutting (see, for example, Figure 8-52 and 8-53).

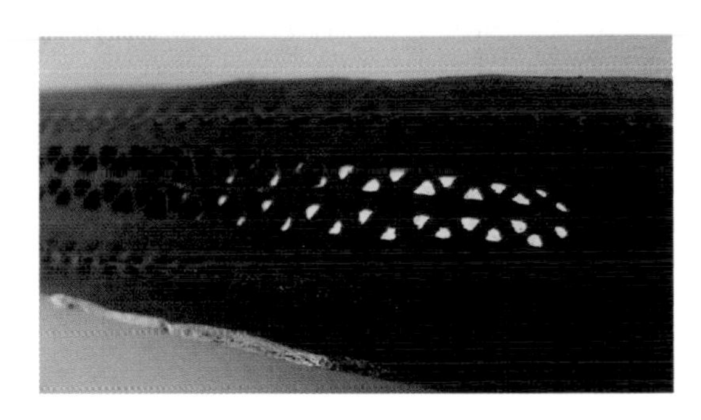

Figure 8-52. *This laser cut leather belt gives the EL tape a much different look*

Figure 8-53. *"Butt Blinkers" by Jen Liu are a wearable signaling mechanism for cyclists made of EL panels cut in the shape of eyes (photographed by Michael Glen)*

EL materials are an attractive lighting option because they provide a large, consistent surface area of light that is quite different from LEDs. EL wire has become a popular costume accessory at various raves, festivals, and party scenes.

While the visual boldness of EL materials is advantageous in terms of visibility, it also means that they need to be managed thoughtfully from an aesthetic perspective. How, through the use of materials and design strategy, can you make the look and feel of electroluminescent materials fit with the intent and vision for *your* project? Figures 8-54 through 8-57 show some examples of EL materials elegantly integrated into wearables.

Figure 8-54. *Diana Eng took EL wire to the runway with her "Fairytale Fashion" project (photographed by Douglas Eng)*

Figure 8-55. *"Electric Parrot Fascinator" by gaïa orain (modeled by Carson Chodos) brings EL wire to life in a colorful headpiece*

Figure 8-56. *"Electric Parrot Fascinator" (detail)*

Figure 8-57. *Syuzi Pakhchyan's "Tron: Quorra Costume" brings movie magic into real life*

When working with EL materials, there are multiple aspects of the system that you need to understand:

Inverters
: Inverters (or drivers), shown in Figure 8-58, are what convert the DC power from the battery to the AC power needed by the EL material. This is an essential part of your EL circuit. The inverter is what drives or lights the EL material. Inverters sometimes have additional functionality built in. They can be sound-activated (AF 831) or can produce a strobe or blinking pattern.

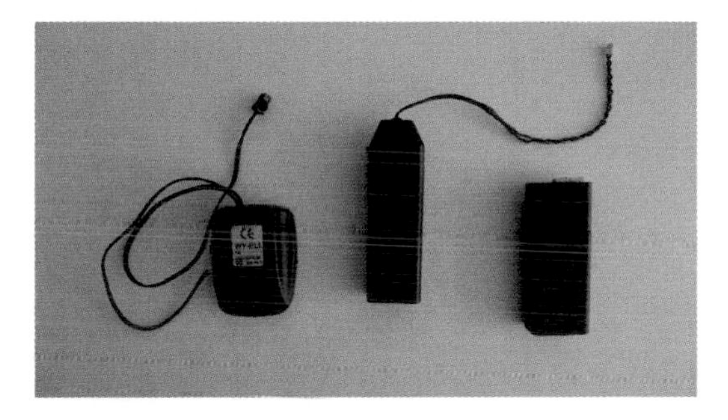

Figure 8-58. *EL inverters (AF 831, AF 317, SF COM-11222)*

Battery holders
: Battery holders are often integrated with the inverter, which helps reduce the bulk of the overall system. Otherwise you'll need to select a battery holder to integrate into your design.

Connectors
: You can solder your own connectors (see Figure 8-59) to EL materials, but it's a bit tricky. If you're a beginner, just stick with parts from the same supplier, and they should be compatible. SparkFun uses JST PH connectors for most of their EL products (like the ones used for 3.7V lithium polymer batteries). Adafruit uses JST SM connectors for their EL products, which are a little different.

Figure 8-59. *Connectors used with EL systems from Adafruit (left) and SparkFun (right)*

Sequencers
: Sequencers enable you to light up multiple EL wires in a sequence. While sequencers can produce complex and interesting effects, they are not necessary in a basic EL circuit.

A simple EL circuit usually consists of the following:

- A power source (in this case, a battery pack)
- An inverter
- An EL material

The type and length or size of the EL material you are working with will determine your inverter and power needs. Read the descriptions of the products you are working with carefully to ensure you are working with compatible parts.

The easiest way to get started is to work with a kit that contains everything you need. SparkFun and Adafruit have some handy starter kits for EL wire (SF RTL-11421, AF 320), tape (AF 637), and panels (AF 628).

Working with AC

*These materials are different in that they work with **alternating current** (or AC), whereas all other projects and materials in this book work with **direct current** (DC). Make sure your batteries are always removed when you are connecting the EL material to the inverter or else you might get shocked!*

If you would like to program your sequencer, SparkFun's EL Sequencer (SF COM-11323) is Arduino-compatible and can individually control up to eight wires. Diana Eng published a fantastic tutorial (*http://bit.ly/UcDzsb*) on the Make Blog about how to use this board in wearables.

Keep in mind that there are also materials that can provide visibility without electronics, namely reflective and glow-in-the-dark materials. See Appendix D for more information.

Experiment: Be Safe, Be Seen

Using one of the tools you've learned about, incorporate light into a piece of clothing for fashion or utility. Think about when it should be lit and when not, and whether it is the user or the wearable that determines changes in state.

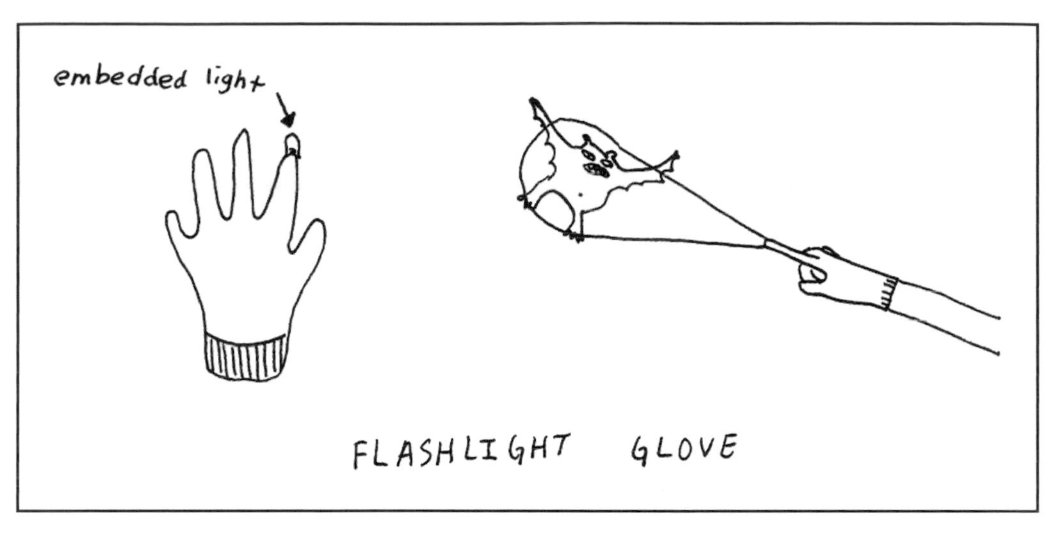

Figure 8-60. *A flashlight glove (illustration by Jen Liu)*

Sound

Sound can be soothing, informative, and even abrasive. How can you get your clothing to speak, sing, or shout? When working with audio for wearables, here are some helpful questions to ask:

- Would you like to make a simple sound, generate a tone, or play an audio file?
- How will the sound be triggered or controlled?
- Where will the sound-emitting device live?
- How loud should the sound be? Is it intended only for the wearer or also for those who are nearby?

With those considerations in mind, let's explore your options for embedding audio close to the skin.

Buzzers

Buzzers are a simple way to provide audio feedback. They are devices that create an audible sound as the result of a electrical signal. There are two types of buzzers that you will encounter: electromagnetic and piezoelectric.

Electromagnetic buzzers create a noise when continuous voltage is applied.

Piezoelectric buzzers require an oscillating signal and can function much like speakers. You'll get to know them in the next section.

3V electromagnetic buzzers (see Figure 8-61) are great standalone actuators and can act as an interesting alternative for LEDs when creating simple analog circuits.

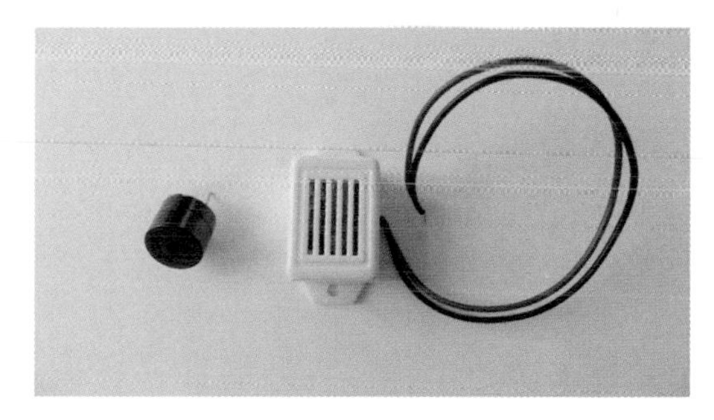

Figure 8-61. *Electromagnetic 3V buzzers—panelmount and with wires*

Be sure to look out for the polarity of these buzzers. The panelmount buzzers usually have a + sign to signify the positive side and with the wired version you can tell by the colors of the wires (red for positive, black for negative).

Simple circuit

A simple circuit can be wired up with a CR2032 3V battery and LilyPad Button board.

Parts and materials:

- 3V buzzer (AF 1536, DK 102-1646-ND, SF COM-07950)
- LilyPad Button Board (SF DEV-08776)
- CR2032 battery (AF 654, DK P189-ND, SF PRT-00338)
- CR2032 battery holder (AF 653, DK BA2032SM-ND, SF DEV-08822)
- Alligator clip test leads (AF 1008, RS 278-1156, SF PRT-11037)

Figure 8-62 shows the circuit, and Figure 8-63 shows the circuit being activated.

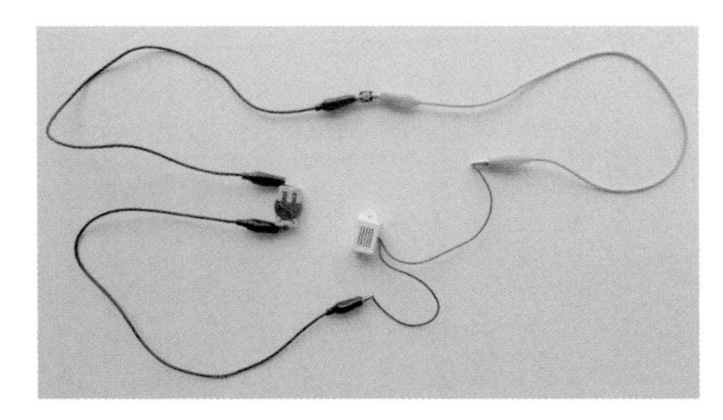

Figure 8-62. *3V buzzer in simple circuit*

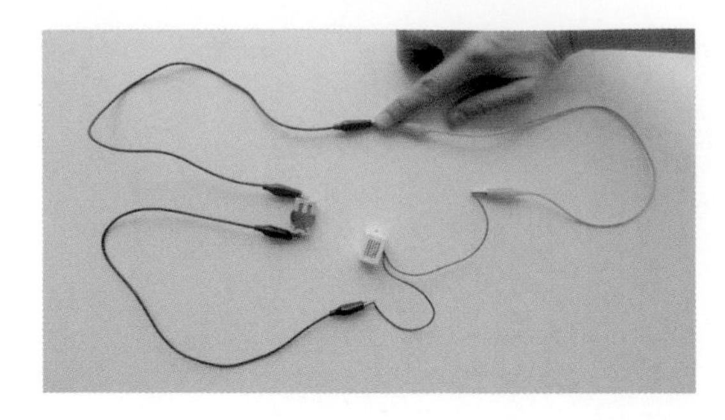

Figure 8-63. *The buzzer will sound when the button is pushed*

Buzzer with microcontroller

These buzzers can also be activated using a microcontroller. Connect the positive side to a digital output pin and connect the negative side to ground. Simply set that digital output pin to "HIGH" and the buzzer will sound.

The connections are shown in Figure 8-64.

Parts and materials:

- LilyPad Arduino Simple (SF DEV-10274)
- 3V buzzer (AF 1536, DK 102-1646-ND, SF COM-07950)
- FTDI board (AF 284, SF DEV-10275)
- USB mini-B cable (AF 899, DK WM5163-ND, RS 55010682, SF CAB-11301)
- Alligator clip test leads (AF 1008, RS 278-1156, SF PRT-11037)

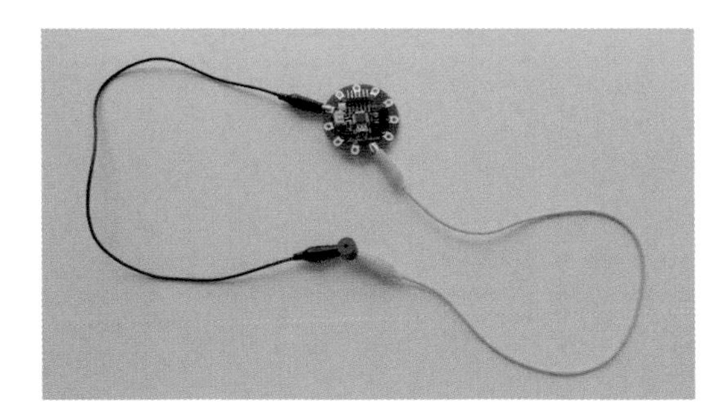

Figure 8-64. *LilyPad Arduino Simple with a panel mount 3V electromagnetic buzzer*

Here is the code:

```
/*
Make: Wearable Electronics
Buzzer example
*/

int buzzerPin = 9;

void setup() {
  pinMode(buzzerPin, OUTPUT);
}

void loop() {
  digitalWrite(buzzerPin, HIGH);
  delay(500);
  digitalWrite(buzzerPin, LOW);
  delay(3000);
}
```

Tones

The simple buzzers you've looked at so far are great for producing a single, simple tone, but if you want to produce a broader range of sounds, you can also generate specific notes using a microcontroller and a speaker (or a piezoelectric buzzer).

Both speakers (see Figure 8-65) and piezoelectric buzzers contain materials that move when voltage is applied. When voltage is applied, the material is in one position, and when it is not, the material is in another position. It is the frequency of switching back and forth between these two positions that moves air in such a way to create different sounds.

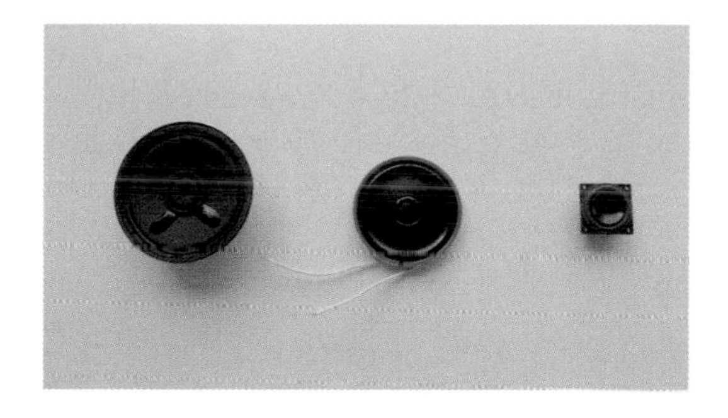

Figure 8-65. *Speakers come in many shapes and sizes*

Figure 8-66. *You can sometimes learn a lot about a speaker by looking at the back of it; this is a 2", 0.5w, 8ohm speaker*

Let's take a look at how you can use the Arduino to produce particular notes.

Circuit

The circuits for connecting a speaker or piezoelectric buzzer to an Arduino are pretty similar. Figure 8-67 shows the circuit layout for connecting a LilyPad Arduino Simple to a speaker, and you can see it assembled with alligator clips in Figure 8-68. Figure 8-69 shows a similar circuit using the LilyPad Buzzer (the assembled circuit with alligator clips is shown in Figure 8-70).

Parts:

- LilyPad Arduino Simple (SF DEV-10274)
- Speaker (SF COM-09151, COM-10722, RTL-10766) and 100Ω resistor *or*
- LilyPad Buzzer (SF DEV-08463)
- FTDI board (AF 284, SF DEV-10275)
- USB mini-B cable (AF 899, DK WM5163-ND, RS 55010682, SF CAB-11301)
- Alligator clip test leads (AF 1008, RS 278-1156, SF PRT-11037)

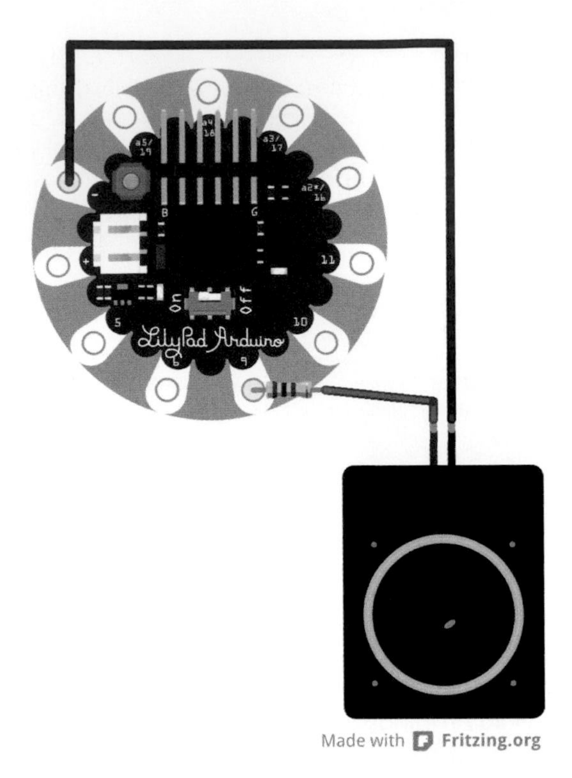

Figure 8-67. *LilyPad Arduino Simple with speaker and 100Ω resistor circuit layout*

Figure 8-68. *LilyPad Arduino Simple with speaker and 100Ω resistor with alligator clips*

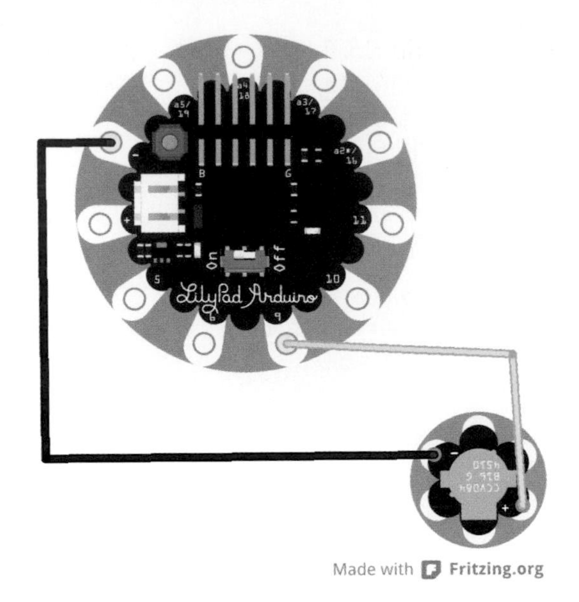

Figure 8-69. *LilyPad Arduino Simple with LilyPad Buzzer circuit layout*

Figure 8-70. *LilyPad Arduino Simple with LilyPad Buzzer connected with alligator clips*

Code

In order to make a note, you need to use the Arduino to turn the pin on and off at a particular frequency. Luckily, there is an Arduino function called `tone()` that handles most of this for you. It looks like this:

```
tone(pin, frequency, duration)
```

Just provide the pin, frequency in hertz, and duration in milliseconds (optional) parameters and the Arduino will generate your desired tone. Try this code as an example:

```
/*
Make: Wearable Electronics
Tone example
*/

int C = 1047;
int D = 1175;
int E = 1319;
int F = 1397;
int G = 1568;
int A = 1760;
int B = 1976;
int c = 2093;

int buzzerPin = 9;

void setup() {
  pinMode (buzzerPin, OUTPUT);
}

void loop() {
  tone(buzzerPin, C, 250);
  delay(300);
  tone(buzzerPin, E, 250);
  delay(300);
  tone(buzzerPin, G, 250);
  delay(300);
  tone(buzzerPin, c, 250);
  delay(300);
  tone(buzzerPin, G, 250);
  delay(300);
  tone(buzzerPin, E, 250);
  delay(300);
  tone(buzzerPin, C, 500);
  delay(1000);
}
```

See also these examples:

- Arduino melody tutorial (*http://arduino.cc/en/Tutorial/Tone*)
- Arduino pitch follower using the `tone()` function (*http://arduino.cc/en/Tutorial/Tone2*)
- Arduino Simple keyboard using the `tone()` function (*http://arduino.cc/en/Tutorial/Tone3*)

Note that there is a second tab in these sketches titled *pitches.h* that defines frequencies for pitches at many octaves (Figure 8-71).

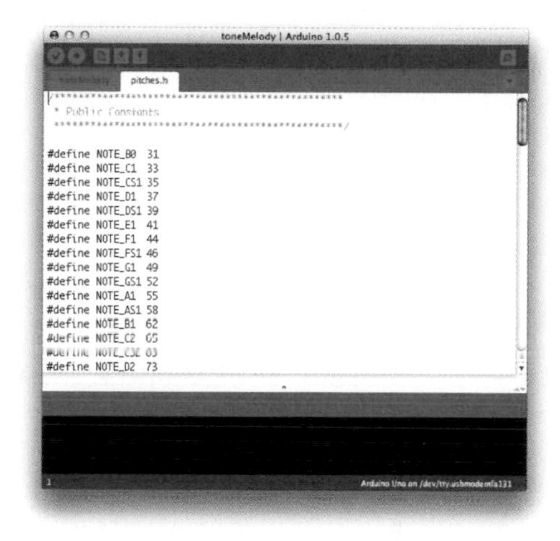

Figure 8-71. *pitches.h tab in the Arduino melody example*

Despite the simplicity of these tones, they can still be combined to create a variety of melodies, sound effects, and feedback noises.

Figure 8-72. *"Pixel Foot" by Ken Leung is a pixel-covered oversized shoe that plays 8-bit music (evocative of 1980s video games) in response to different foot movements, such as stomping, jumping, or kicking*

See also:

- *Arduino Cookbook*, "Chapter 9: Audio Output"

Audio Files

While the ability to produce tones with the Arduino is useful, being able to play digital audio files greatly expands your project's horizons.

There are a variety of Arduino-compatible tools that enable the playback of digital audio files, such as the Adafruit Wave Shield (AF 94) and the SparkFun MP3 Shield (SF DEV-10628). These shields are intended to sit atop an Arduino Uno, which makes for a bulky solution that is not particularly wearables friendly.

The LilyPad MP3 board (SF DEV-11013, shown in Figures 8-73 and 8-74) is a wearable alternative to the MP3 shield that combines a LilyPad Arduino with an assortment of useful audio tools in one slim package. A microSD card holder enables storage of audio files. The ATmega 328 processor is Arduino-compatible, so this board can be easily programmed, and there is no need for an additional Arduino board. An onboard amplifier chip provides increased volume and prevents the need for a bulky additional circuit. A mini headphone jack as well as left and right speaker connection pins expand the options for possible audio output devices.

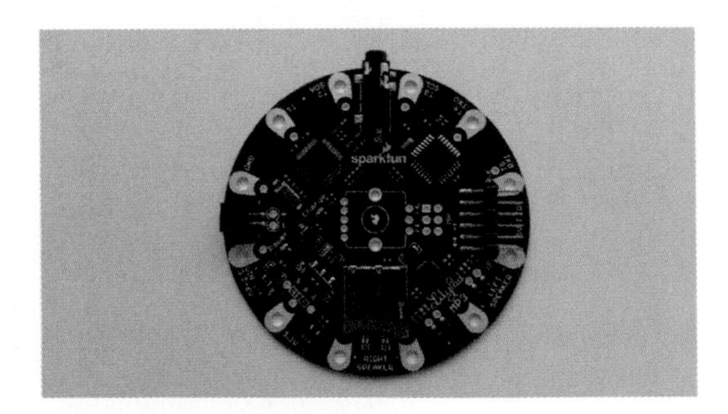

Figure 8-73. *LilyPad MP3*

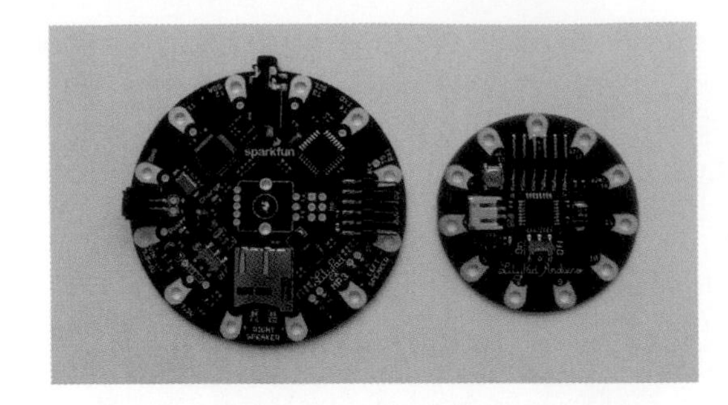

Figure 8-74. *The LilyPad MP3 is a bit bigger than a standard LilyPad board*

If you'd like to trigger audio files without the fuss of programming, the LilyPad MP3 ships with a test sketch loaded on it that will play back five different audio files when designated trigger pins (marked T1, T2, etc.) are connected to ground. Let's get that circuit up and running.

Parts:

- LilyPad MP3 (SF DEV-11013)
- MicroSD card
- Speaker (SF COM-09151, COM-10722, RTL-10766) and 100Ω resistor *or* LilyPad Buzzer (SF DEV-08463)
- Alligator clip test leads (AF 1008, RS 278-1156, SF PRT-11037)
- 3.7V lithium-ion polymer rechargeable battery (AF 258, SF PRT-00339)

First, you will need to load audio files onto your microSD card. Connect your microSD card to your computer using a card reader and transfer your audio files onto the microSD card. The LilyPad MP3 will read a number of audio file types, including MP3 and WAV. Just be sure to change the filenames so that the first character of each is a number from 1 to 5. It doesn't matter what the remaining characters in the filenames are. I like to keep my filenames simple: *1.mp3*, *2.mp3*, and so on.

Once you've prepared the microSD card, load it into the slot on the LilyPad MP3, as shown in Figure 8-75.

Figure 8-75. *Insert a microSD card loaded with audio files into the slot*

Next, use a red alligator clip to connect the positive terminal of the speaker to "Right Speaker +" pin and a black alligator clip to connect the negative terminal of the speaker to the "Right Speaker –" pin (see Figure 8-76).

Figure 8-76. *Speaker connections*

Connect a black alligator clip to ground (GND) and leave the other side unconnected for now, as shown in Figure 8-77.

Figure 8-77. *Black alligator clip to ground (GND)*

Plug a LiPo battery into the JST connector (see Figure 8-78).

Figure 8-78. *Battery connected*

Move the power switch to the "ON" position. The Power LED should turn on as shown in Figure 8-79. Figure 8-80 shows the final circuit.

Figure 8-79. *LilyPad MP3 powered "ON"*

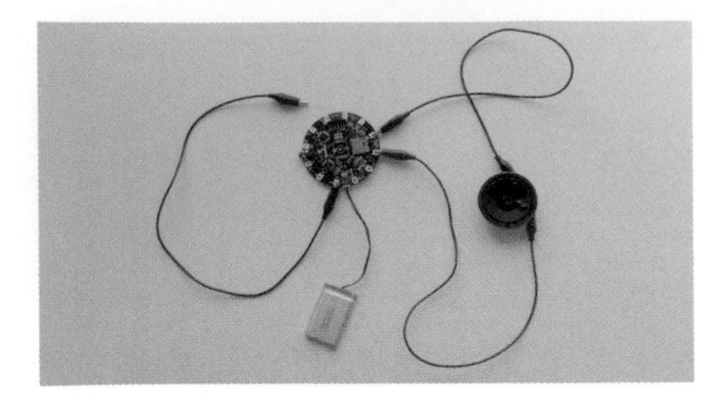

Figure 8-80. *Completed circuit*

Now your circuit is ready to go! Touch the free end of the black alligator clip to pins T1-T5 to play the corresponding audio files as shown in Figure 8-81.

Figure 8-81. *Triggering audio file #4*

With your knowledge of how to make creative switches from Chapter 3, you know that it is possible to trigger these audio files in unexpected and delightful ways.

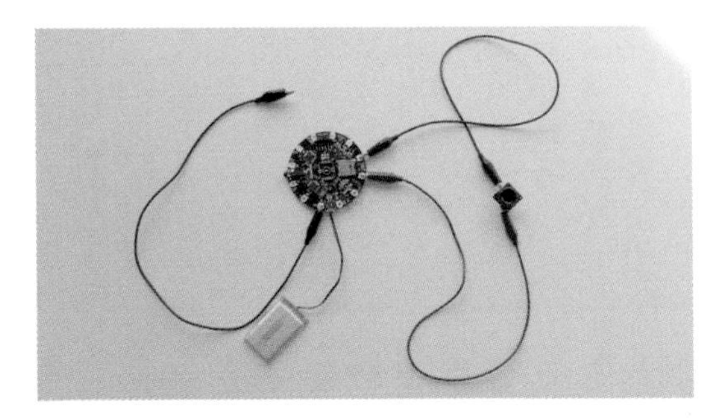

Figure 8-82. *Try different types of speakers to hear the difference in volume and sound quality*

You can also program the LilyPad MP3 to act as an MP3 player! For instructions on how to do this, check out SparkFun's Getting Started with LilyPad MP3 (*https://learn.sparkfun.com/tutorials/getting-started-with-the-lilypad-mp3-player*) tutorial.

Audio file playback can be used to create interesting wearables. There's a lot to consider in terms of what the content is, what triggers the audio, and where it is heard. It's also worth considering whether the audio is meant to be heard only by the wearer or if it is also intended for others who are nearby.

"Bio Circuit" by Dana Ramler and Holly Schmidt (Figure 8-83) is a vest that generates a soundscape in response to the wearer's heart rate. The garment is designed so that the sound is played back to the wearer privately.

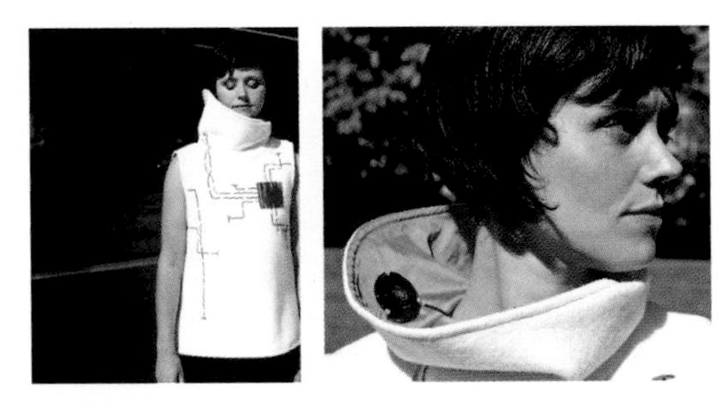

Figure 8-83. *"Bio Circuit" by Dana Ramler and Holly Schmidt*

"Yuga" by Teresa Almeida (Figure 8-84) is a pair of wearable devices that play mood sounds meant to engage people in the immediate vicinity of the wearer.

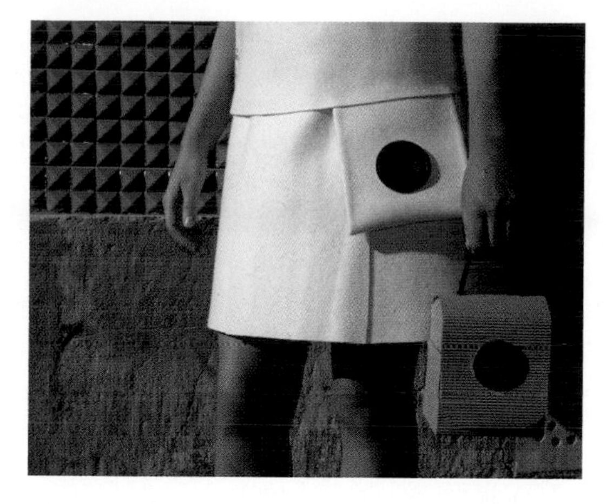

Figure 8-84. *"Yuga" by Teresa Almeida (photographed Pietro Romani)*

"Small Talk Destroyer" by Mitch McGooey (Figure 8-85) is a necktie that plays back prerecorded small talk when the wearer shakes the hand of a new acquaintance.

Figure 8-85. *"Small Talk Destroyer" by Mitch McGooey*

Figure 8-86. *"Sock Hop Socks" (illustration by Jen Liu)*

Experiment: Wearable Instrument

Now that you know how to generate tones using the Arduino, use your knowledge of sensors and wearable construction techniques to create a body-based instrument with an unusual interface.

Motion

Making things move can be an enticing prospect. It is also a challenging one in the dynamic arena of the human form. From the tiny buzz of a vibration motor to the sharp and precise movements of a servo to the significant physical transformations created by a gearhead motor, this section covers how to use motors (see Figure 8-87) to accomplish a range of movement possibilities.

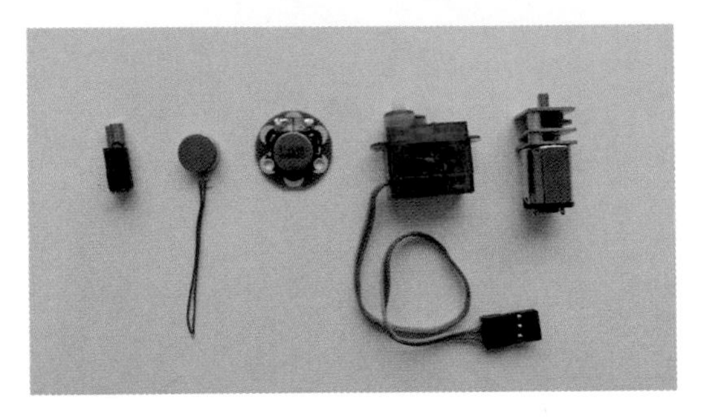

Figure 8-87. *Small motors well suited for wearable applications (left to right: vibration motor (exposed), vibration motor (enclosed), LilyPad vibe board, microservo, and a small gearhead motor)*

Vibrating Motors

Vibrational feedback can be powerful, subtle, and even seductive. It can simulate a stroke, a tap, or a tickle. It holds the potential to be perceived only by the wearer and is ideal for situations that warrant privacy and discretion or situations where it is inconvenient or impossible for the wearer to see or hear feedback.

Vibrating motors are basically DC motors with a weighted head (Figure 8-88) attached to the shaft. As the motor spins, the weight spins, thus causing the motor to rock back and forth. Many vibrating motors come with their weighted head exposed (Figure 8-89). This can be a bit problematic if you're incorporating the motor into a garment with folds of fabric or other intrusions that can interfere with the spinning of a head. The advantage of small, open vibrating motors (the type often found in cell phones or pagers) is that they are often available at surplus stores for very cheap. When using them, be sure to build in protection so the head can spin freely. The leads also tend to be a bit delicate, so it's worth using heat shrink tubing to reinforce your connections.

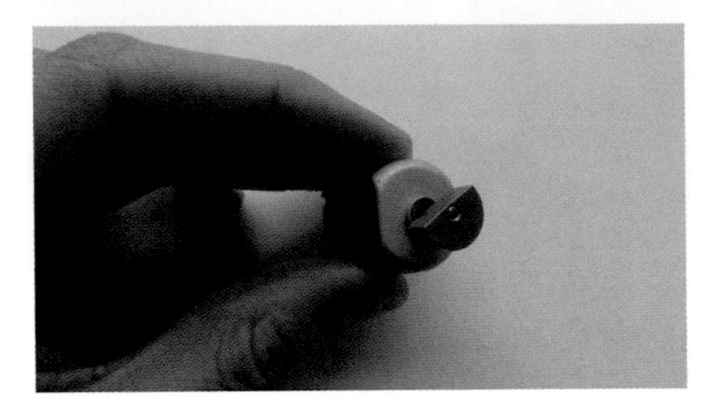

Figure 8-88. *A weighted head causes the DC motor to shake as it spins*

Figure 8-89. *Vibrating motors with exposed heads*

There are also completely enclosed small, flat vibrating motors, sometimes called *pancake motors* (see Figure 8-90). These are well-suited for wearable applications and very easy to work with. This is the same kind used on the LilyPad Vibe board (SF DEV-11008), but you can also purchase the motor on its own (AF 1201, SF 1201) and incorporate it into your project as you like.

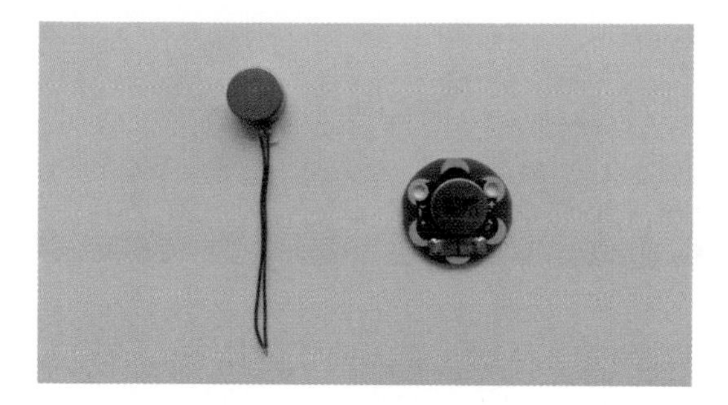

Figure 8-90. *Pancake vibrating motors*

These motors can be directly connected to either a digital or analog output pin on the Arduino, depending on whether you want to control the intensity of the vibration. Simply connect one end of the motor to the output pin and the other to ground. The 40 mA provided by the Arduino output pin is plenty to get these motors shimmying, but if you'd like a more intense vibrational kick, you just need to supply them with additional current. To learn how to do this, check out the discussion of transistors in "Gearhead Motors" on page 177.

Figure 8-91 shows the circuit layout, and you can see the assembled circuit in Figure 8-92.

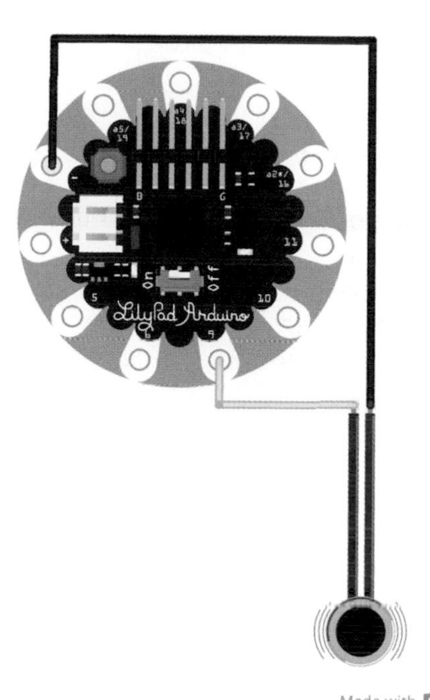

Figure 8-91. *LilyPad Arduino Simple with LilyPad Vibe board circuit layout*

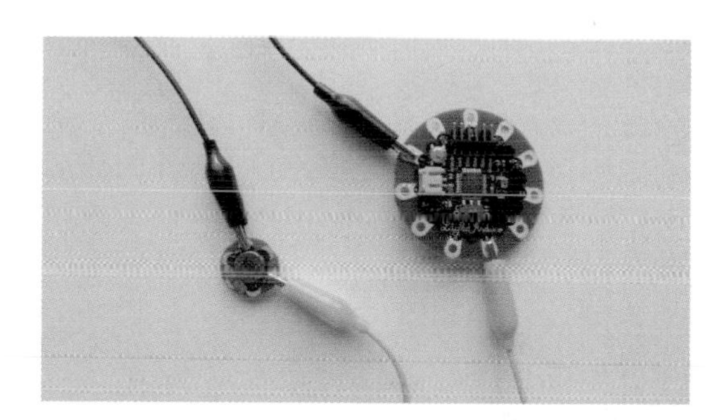

Figure 8-92. *LilyPad Arduino Simple with LilyPad Vibe board connected with alligator clips*

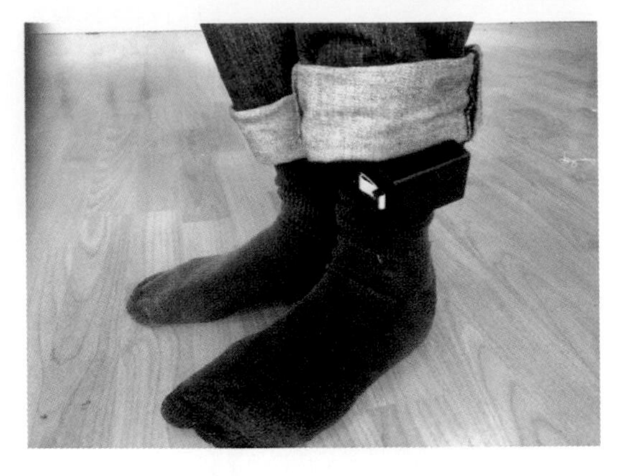

Figure 8-93. *"North Paw" by Eric Boyd is a direction-signaling ankle bracelet containing eight motors that vibrate when that side of the body is facing north*

Servo Motors

Sometimes you want to use motors to accomplish precise movements. This could be for functional purposes, such as opening and closing a pocket, or for aesthetic purposes, such as the movement of materials to create a dynamically shifting design.

Servo motors (Figure 8-94) are capable of accomplishing small, discrete movements. They are extremely precise in their position and most often have a turning range of 180 degrees, though there are 360 degree models available. A servo motor can be told to turn to any location within its potential range of movement.

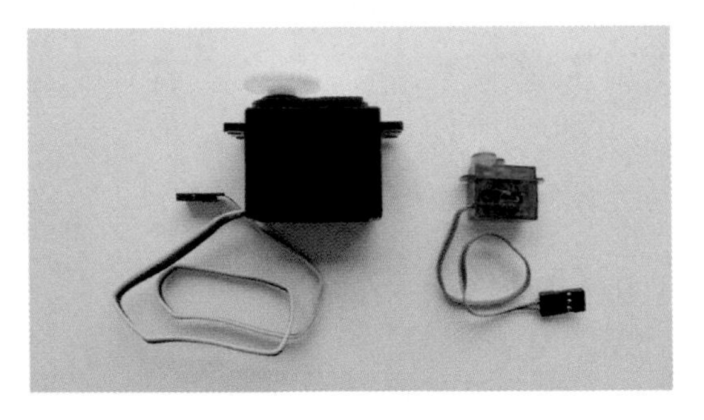

Figure 8-94. *A medium servo and micro servo*

Microservos are miniature servo motors that are useful for wearables because they are small and lightweight. As with any motor, it is important to pay attention to the power requirements of the particular model you are working with. Many servos need 5V to run, in which case you will need to make sure you have a 5V power supply included in your circuit.

There are a few microservos (such as AF 169, shown in Figure 8-95) that will work with as little as 3V, which is helpful if you are using a 3.7V lithium polymer battery.

Figure 8-95. *This microservo (AF 169) will run on 3-6V*

Servos usually come with a number of attachments (Figures 8-96 and 8-97). These can be screwed directly to the shaft to provide leverage, enable attachment to other materials, or for mechanical purposes.

Figure 8-96. *Servo attachments include arms, propellers, and wheels*

Figure 8-97. *Small servo with propeller attached*

A servo has three connections: power, ground, and *signal*. The servo cable is usually terminated with a female header. You can either insert hookup wire to make temporary connections, as shown in Figure 8-98, or snip off the header to access the wires for soldering or sewing.

Figure 8-99 shows a circuit layout for use with a servo, and you can see this circuit build with alligator clips in Figure 8-100.

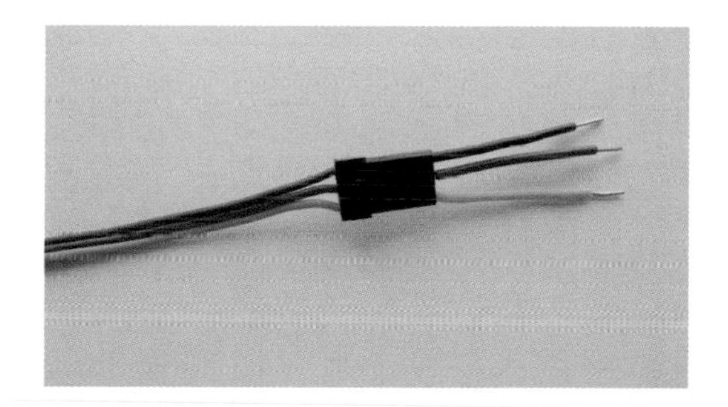

Figure 8-98. *Servo cable with hookup wires*

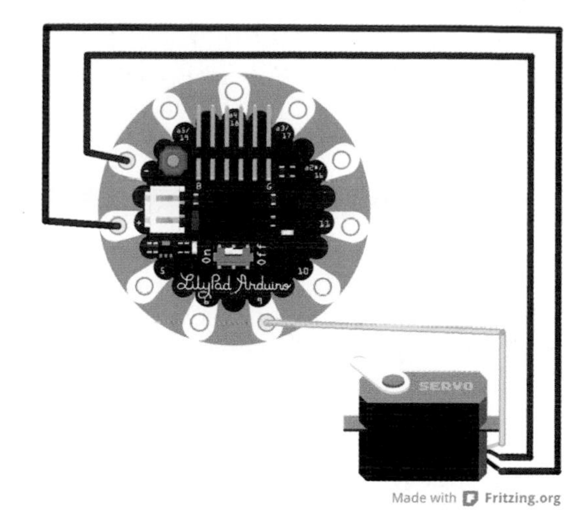

Figure 8-99. *Servo circuit layout*

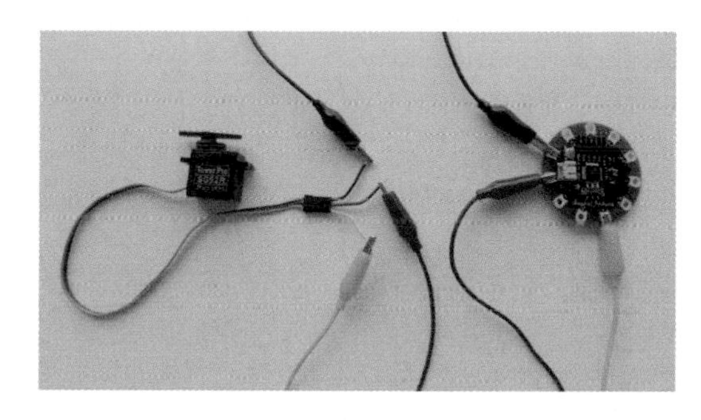

Figure 8-100. *Servo circuit with alligator clip connections*

Most Arduinos can control up to 12 servos simultaneously. Arduino even has a built in Servo library. Here are some commands that are useful to know when working with the library:

`#include <Servo.h>`
: This includes the servo library into your code, so that it's incorporated into your code when it's compiled.

`Servo mrSpinny`
: This declares a variable name for the particular servo you are working with. In this case, it is `mrSpinny`. But it could be `myFavoriteServo`, `servo`, or even `Bob`. You'll see in the following commands that it is `mrSpinny` followed by a

period followed by the command. `mrSpinny` would be replaced with whatever variable you've declared.

`mrSpinny.attach(pin)`
: This declares which pin the servo will be connected to.

`mrSpinny.write(angle)`
: The angle is the position between 0 and 180 that you would like the servo to turn to. Keep in mind that it takes time for the servo to turn, so you should always include a delay between `.write` commands so that it has adequate time to turn.

With the circuit complete and this knowledge in hand, you can go ahead and program the Arduino to control the servo! Here's an example:

```
/*
Make: Wearable Electronics
 Servo example
 */

#include <Servo.h>

// name your servo
Servo mrSpinny;
int servoPin = 9;

void setup()
{
  // set the servo pin
  mrSpinny.attach(servoPin);
}

void loop()
{
  // turn to 0 degree position
  mrSpinny.write(0);
  // wait 1000 milliseconds
  delay(1000);
  mrSpinny.write(45);
  delay(300);
  mrSpinny.write(90);
  delay(300);
  mrSpinny.write(135);
  delay(300);
  mrSpinny.write(180);
  delay(1000);
}
```

See also:

- Arduino Sweep tutorial (*http://arduino.cc/en/Tutorial/Sweep*)
- Arduino Knob tutorial (*http://arduino.cc/en/Tutorial/Knob*)

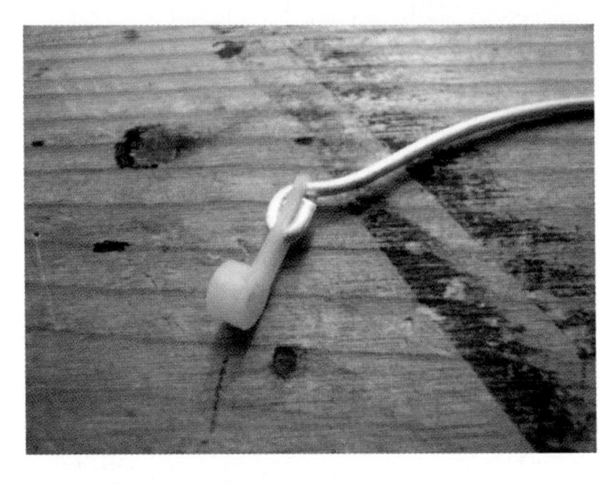

Figure 8-101. *"Soft Cyborg" by Rachael Kess uses extended servo motors to animate the eyelids of a felt mask and make it blink*

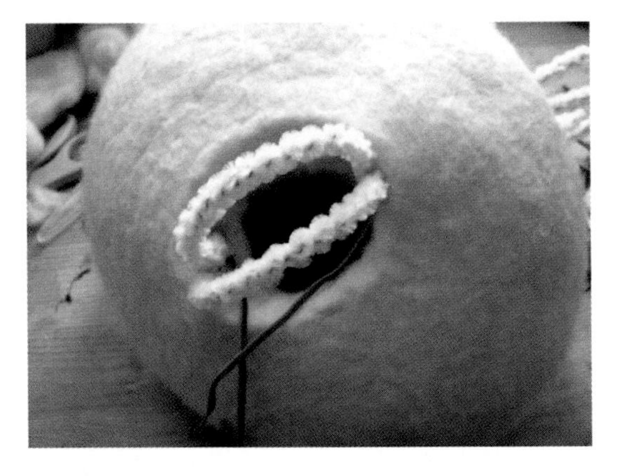

Figure 8-102. *A pipecleaner frame for the eyelids is attached to the servo motor's propeller*

Figure 8-103. *Eyelids covered with felt*

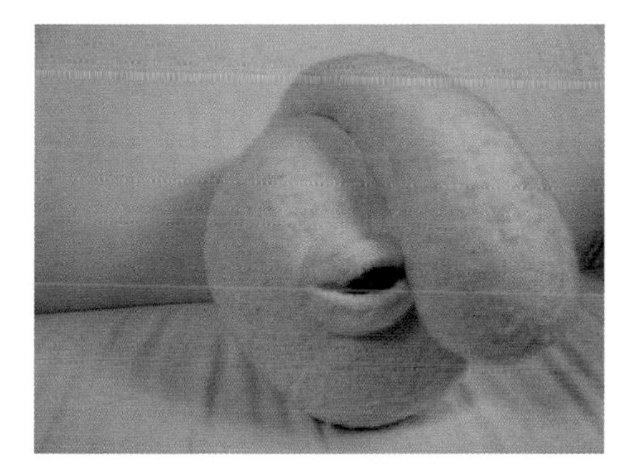

Figure 8-104. *Completed mask*

Figure 8-105. *Mask in use in performance*

Gearhead Motors

A DC motor spins freely when voltage is applied. DC motors usually spin quite quickly. A gearhead motor, or gear motor (Figure 8-106), is a DC motor augmented with a set of gears that reduce the number of revolutions per minute (RPMs). They tend to be on the larger side, though there are some smaller ones (SF ROB-08911) that are nice to work with in wearables (see Figure 8-107). A gearhead motor tends to be much stronger than a typical servo motor.

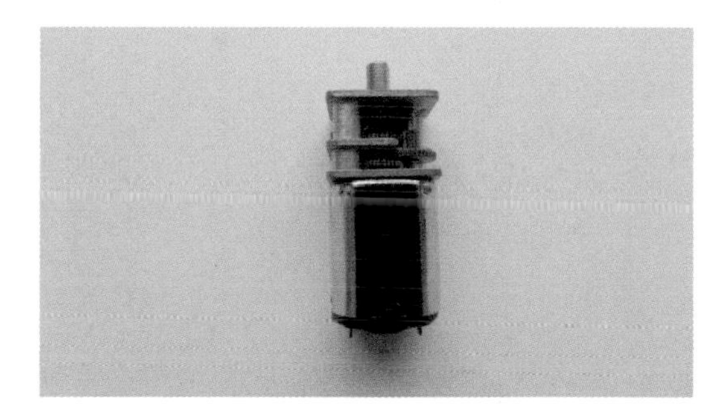

Figure 8-106. *Gearhead motor*

Figure 8-107. *A very large and very small gearhead motor*

Motors such as this one often require more current than the 40 mA that an output pin on the Arduino can provide. A *transistor* is a component that allows a small amount of current to *trigger* a device that requires a larger amount of current. A transistor circuit can enable an Arduino to control a motor that needs more than 40mA of current.

The servo circuit shown in Figure 8-99 did not require a transistor because the servo is powered from the + pin, which connects directly to the power source.

As with most electronic components, there are many types of transistors to choose from. Let's take a look at two commonly available transistors that work well with the types of circuits you might encounter in wearables (Figure 8-108).

Figure 8-108. *NPN Bipolar Transistor (PN2222) and TIP120 Power Darlington Transistor*

The PN2222 is a medium-power transistor that that can switch currents up to 500mA. The TIP 120 comes in a slightly larger packaging. It is a medium- to high-power transistor that can switch currents up to 5A.

The transistors that are shown in this chapter (NPN transistors) have three pins (Figure 8-109). Their functions are as follows:

Base
: This is what gets connected to the microcontroller output pin

Collector
: The collector is connected to the power source, often with the load (in this case, the motor) in series

Emitter
: The emitter is what gets connected to ground

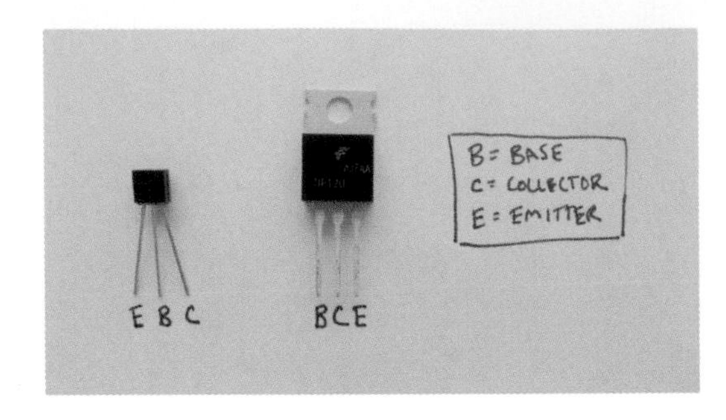

Figure 8-109. *Transistor pinouts*

When a small amount of electricity is applied to the base, a larger amount of electricity flows between the collector and the emitter.

In practice, a circuit might look something like Figure 8-110.

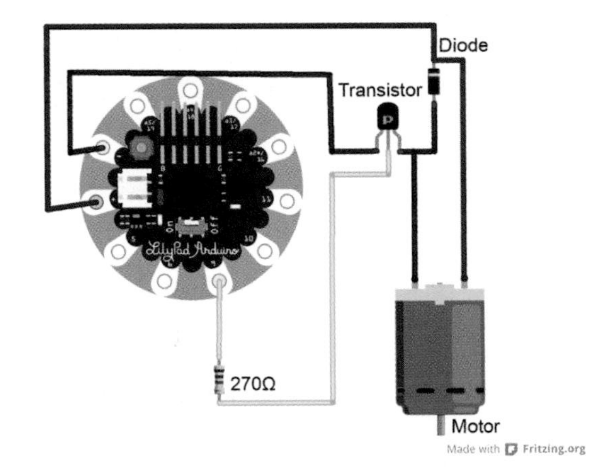

Figure 8-110. *DC motor circuit layout*

Notice that in this circuit you're using a new component called a *diode* (Figure 8-111).

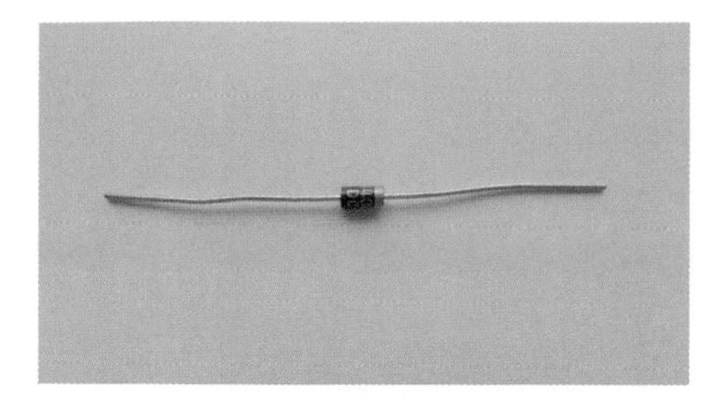

Figure 8-111. *Diode*

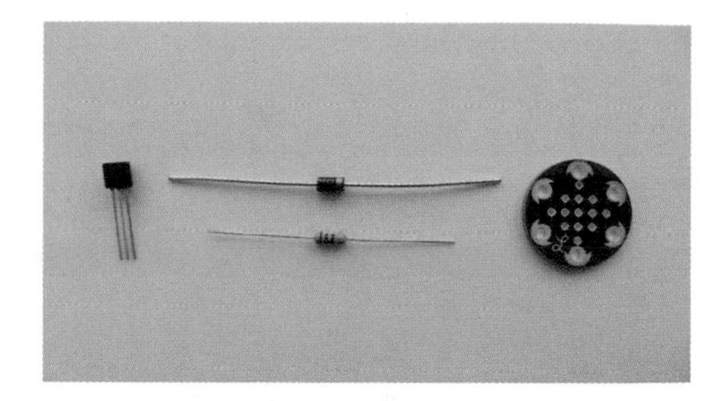

Figure 8-112. *Parts*

A diode is a component that allows current to flow only in one direction. You have previously encountered diodes in the form of light-emitting diodes (LEDs). The diode in this circuit is similar except it does not emit light—it simply limits the flow of the electricity to one direction.

The purpose of this diode is to prevent *blowback voltage*. When voltage is supplied to a motor, it turns. This relationship also works in reverse. If you turn a motor, it can actually function as a generator and produce voltage. Should that happen accidentally, the diode in this circuit prevents the voltage from traveling back to, and damaging the microcontroller.

What's the best way to integrate this transistor circuit into the parts you've been working with? It just so happens that there are LilyPad Protoboards that fit this purpose well. Let's use the small size to assemble these through-hole parts into a compact bundle that's easy to connect to your other LilyPad components.

Parts, as shown in Figure 8-112:

- LilyPad Protoboard Small (SF DEV-09102)
- PN2222 transistor (AF 756)
- Diode (AF 755 or SF COM-10926)
- 270Ω resistor

Tools:

- Solder and soldering iron
- Helping hands
- Ruler
- Knife
- Multimeter
- Needle-nose pliers

The protoboard transistor circuit you'll be creating looks like Figure 8-113.

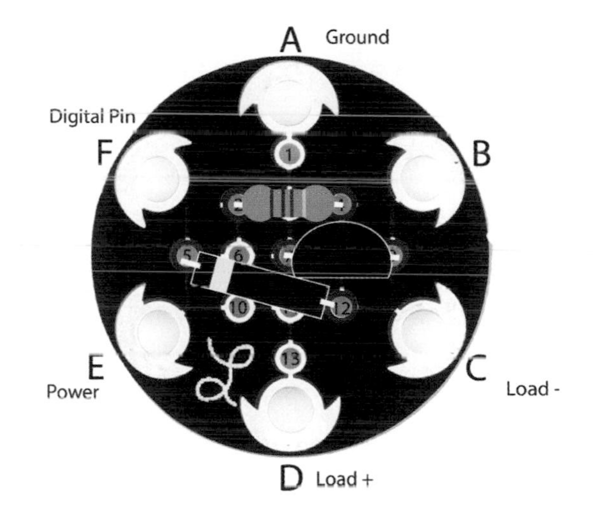

Figure 8-113. *Protoboard transistor circuit*

A LilyPad Protoboard comes with all of its pins connected together at first, so rather than making connections, you'll be breaking the ones you don't need. If you're at the side of the board with the "L" on it (which I will refer to as the front), you can see that these connections are close to the surface (Figure 8-114). You will be using a ruler and knife to break some of these connections. The back is shown in Figure 8-115.

Figure 8-114. *LilyPad Protoboard Small; I refer to the side with the "L" as the front*

Figure 8-115. *LilyPad Protoboard Small; I refer to the side without the "L" as the back*

Despite the fact that these protoboards are small, there are a number of connections that you will be working with within them. Let's name the holes so that you can have a good sense of what goes where. The sew tabs will be represented by letters and the interior holes by numbers. Throughout these instructions, you can refer to Figure 8-116 for reference. Figure 8-117 shows the labels from the back side.

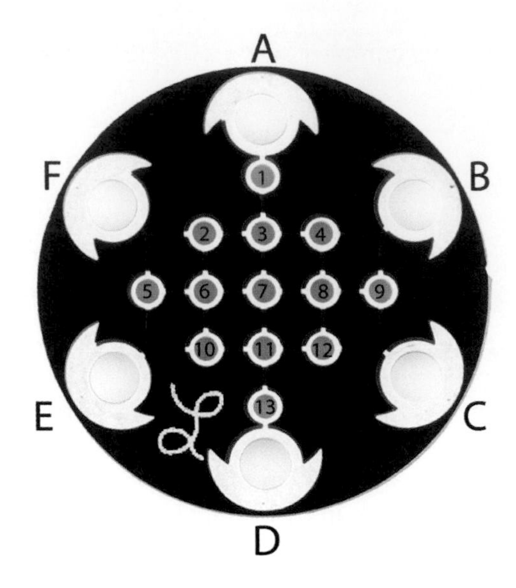

Figure 8-116. *Hole and pin numbers (front)*

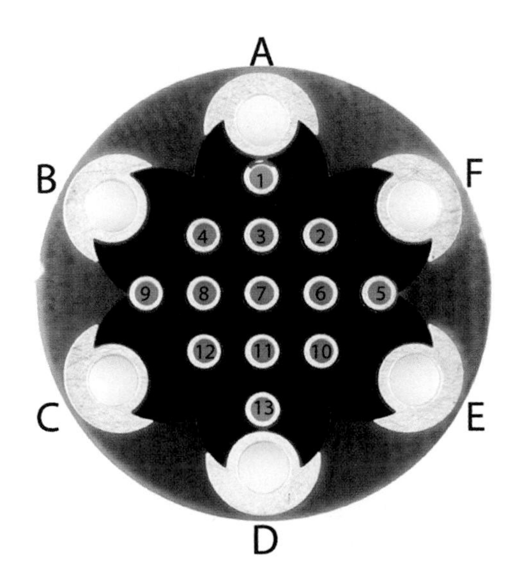

Figure 8-117. *Hole and pin numbers (back)*

In order to prepare the board for your transistor circuit, you will need to make some cuts to disconnect some of the holes. Keeping the "L" at the lower left, line up the ruler so that the edge falls between the column that contains "3," "7," and "11" and the column that contains "4," "8," and "12," as shown in Figure 8-118.

Figure 8-118. *Cutting the traces*

Use a knife to carefully and fully score the board to cut the underlying copper trace (Figure 8-119). This may take two or three cuts, depending on the sharpness of the knife.

Figure 8-119. *The resulting cut*

Using a multimeter set to the continuity setting, test each row on either size of the cut to see if the pins are now disconnected (Figure 8-120). When testing the disconnected pins, the multimeter should not beep.

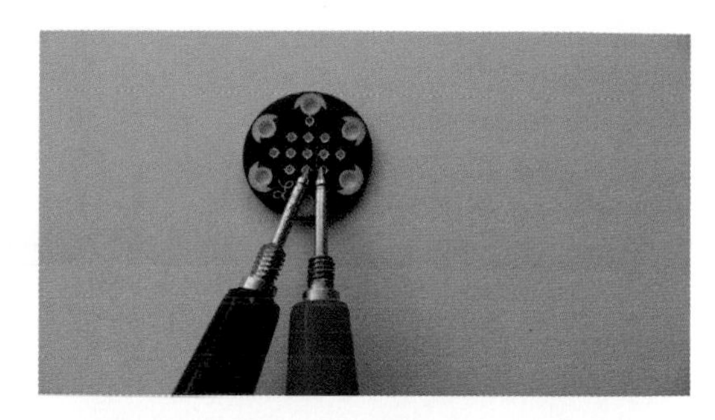

Figure 8-120. *Testing with a multimeter to see if the pins are disconnected*

Now that you know how to make a successful cut, you can go ahead and do the rest. Make the remaining cuts according to Figure 8-121. Figure 8-122 shows the resulting zones that remain connected.

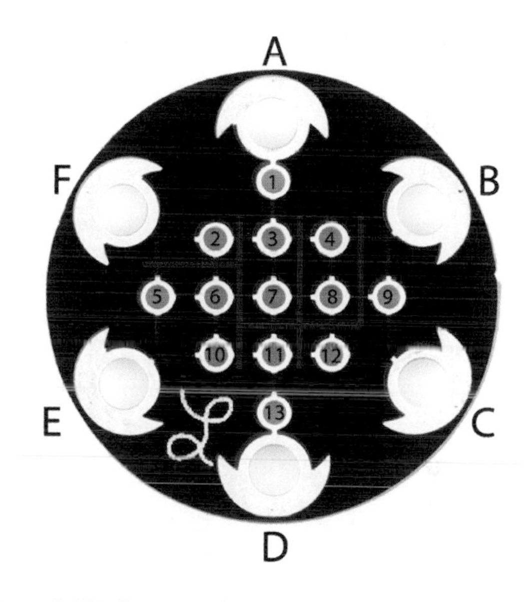

Figure 8-121. *Diagram of cuts*

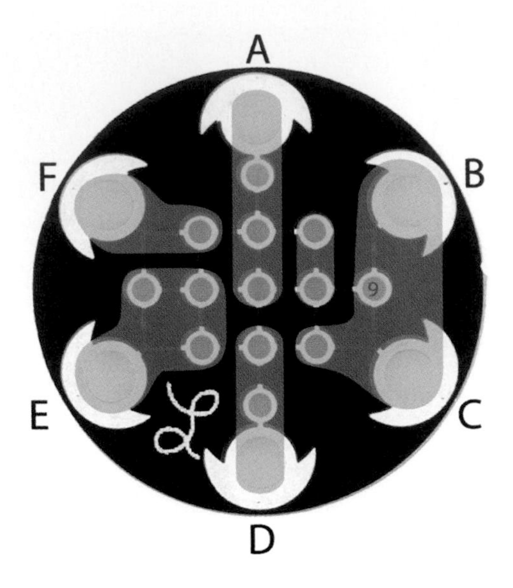

Figure 8-122. *Zones of connectivity*

If you find that a cut isn't deep enough, you can score it with a knife again. If you accidentally cut something you shouldn't, it's possible that you can later go back and use a jumper wire to repair that connection.

Once all of your scores are complete, use your multimeter to double-check that all of your cuts are deep enough. When you are done, your board should look like Figure 8-123.

Figure 8-123. *LilyPad Small Protoboard with completed cuts*

Now it's time to start assembling the circuit. The diagram in Figure 8-124 shows you where the components will be placed on the board.

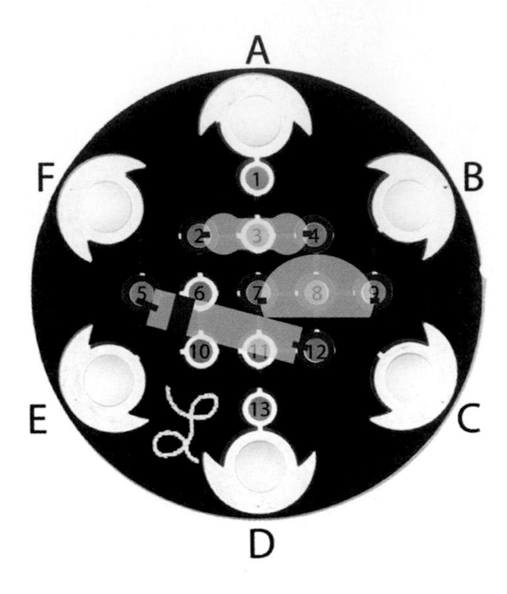

Figure 8-124. *Component placement*

Figures 8-125 through 8-129 walk you through the assembly of the board. First, you will add the resistor.

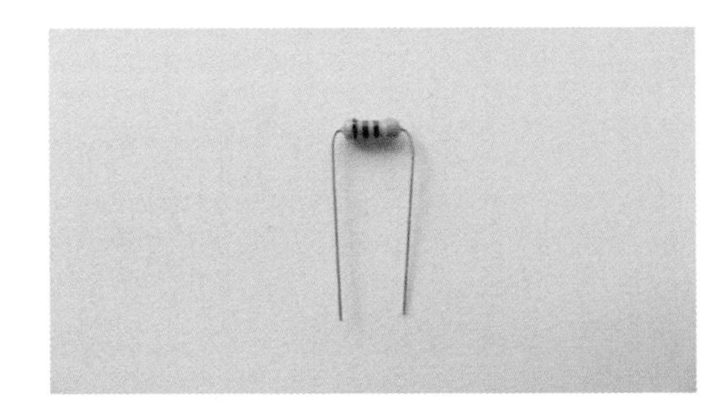

Figure 8-125. *Take the resistor and use the needle-nose pliers to bend the legs at a 90-degree angle*

Figure 8-126. *Insert the legs into hole "2" and hole "4," then pull them through so the resistor sits close to the board*

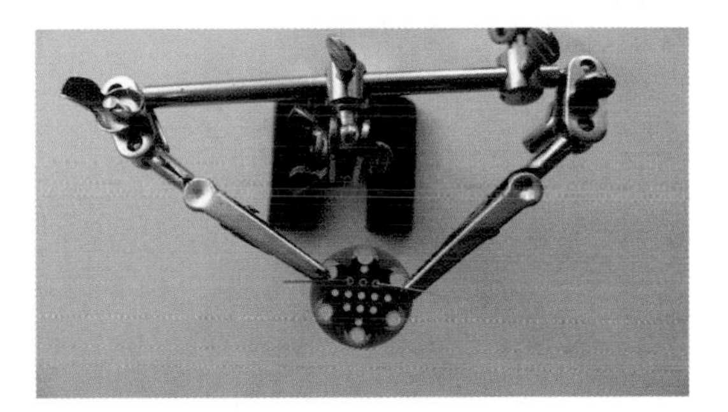

Figure 8-127. *Flip the board over and spread the legs slightly so the resistor stays in place*

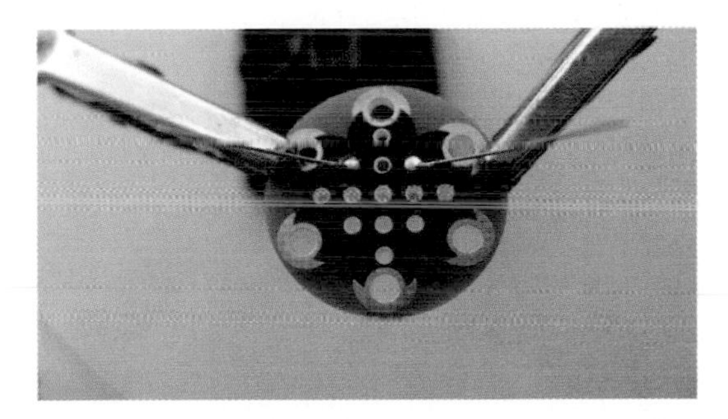

Figure 8-128. *Solder the resistor legs*

Figure 8-129. *Snip the excess*

Next, let's add the transistor. Orient the transistor so the flat part is facing you. Spread the legs of the transistor slightly and insert them into holes "7," "8," and "9," as shown in Figures 8-130 and 8-131.

Figure 8-130. *With the flat side of the transistor facing the "L" on the board, insert the legs into "7," "8," and "9"*

Figure 8-131. *Push the legs through so that the head of the transistor sits close to the surface of the board*

Flip the board over, secure in place with a set of helping hands, and solder the three connections.

Once the soldering is complete (Figure 8-132), snip off the remainder of the legs.

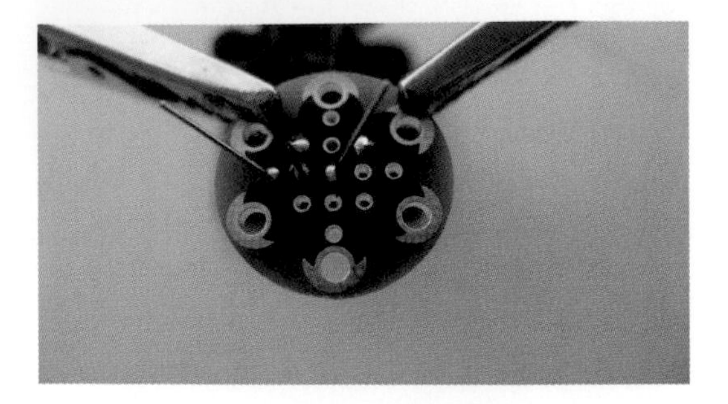

Figure 8-132. *Transistor connections soldered*

Finally, it is time to add the diode. Bend the diode legs to a 90-degree angle, as shown in Figure 8-133.

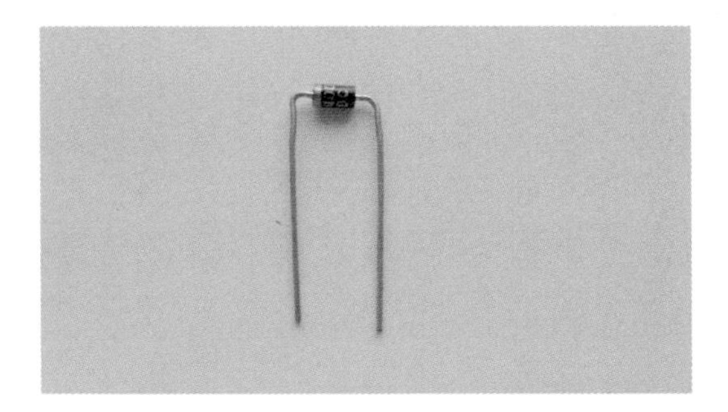

Figure 8-133. *Diode with bent legs*

Place the diode on the board so that the leg close to the stripe is in hole "5" and the other leg is in hole "12" (Figure 8-134). Solder it into place as shown in Figure 8-135.

Figure 8-134. *Diode with legs in holes "5" and "12"; be sure to pay attention to the orientation of the diode!*

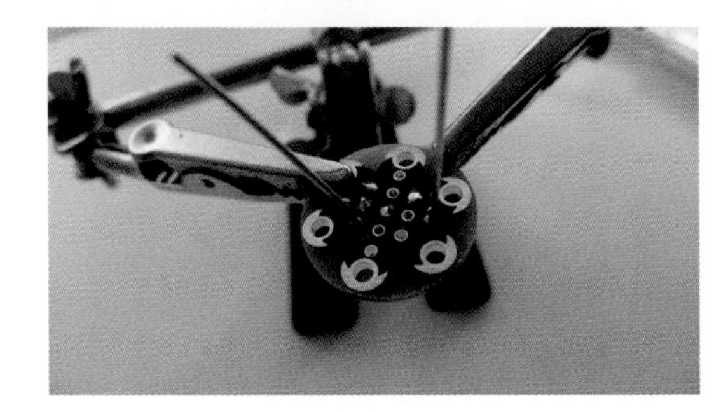

Figure 8-135. *Flip the board and solder the legs in place.*

Snip the excess. The completed back of the board should look like Figure 8-136.

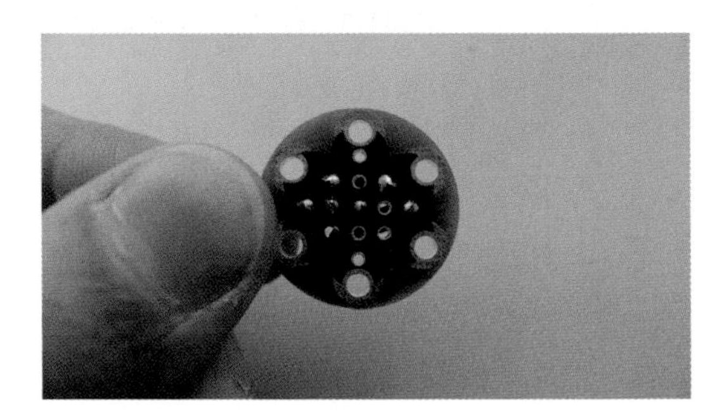

Figure 8-136. *Back of completed board*

And the completed front of the board should look like Figure 8-137.

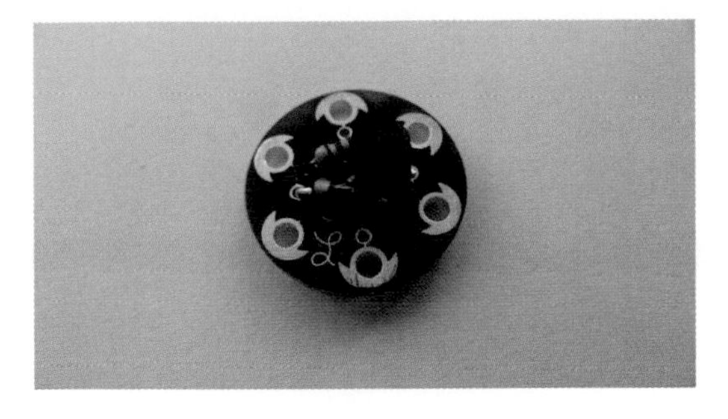

Figure 8-137. *Front of completed board*

See Figures 8-124 and 8-113 for the component placement.

Your transistor circuit is now complete! The functions of the pins on your new transistor module are as follows:

A	Ground
B	(None)
C	Load negative (-)
D	Load positive (+)
E	Power
F	Digital pin

The "load" is whatever it is that you are controlling with a transistor. In this case, it is the gearhead motor.

This board can now be added to a circuit with your LilyPad Arduino Simple to control a gearhead motor. Figure 8-138 shows the circuit diagram. Figure 8-139 shows the circuit assembled with alligator clips.

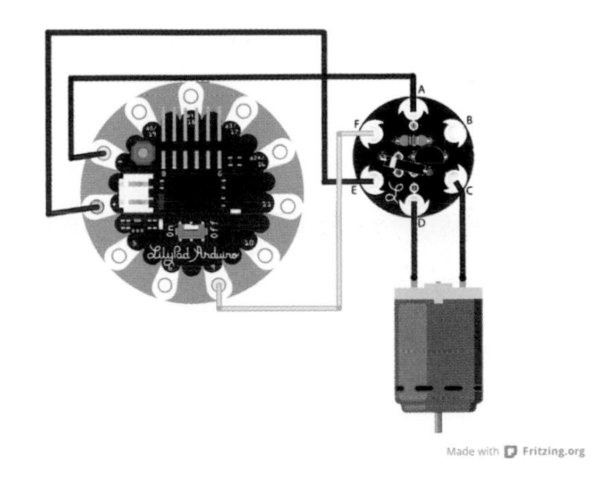

Figure 8-138. *Circuit diagram*

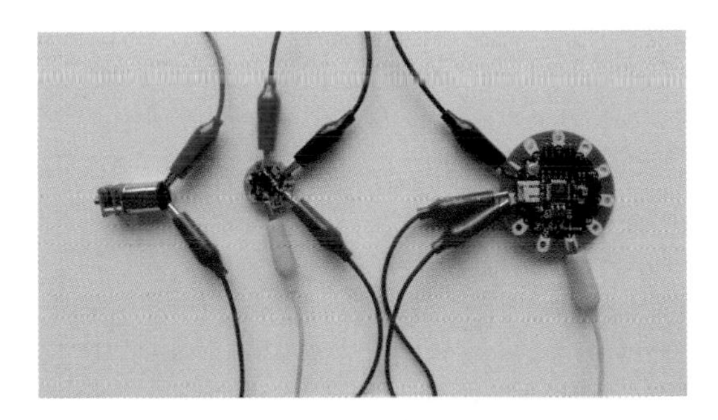

Figure 8-139. *Gearhead motor alligator clip circuit with transistor*

Once your circuit is connected, you can program the Arduino to control the motor. To simply turn the motor on and off, you can set pin 9 as a digital output:

```
/*
Make: Wearable Electronics
 Gearhead Motor Digital Example
 */

int motorPin = 9;

void setup(){
  pinMode(motorPin, OUTPUT);
}

void loop (){
  // turn motor on
  digitalWrite(motorPin, HIGH);
```

```
    delay(5000);
    // turn motor off
    digitalWrite(motorPin, LOW);
    delay(1000);
  }
```

If you'd like a bit more control over the speed, make use of the `analogWrite()` function. Here's an example:

```
/*
Make: Wearable Electronics
 Gearhead Motor Analog Example
  */

int motorPin = 9;

void setup(){
}

void loop (){
  // turn motor off
  analogWrite(motorPin, 0);
  delay(500);
  // spin motor slowly
  analogWrite(motorPin, 100);
  delay(5000);
  // turn motor off
  analogWrite(motorPin, 0);
  delay(500);
  // spin motor at full speed
  analogWrite(motorPin, 255);
  delay(5000);
}
```

The transistor board you created can also be used with vibrating motors to power a more intense vibration, as shown in Figures 8-140 and 8-141.

Once you're able to get a gearhead motor moving, then you have to figure out what to do with it! Here are some examples of gearhead motors at work in wearables.

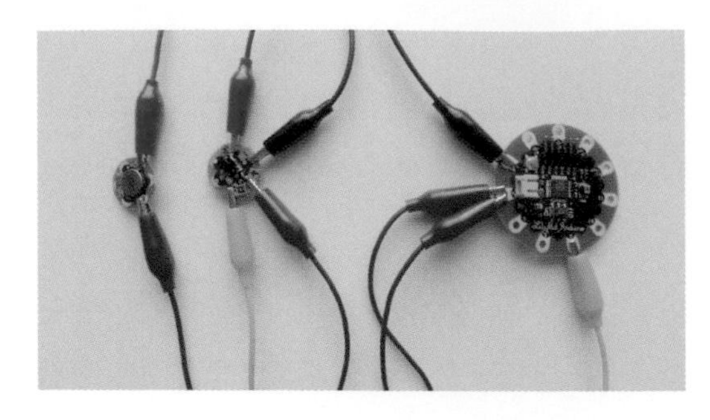

Figure 8-140. *LilyPad Vibe motor with transistor*

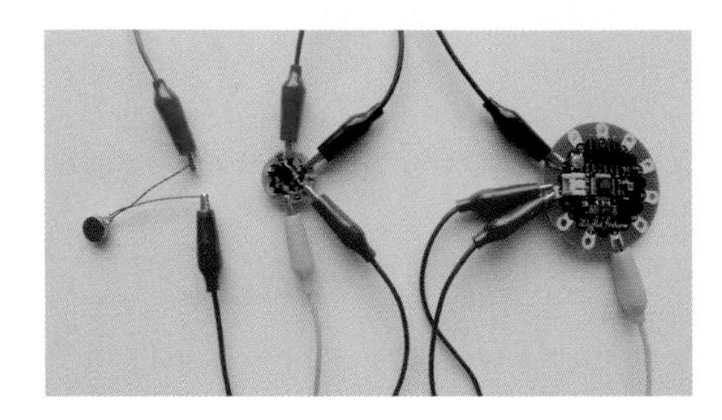

Figure 8-141. *Pancake vibe motor with transistor*

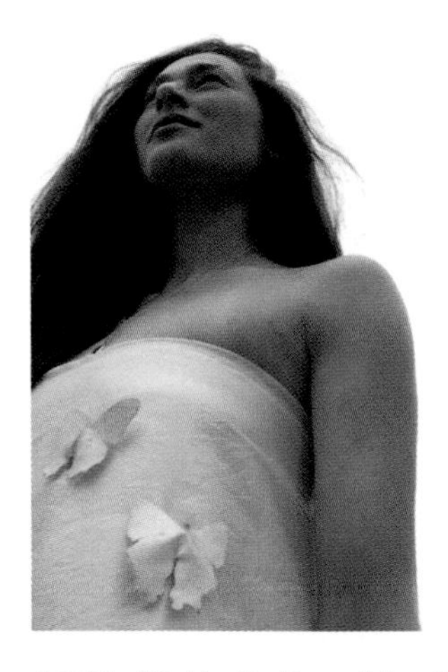

Figure 8-142. *"Butterfly Dress" by Alexander Reeder uses micro metal gearmotors, with a custom attachment for the shaft, to flap the wings on these wearable butterflies*

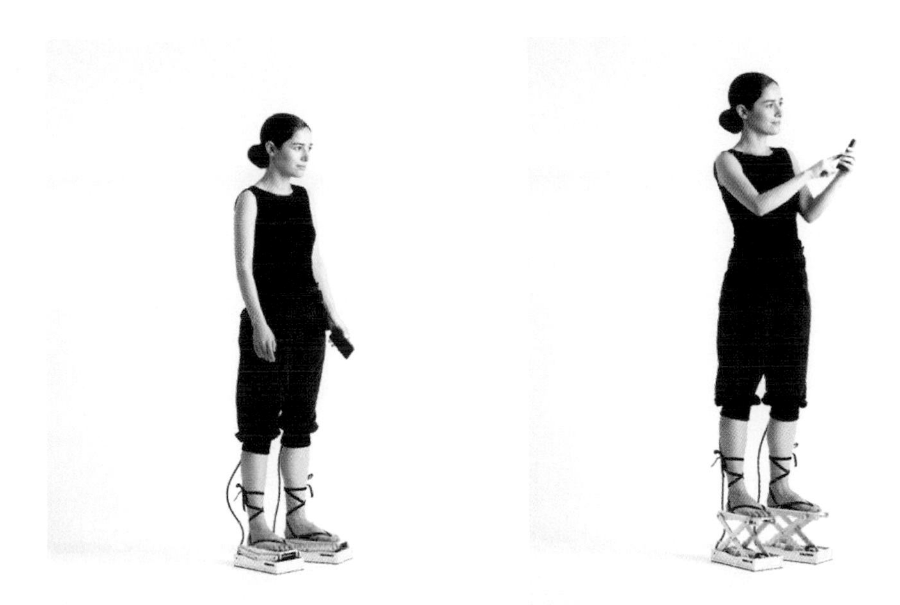

Figure 8-143. *"Short ++" by Adı Marom uses a heavy-duty gearhead motor to physically adjust the height of the wearer (photographed by Charlie Wan)*

When working with motors, a knowledge of mechanics can be a tremendous asset. Check out *Making Things Move* by Dustyn Roberts (McGraw-Hill) to learn how to prototype mechanical systems.

Experiment: Shake, Spin, or Shimmy

Create a piece of clothing that moves in response to stimuli. Think about where the motor will sit, what and how it will move, and how your material design can best support its movements.

Figure 8-144. *"Here-I-Am Hat" (illustration by Jen Liu)*

Temperature

Because clothing is often used to provide heat and protection, it's no surprise that wearable technology designers are often interested in working with actuators that provide a heating and cooling effect.

Keep in mind that these are usually higher current devices, so the LilyPad transistor board you created in the last section will come in handy.

Fans

Thanks to the temperature needs of desktop and laptop computers, there are a significant number of small 5V fans that are readily available for an equally small price, such as the one shown in Figure 8-145.

Figure 8-145. *Small fan*

By taking a closer look at the label on the fan, you can learn about its power needs (Figure 8-146).

Figure 8-146. *This is a 5V, 200mA fan*

Heatit

It's important to know how transistors work, but it is also helpful if you can pack up that functionality into a smaller package. At the time of this writing, there is a new tool being developed called Heatit. Heatit is an open source electronics platform (based on Arduino) that offers precise high current output in a small convenient package. Its output pins are able to supply up to 500mA, making it well suited to control motors, heating pads, fans, shape-memory alloys, and other higher current actuators.

Here's the Heatit board by the Heatit Team (photographed by Eszter Ozsvald):

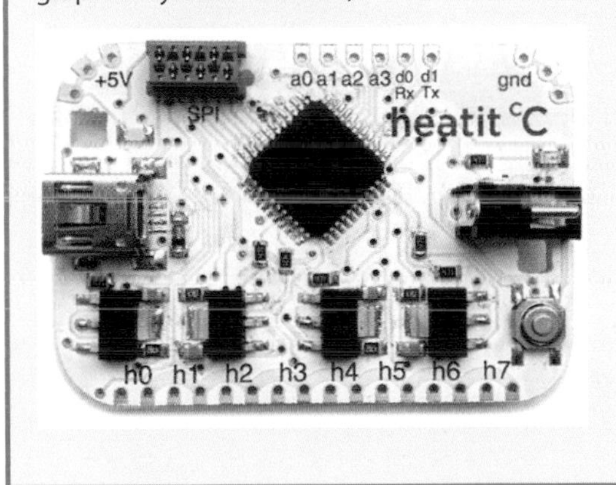

200mA is far beyond the 40mA that an Arduino output pin supplies. This is another situation in which you can make use a transistor, as shown in Figure 8-147. One way to do this is to use the protoboard transistor circuit you created in the last section and swap out the motor for the fan as shown in Figure 8-148.

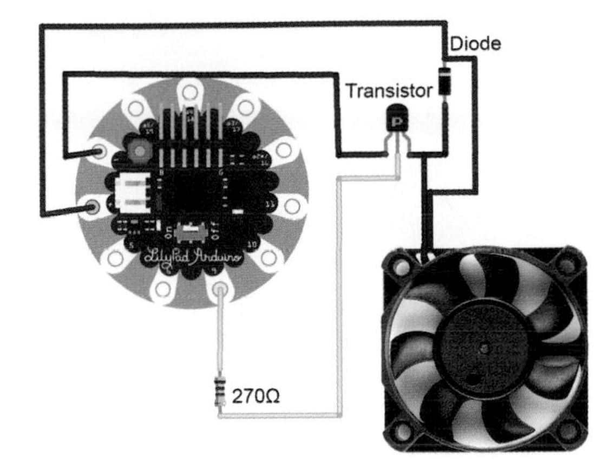

Figure 8-147. *Transistor and fan circuit diagram*

Figure 8-148. *Transistor proto board circuit and fan diagram*

While this type of fan is still a bit bulky to incorporate into clothing, it can have some fun, cooling results.

Keep in mind that when incorporating this into a garment, it is helpful to use a stiffer, thicker material that can provide proper support for the fan (for example, Figure 8-149).

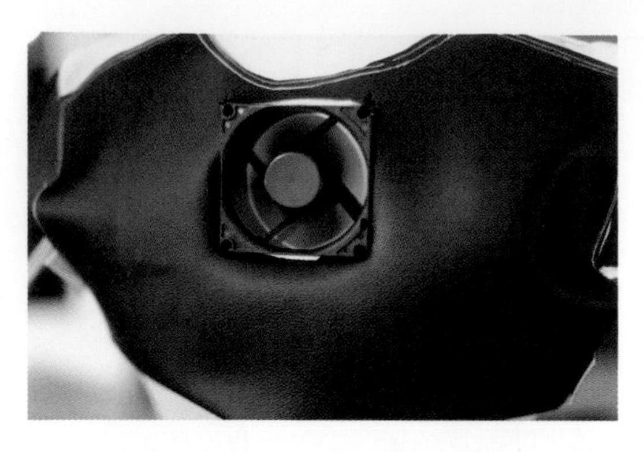

Figure 8-149. *The Cool Suit by Yuxi Wang and Robert Tu*

Heat

From keeping hands toasty on a winter day to providing a slow-growing warming sensation over your heart when someone is thinking of you, heat can provide both physical comfort and even elicit an emotional response when handled in an interesting way (Figure 8-150). Keep in mind that heating pads are slow and subtle actuators. Your interaction scenario should be designed accordingly.

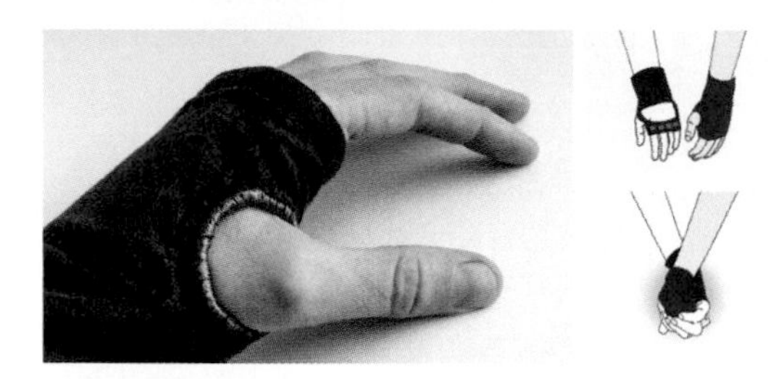

Figure 8-150. *"The CoDependent Gloves" by Fiona Carswell provide warmth when two people hold hands*

Electric heating pads (AF 1481, SF COM-11288) are thin and flexible (Figure 8-151), making them easy to integrate into clothing (see Figure 8-152). Like fans, heating pads are higher current actuators and will require the use of a transistor. Swap a heating pad into the circuit you used for the fan and the gearhead motor and you'll be good to go!

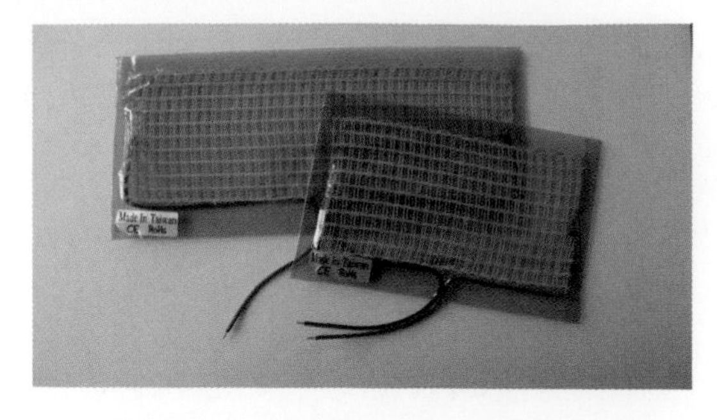

Figure 8-151. *Heating pads*

Figure 8-152. *A pocket can be a great way to hold a heating pad in place*

Experiment: It's Gettin' Hot in Here

Using a fan, heating pad, or both, create a climate-controlled wearable that responds to the current temperature. Refer back to Chapter 7 for more information on how to sense temperature.

Conclusion

As you can see, there's no end to the ways you can use actuators to make things happen when creating wearable electronics. Now that you know how to build a full interactive system that lives on the body, it's time to move beyond the bodysphere and out into the rest of the world. Next up: wireless wearables!

Figure 8-153. *"Fan Suit" (illustration by Jen Liu)*

9 Wireless

So far, you've used a variety of materials, tools, and components to create interactive systems that reside on the body. But what if you want to design wearable systems that communicate beyond the body?

What if you want to use gestures, biometric data, or body language to control what happens on a screen? Or log body-generated data to a shared database? Or send a signal from one wearable to another?

While communication between interactive systems can easily be accomplished with wires, this is not terribly practical in the wearable context. Wires physically tether the wearable to whatever external system it is communicating with. Who wants to get tangled up in wires when they're going for a run, bustin' a move, or just walking around the house?

In this chapter, you'll explore some introductory options for wireless wearable communications. There are many ways in which wireless communication can be accomplished, and here you'll focus on a few simple ones to get you started.

Figure 9-1. *Robert Tu's "MeU" is a flexible LED matrix that is worn on the body to display information; it is a modular system of flexible 8 x 8 LED matrices controlled by a smartphone via Bluetooth (photographed by Robert Tu and Gordon Pietzsch)*

Bluetooth

Bluetooth is a convenient communication protocol to work with because of its ubiquity. You use it for your wireless headsets, your keyboards, and your mice. It's quite likely that your laptop, smartphone, and tablet all already have Bluetooth built in. Even if you have an older computer, you can get a USB Bluetooth dongle for under $20 these days (SF WRL-09434). Bluetooth can reduce your expenses and setup time because one side of the communication is already taken care of for you.

There are many types of Bluetooth radios available for Arduino. For the example in this chapter, you will use the Bluetooth Mate Silver from SparkFun (Figure 9-2). It's similar to their BlueSMiRF modem, but the pins have been arranged so that they connect easily with a LilyPad Arduino and a few other Arduino boards. This device runs off 3.3-6V and consumes an average of 25mA, making it a fairly low-power device. For wearables, low power is advantageous because it means a smaller battery pack will last a lot longer.

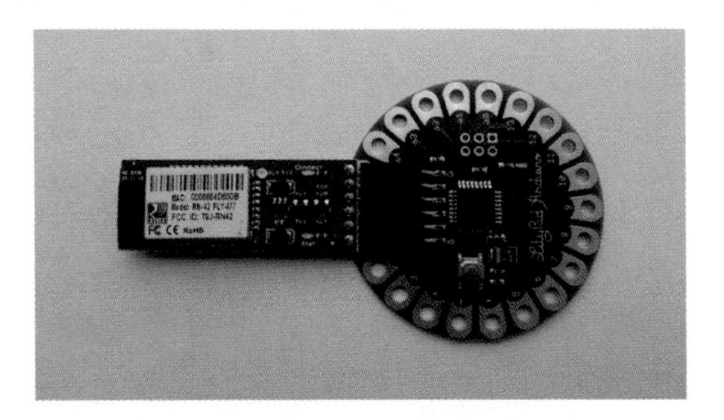

Figure 9-2. *Bluetooth Mate Silver connected to LilyPad Arduino*

The actual chip on the board is Roving Network's RN-42. While this radio is capable of being customized with a variety of configurations, it is also possible to accomplish a lot with the default settings.

The datasheet for the RN-42 claims that it can operate at a range of up to 60 feet (20 meters) distance, though keep in mind that this is only in optimal conditions, typically line-of-sight on raised poles with no obstructions or radio interference. If you're looking for a greater range, consider upgrading to the Bluetooth Mate Gold, which has a RN-41 onboard with a range up to 330 feet (100 m) distance.

Experiment: Communicating with Bluetooth

In this example, you will learn how to communicate sensor data wirelessly from a wearable circuit to a nearby computer via Bluetooth.

Parts and materials:

- LilyPad Arduino 328 (SF DEV-09266)
- LilyPad Light Sensor (SF DEV-08464)
- LilyPad Simple Power (SF DEV-10085)
- FTDI board (AF 284, SF DEV-10275)
- Bluetooth Mate Silver (SF WRL-10393)
- 6-pin set of right-angle female headers with 0.1" (2.54mm) spacing (AF 1542, SF PRT-09429)
- USB mini-B cable (AF 899, DK WM5163-ND, RS 55010682, SF CAB-11301)
- Alligator clip test leads (AF 1008, RS 278-1156, SF PRT-11037)
- 3.7V lithium-ion polymer rechargeable battery (AF 258, SF PRT-00339)

Tools:

- Soldering iron
- Solder
- Bluetooth-enabled computer

The Bluetooth Mate Silver is not compatible with the LilyPad Arduino Simple Board.

Prepare the LilyPad Simple Power board

The LilyPad Simple Power board provides a location for a resistor in case you want to modify the voltage of the battery. In this case, you do *not*, so before using this board it is necessary to bridge this connection. Using a small bit of solder, close the gap between these two solder pads, as shown in Figure 9-3.

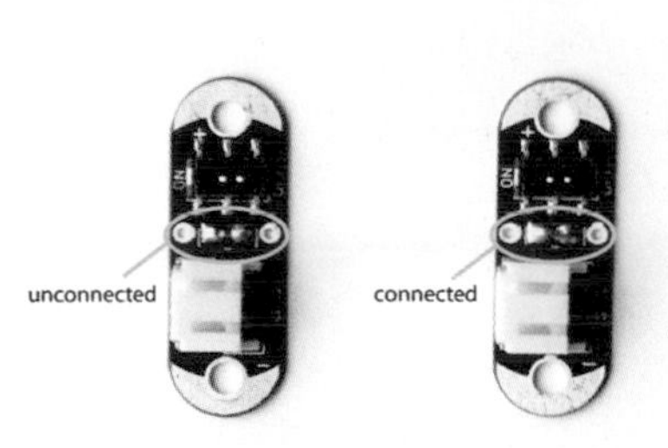

Figure 9-3. *LilyPad Simple Power Board with solder bridge*

Solder headers to the Bluetooth Mate

When you first get the radio, you will find that there are no headers connected to board, as shown in Figure 9-4. Without headers, you won't be able to plug the radio into anything else.

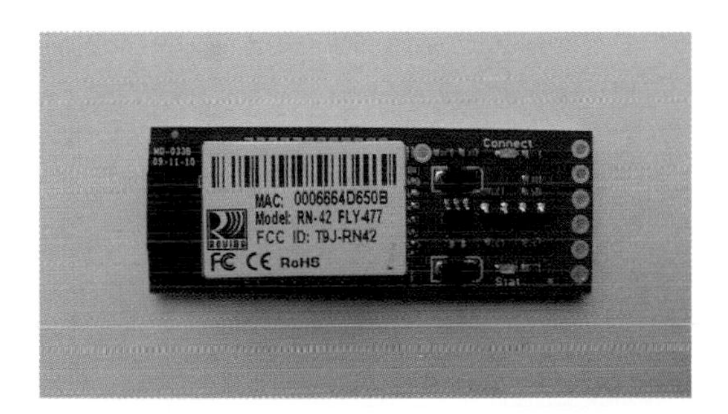

Figure 9-4. *Bluetooth Mate Silver, without headers*

Take a 6-pin set of right angle female headers. Place them so that they pass from underneath the board and point up and out the holes on the side of the board that holds the radio and other components (Figure 9-5).

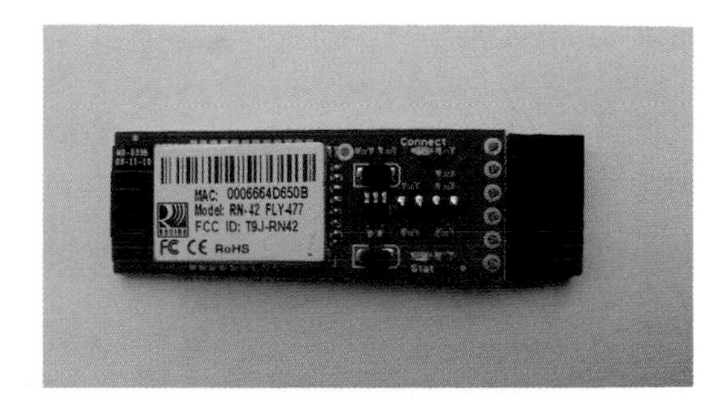

Figure 9-5. *Bluetooth Mate Silver, headers in place*

Solder them in place as shown in Figures 9-6 and 9-7.

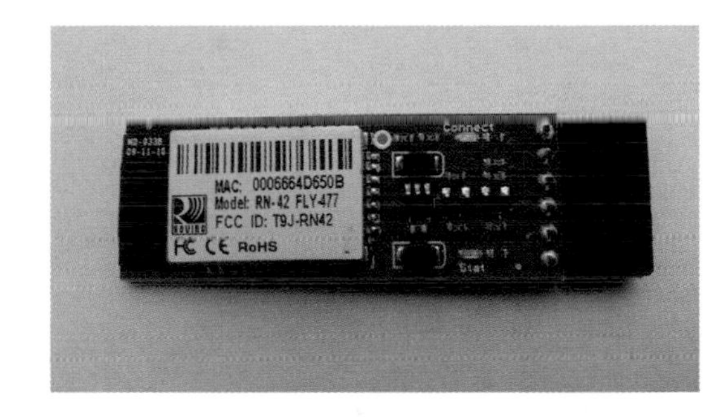

Figure 9-6. *Bluetooth Mate Silver, headers soldered*

Figure 9-7. *Bluetooth Mate Silver, back*

Program the LilyPad

Connect the LilyPad Arduino to the FTDI board as shown in Figure 9-8. Then connect the FTDI board to your computer with a USB miniB cable.

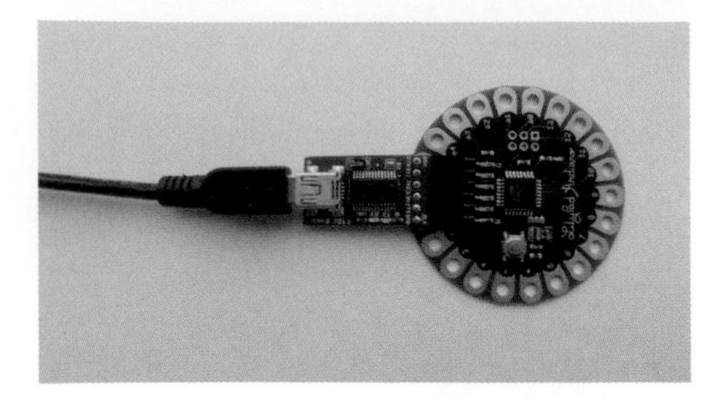

Figure 9-8. *LilyPad Arduino with FTDI programming*

Open Arduino. Create and run the following sketch:

```
/*
Make: Wearable Electronics
 Bluetooth Pairing example
 */

void setup() {
  // Initialize serial communication at 115200
  // bits per second. This is the default speed
  // of communication for the RN-42.
  Serial.begin(115200);
}


void loop() {
  // Leave the loop empty. You're just looking
  // to make contact.
}
```

Upload it to your LilyPad Arduino.

Prepairing to pair

Once the program has been uploaded successfully, you can prepare your Bluetooth circuit to be paired.

Disconnect the FTDI board and then make the connections shown in Figure 9-9.

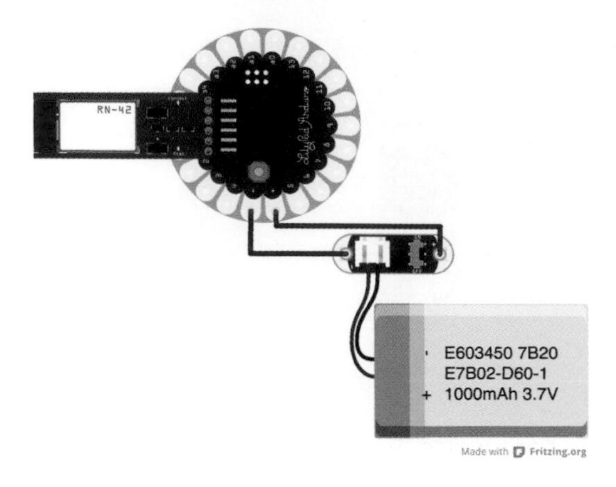

Figure 9-9. *LilyPad Arduino with Bluetooth Mate Silver and LilyPad Simple Power board*

Turn the switch on the LilyPad Simple Power board to ON. The STAT LED on the Bluetooth Mate will begin blinking red.

Pairing on a Mac

On your computer, go to the apple in the upper-left corner, then System Preferences, then Bluetooth.

First, make sure Bluetooth is on.

Next choose "Set Up New Device":

This will open the Bluetooth Setup Assistant. Wait while nearby devices are detected. You are looking for a device whose name starts with "RN42-":

Once the device appears, select the device name and then click Continue. You will then likely get an error screen that looks like this:

Click "Passcode Options." Then choose "Do not use a passcode with this device":

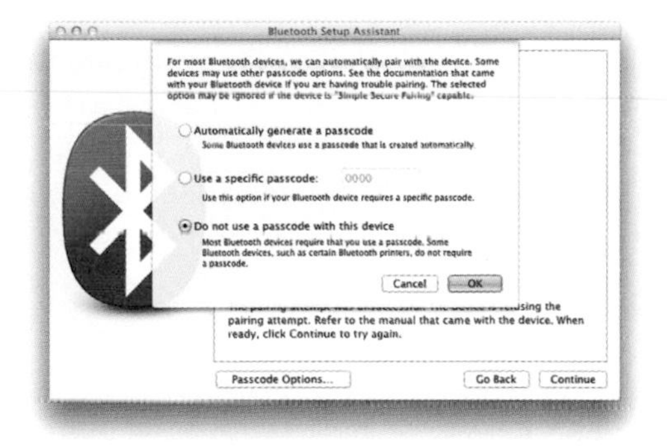

Click OK and then you should see this screen:

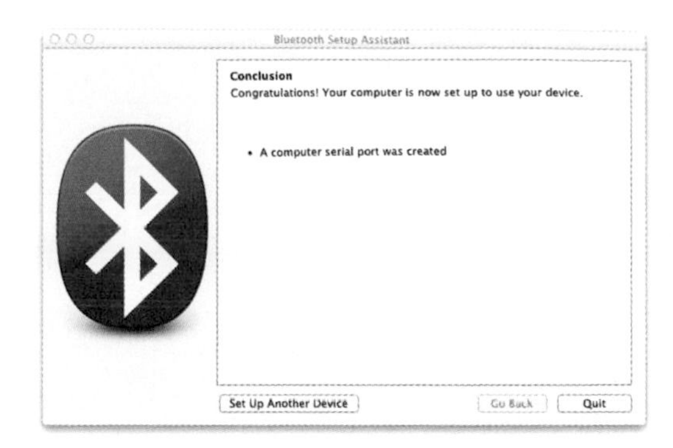

The Bluetooth Mate is now paired with your computer! Go ahead and click Quit. Now you're ready to go!

Pairing on a Windows machine

On Windows 7, locate the Bluetooth icon in the System Tray. Right-click it, and choose "Add a Device." It may take a minute or two before all the nearby devices appear. Locate the one named *RN42-XXXX*, where *XXXX* is some sequence of numbers and letters:

Select the RN42 device, then click Next. When prompted to choose a pairing code, select "Pair without using a code":

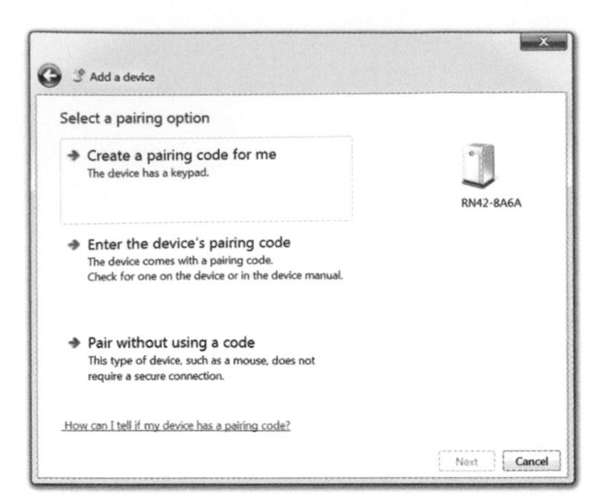

When Windows is done adding the device, it will show the message, "This device has been successfully added to this computer." The Bluetooth Mate is now paired with your computer!

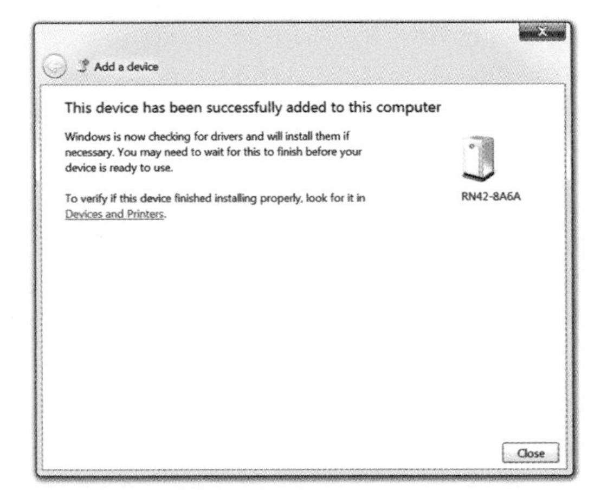

Sending light sensor data

Now that the radio is able to connect to the computer, you can try to transmit some data.

Turn off the circuit with the switch on the LilyPad Simple Power board. Disconnect the Bluetooth Mate, and make the connections shown in Figures 9-10 and 9-11.

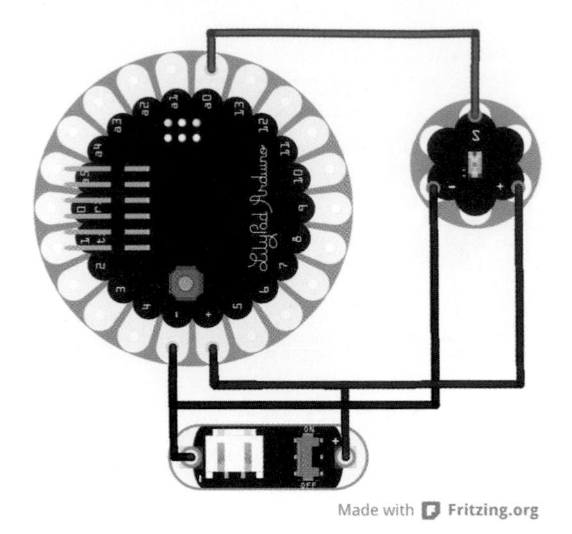

Figure 9-10. *Circuit layout for LilyPad Arduino with Light Sensor and battery*

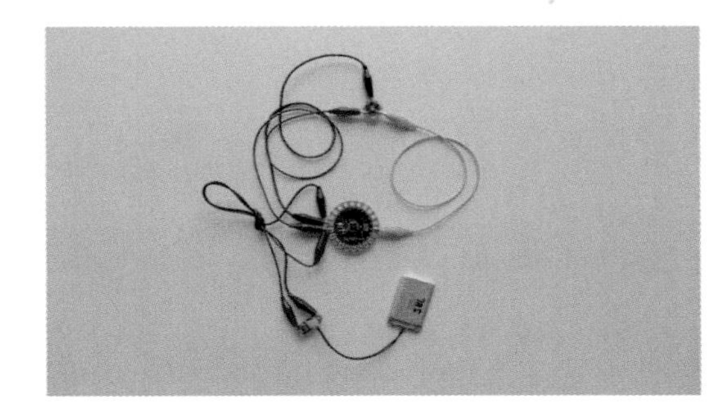

Figure 9-11. *LilyPad Arduino with Light Sensor and battery*

With the circuit ready, you can now update the Arduino program. Connect the FTDI board to the LilyPad Arduino and connect the FTDI board to your computer. Upload the following sketch to the Arduino board:

```
/*
Make: Wearable Electronics
 Bluetooth Light Sensor example
 */

//initialize variables
int lightSensorPin = A0;
int lightSensorValue = 0;
```

```
void setup() {
  // Initialize serial communication at 115200
  // bits per second. This is the default speed
  // of communication for the RN-42.
  Serial.begin(115200);
}

void loop() {
  // read the light sensor value
  int lightSensorValue =
  analogRead(lightSensorPin);
  // print the value of the light sensor
  Serial.println(lightSensorValue);
  // add a delay between readings so as not
  // to lock the radio with data overflow
  delay(200);
}
```

Once the sketch is uploaded, open the Serial Monitor and change the baud rate to 115200, as shown in Figure 9-12.

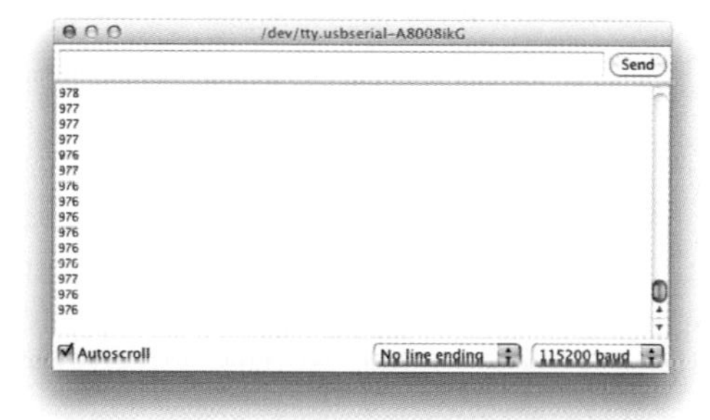

Figure 9-12. *Changing baud rate to 115200*

Just like the example in "Analog Input" on page 108, you should be able to see the light sensor readings on screen. Now that you know the data is being transmitted properly over a USB cable, let's move on to wireless!

Disconnect the FTDI board and reconnect the Bluetooth Mate, as shown in Figure 9-13.

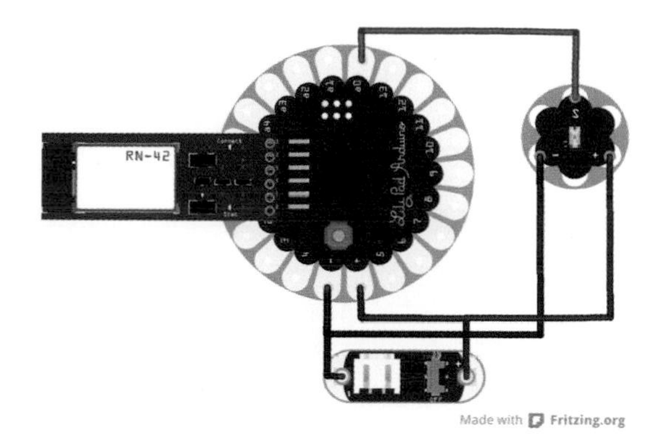

Figure 9-13. *LilyPad Arduino with Light Sensor, battery, and Bluetooth Mate Silver*

Turn the power for the circuit back on. Back on your computer choose Tools → Serial Port in Arduino, and set the Serial Port to the serial port corresponding to your Bluetooth Mate. This will be a Bluetooth serial port, which will appear as a numbered COM port in Windows, and something like */dev/tty.RN42-XXXX* on Mac, as shown in Figure 9-14.

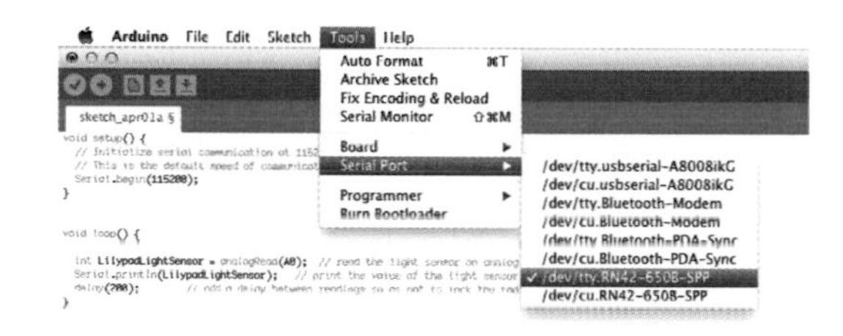

Figure 9-14. *Choosing the Bluetooth serial port*

Open the Serial Monitor (Figure 9-15). Make sure the baud rate is still at 115200. You should now be seeing data in the Serial Monitor!

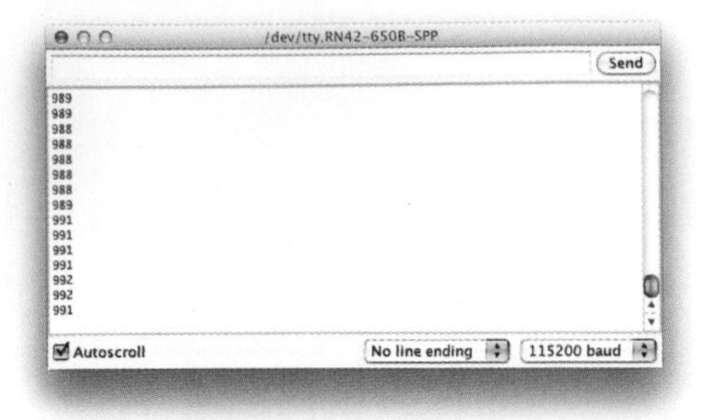

Figure 9-15. *Watching the data scroll by*

You will also notice that the Stat LED on the Bluetooth Mate is no longer blinking and the Connect LED is now lit. This means that the radio is connected to your computer.

If you are not able to see the data coming through, check that:

- The wiring is correct
- The right program is loaded onto the Arduino
- The radio is paired with the computer
- The battery is charged

Once your Bluetooth circuit is working, carefully pick it up and move it around the room to areas with different lighting conditions. Notice how the values in the Arduino Serial Monitor change in response! Put it next to the window, under the table, or hold it with you as you spin. You can even take it on a walk and see how far you can get before losing contact.

Congratulations! You're communicating wirelessly!

Hello XBees

There are many wireless-communication options besides Bluetooth. For the remaining examples, you will work with XBees. XBees is a brand of radio transceiver that includes many types: point-to-point, 802.15.4, mesh networking ZigBee, and Internet-ready WiFi to name a few. There are many ways in which XBees can be used (in pairs, multiples, or networks, with and without a microcontroller). In this chapter, I cover two simple examples using the most basic setup—XBee 802.15.4 radios in pairs.

Figure 9-16. *XBee radio*

Configuring XBees

To enable your two XBees to communicate, you first need to configure them. To do this, you'll be using CoolTerm (*http://freeware.the-meiers.org/*), a terminal program that allows you to communicate with hardware connected to your serial port.

These radios are *configured* with AT commands rather than *programmed* like a microcontroller. Think of it as determining settings on a control panel. There are a limited number of settings that you can customize with your selection, but you can't invent new settings. For the configuration process you'll be using the following items.

Parts and materials:

- (2) XBee Series 1 (802.15.4) with trace or wire antenna (AF 128 or SF WRL-11215)
- (1) LilyPad XBee (SF DEV-08937)
- (1) set of 6-pin right-angle male headers with 0.1″ (2.54mm) spacing (AF 1540, SF PRT-00553)
- (1) FTDI board (AF 284, SF DEV-10275)
- (1) USB mini-B cable (AF 899, DK WM5163-ND, RS 55010682, SF CAB-11301)

Tools:

- Soldering iron and solder
- (1) computer with CoolTerm installed

With the Bluetooth example, you only needed one radio because you were communicating with devices (in particular, a computer) that already have a Bluetooth radio installed. For these XBee examples, you will need two radios, as shown in Figure 9-17. It takes two to tango, and to have a conversation.

Figure 9-17. *Two XBees*

If you haven't already installed FTDI drivers for programming the LilyPad Arduino or other devices, you will need to do that now. See "Software" on page 95 for details.

In order to use the LilyPad XBee to configure XBee radios, you will need to solder a 6-pin set of right-angle male headers onto the board (visible in Figure 9-18). This is where you will connect the FTDI board.

Figure 9-18. *LilyPad XBee (front and back) with 6-pin right-angle male headers*

Use some tape and a marker to label the XBees so that it is easy to identify which is which. Here you can see that the XBees have been labeled "A" and "B" for easy reference, as shown in Figure 9-19.

Figure 9-19. *Two Labeled XBees*

Now you're ready to configure the XBees. Table 9-1 shows the configurations that you will use.

Table 9-1. XBee configurations

AT Command	XBee "A"	XBee "B"	Function
ATRE	—	—	Factory reset (erases previous settings)
ATID	B0D1	B0D1	Sets PAN ID (same for both XBees)
ATMY	A	B	Sets individual XBee IDs
ATDL	B	A	Sets Destination Low address
ATWR	—	—	Saves settings to firmware so they are retained when the radio loses power

XBee Explorer

For configuring the radios, you can also use an XBee Explorer:

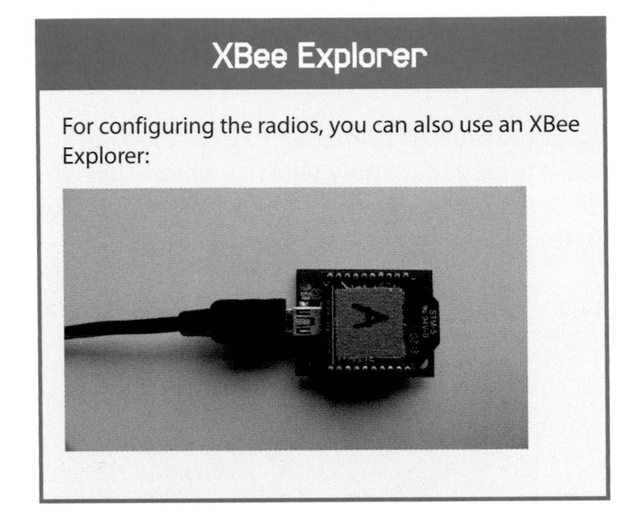

Configure XBee "A"

Place XBee "A" on the LilyPad XBee board, using the visible guidelines inscribed on the board. Ensure that the pins are properly lined up and that the XBee is not plugged in upside down (see Figure 9-20).

Connect the same FTDI board that you use with your LilyPad Arduino to the male headers on the LilyPad XBee board and plug in the USB cable to both the board and your computer, as shown in Figure 9-21.

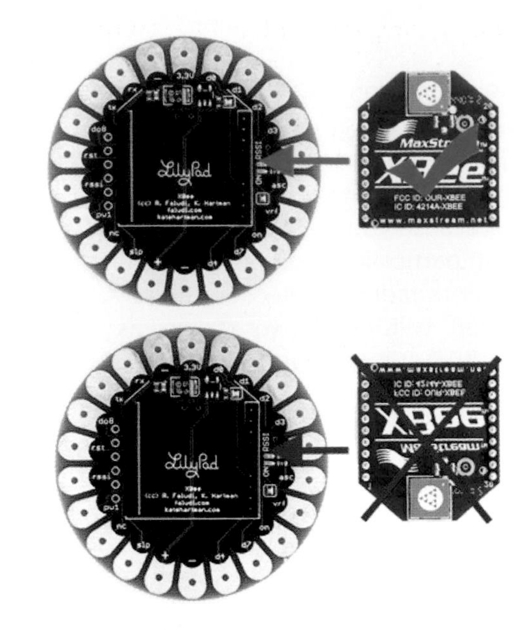

Figure 9-20. *Correct and incorrect ways to connect the XBee*

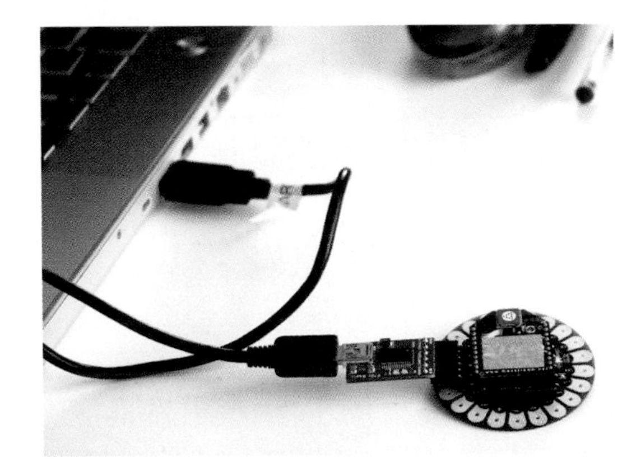

Figure 9-21. *FTDI connection to computer*

XBee radios are configured using commands called *AT Commands*. The configurations you will be using can be found in Table 9-1. Let's start by configuring XBee "A."

To get started, open CoolTerm. In the Menu bar, select Connection → Options. Under the Serial Port tab on the left, first select the serial port. The one you are looking for will have the same name as when you select the serial port in Arduino. On Mac, the name will start with "usbserial-". On Windows, it will be a numbered COM port. Once you've set the serial port, select a baud rate of *9600* (see Figure 9-22).

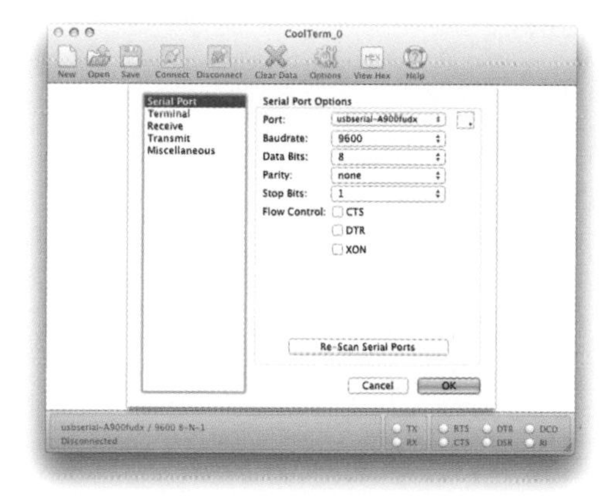

Figure 9-22. *Selecting the serial port and baud rate*

Then, under the Terminal tab, turn Local Echo on by checking the box (Figure 9-23). This will allow you to see what you are typing. Finally, click OK.

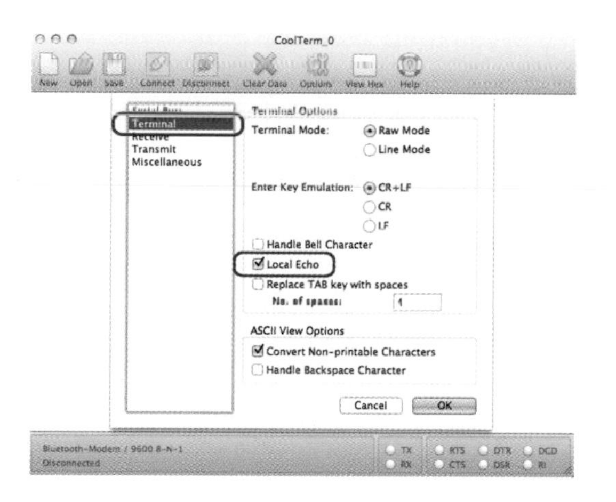

Figure 9-23. *Turning on Local Echo*

Once your settings have been adjusted, click Connect. Now you're ready to start configuring!

Before you can execute AT commands, you must put the XBee into *Command Mode*.

Type "+++" into the CoolTerm window but do *not* press Enter. The radio responds with an OK message, showing that it is in Command Mode.

Staying in Command Mode

If you accidentally press Enter, wait several seconds and try again without pressing Enter.

By default an XBee will fall out of Command Mode in 10 seconds, so try not to leave the window idle. If you aren't getting an OK in response, type in "+++" again to re-enter Command Mode.

Once you've successfully entered Command Mode, you can type your AT commands.

To check an AT setting, you can just type the AT command, press Enter, and it will return the current setting. For example, you can see the XBee's address by typing ATMY and pressing Enter.

To change the setting, type the AT command followed by the new setting. Then press Enter.

Let's get started:

1. First type "ATRE" then press Enter. This will wipe any previous settings from the radio and do a factory reset. You should then receive an *OK* message. If not, make sure you've pressed Enter, and also that you're still in Command Mode.

2. Next, type "ATIDB0D1" then press Enter. This sets the PAN ID, which is the channel on which the two radios will communicate. You will also use the same PAN ID for radio "B".

PAN IDs

In order for two XBees to communicate with each other, they must have the same PAN ID. If you are in an environment where many pairs of XBees are in use, each pair must use a different PAN ID so that they do not interfere with each other.

3. Type "ATMYA" then press Enter. This will set the identity of this radio to "A."
4. Type "ATDLB" then press Enter. This sets the address of the radio this radio talking to radio "B."
5. Finally, type "ATWR" then press Enter. This saves these settings to the radio's firmware so that it will retain the information even if the radio loses power and restarts. This is an important step. If you don't do this, all settings will be deleted when the XBee is unplugged.

If all goes well, your screen should look like this:

```
+++OK
ATRE
OK
ATIDB0D1
OK
ATMYA
OK
ATDLB
OK
ATWR
OK
```

Once this is complete, click the Disconnect button in CoolTerm to end the session.

Unplug your XBee and then carefully plug it back in again. Now that the XBee has been through a power cycle, you'll double-check it to make sure all of your configurations are still intact.

Click Connect. Enter Command Mode (type +++ and wait) and then type each AT command without specifying a new setting, followed by Enter to check the settings:

- ATID <ENTER>
- ATMY <ENTER>
- ATDL <ENTER>

Do *not* type ATRE, as it will erase the settings you just created.

It should look like the following:

```
+++OK
OK
ATID
B0D1
ATMY
A
ATDL
B
```

Configure XBee "B"

Now that you know XBee "A" is properly configured, click "Disconnect." Unplug the USB from the computer. Remove the XBee "A" from the LilyPad XBee board and plug the XBee "B" into the board. Then plug the USB cable back into your computer.

In CoolTerm, select the "Connect" icon in the Menu bar. From here, you will continue with the same steps as with the XBee "A," except that you will use the parameters for XBee "B" listed in the chart in Table 9-1.

The configuration process is shown in the following listing:

```
+++OK
ATRE
OK
ATIDB0D1
```

```
OK
ATMYB
OK
ATDLA
OK
ATWR
OK
```

And be sure to double-check them when you're done, like you did at the end of the preceding section:

```
+++OK
OK
ATID
B0D1
ATMY
B
ATDL
A
```

Experiment: Chat Test

Now that the XBees are configured, they are ready to communicate with each other. If you have two computers and two FTDI Breakout boards available, you can test your "chat" abilities between the two XBees.

Parts and materials:

- (2) XBee Series 1 (802.15.4) with trace or wire antenna (AF 128 or SF WRL 11215)
- (2) LilyPad XBee (SF DEV-08937) with right angle male headers soldered on (AF 1540, SF PRT-00553)
- (2) FTDI board (AF 284, SF DEV-10275)
- (2) USB mini-B cable (AF 899, DK WM5163-ND, RS 55010682, SF CAB-11301)

Tools:

- (2) computers with CoolTerm installed

Note the addition of an extra computer and an extra LilyPad XBee setup. This is needed because you will be using both radios at the same time.

You can also use two instances of CoolTerm on one computer for the chat test, but it's a little less confusing and a little more fun to use two.

Plug the XBees into the LilyPad XBee boards (Figure 9-24). Then plug the FTDI Breakout Boards into the LilyPad XBee boards, and connect them to the computers using mini-USB cables.

Figure 9-24. *Two labeled XBees on LilyPad XBees*

On each computer, open CoolTerm if it is not already open. Double-check your settings. Then select Connect in the menu bar. Once you are connected, try typing a message into the XBee "A" CoolTerm window as shown in Figure 9-25.

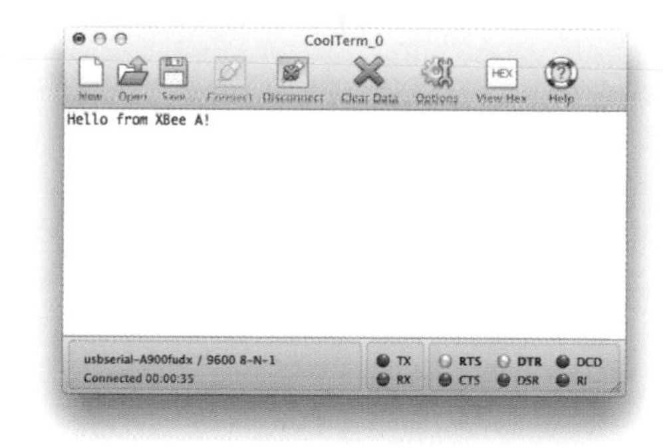

Figure 9-25. *Sending a message from XBee "A" to XBee "B"*

This message will then appear in the CoolTerm window connected to XBee "B". Now try sending an message from XBee "B" to XBee "A".

If your messages aren't coming through in either direction, double-check that each XBee is properly configured and that both CoolTerm windows are in "Connect" mode.

Once you've confirmed that messages can be sent and received in both directions, you can carry on a fully wireless conversation using XBee radios!

Experiment: XBee and Arduino

Now that you have the radios configured, you can start putting them to use.

This example is quite similar to the Bluetooth one in that you'll be using XBees to wirelessly send Arduino sensor data to a nearby computer. There are a few differences, however. First, you will need to connect an XBee radio to the computer as well as the Arduino, given that most computers do not have an 802.15.4 radio built into them. Second, because you have already configured these radios to communicate with each other, there will be no need for a pairing process.

For this example, you'll be using an Arduino Fio. While not intended as a wearable or e-textile tool, the Arduino Fio is a useful and compact version of the Arduino to work with when your circuit includes both an Arduino and an XBee. It features an XBee footprint with XBee headers, a JST connector for a battery, and a USB mini connector for charging the battery. Having all of these features on a single slim board can really reduce bulk. The Arduino Fio fits nicely into a preexisting or custom-built pocket.

Parts and materials:

- (1) Arduino Fio (SF DEV-10116)
- (1) LilyPad XBee (SF DEV-08937) with right-angle male headers soldered on (AF 1540, SF PRT-00553)
- (2) XBee Series 1 (802.15.4) with trace or wire antenna (AF 128 or SF WRL-11215), configured as you did in the previous example
- (1) LilyPad Light Sensor (SF DEV-08464)
- (1) FTDI board (AF 284, SF DEV-10275)
- (1) USB mini-B cable (AF 899, DK WM5163-ND, RS 55010682, SF CAB-11301)
- (1) 3.7V lithium-ion polymer rechargeable battery (AF 258, SF PRT-00339)
- (1) 6-pin set of right-angle male headers with 0.1" (2.54mm) spacing (AF 1540, SF PRT-00553)
- Alligator clip test leads (AF 1008, RS 278-1156, SF PRT-11037)

Tools:

- Soldering iron and solder

You can also use a LilyPad Arduino with a LilyPad XBee for this example, though it will require a few more connections:

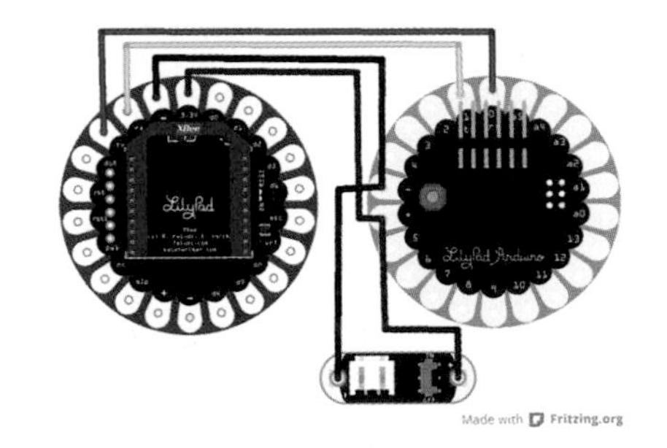

Solder FTDI headers

The USB mini connector on the board is for charging only. In order to program this Arduino, you will need to connect headers for an FTDI board. I find that right-angle headers on the XBee side of the board facing inward (Figure 9-26) provide appropriate access while still maintaining the Fio's slim profile.

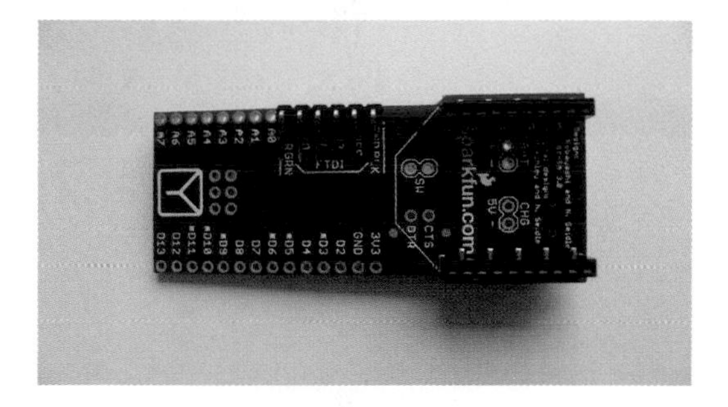

Figure 9-26. *Arduino Fio with headers soldered*

Connect the Light Sensor

Now for the rest of the circuit. You need to make the connections shown in Figure 9-27.

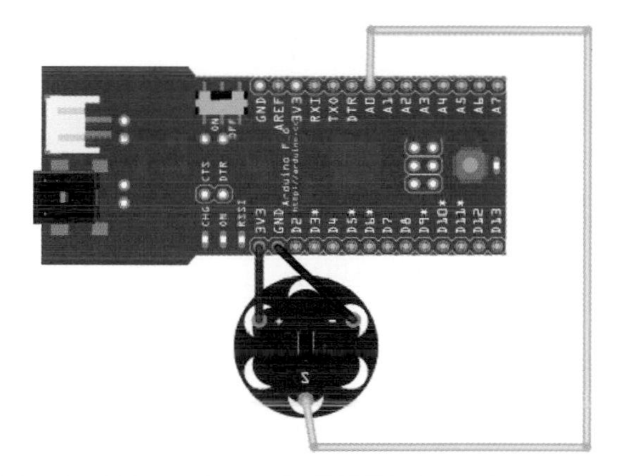

Figure 9-27. *Arduino Fio with Light Sensor*

The tradeoff for Fio's small size is that you need to make some choices as to how you would like to accomplish your connections to other components. It is possible to solder male headers to the pins so it can be plugged into a breadboard, or female headers so that connections can be made with hookup wire. But because this is intended for a wearable context, you're going to avoid both of these bulkier approaches.

Instead you can create these connections either with delicately placed alligator clips or wires soldered directly to the board. Keep in mind that the Fio is not intended for use with alligator clips, so you should be careful that they don't slip and create a short circuit. If you are using wire, three-stranded ribbon cable is a nice solution. If you're really ambitious, it is possible to make connections with conductive thread, though that requires a delicate needle, a thin thread, and a good command of your stitch.

Program the Arduino

To program the board, connect an FTDI board to the newly soldered headers (Figure 9-28). Use a USB mini-B cable to connect the FTDI board to your computer.

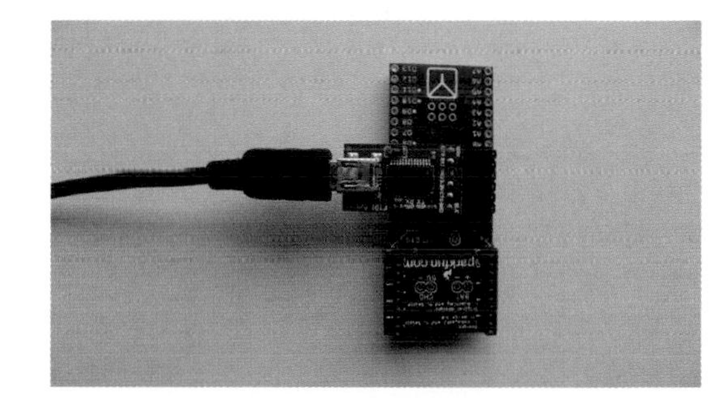

Figure 9-28. *Arduino Fio ready to be programmed.*

Open Arduino. Enter in the following program:

```
/*
Make: Wearable Electronics
 XBee Arduino example
 */

// initialize variables
int lightSensorPin = A0;
int lightSensorValue = 0;

void setup() {
  // initialize serial communication
  Serial.begin(9600);
}

void loop() {
```

```
    // read the light sensor value
    int lightSensorValue =
    analogRead(lightSensorPin);
    // print the value of the light sensor
    Serial.println(lightSensorValue);
    delay(100);
  }
```

Under Tools → Board, select Arduino Fio. Then, under Serial Port, select the appropriate port for your Arduino.

Upload your program. Open the Serial Monitor. You should see the light sensor data appear on screen.

Prepare the circuit

Disconnect the Arduino from the computer.

Connect XBee "A" to the Fio by seating it in the parallel rows of female headers on the back of the board. Be sure that the radio is oriented correctly. The narrowed end should point toward the center of the board, not the edge.

Connect a battery to the JST connector on the Arduino Fio board (see Figure 9-29).

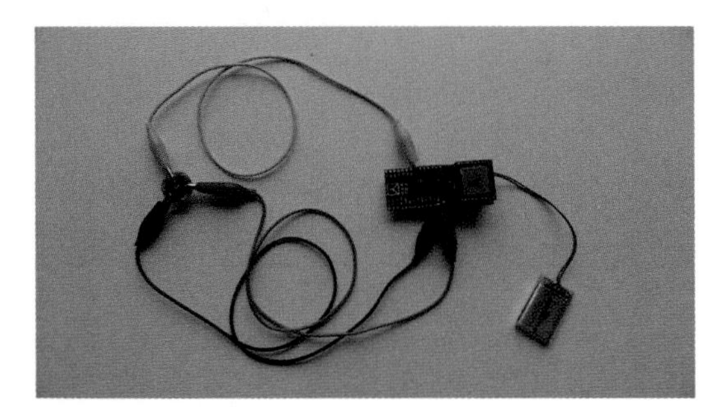

Figure 9-29. *Arduino Fio with Light Sensor connected via alligator clips*

Prepare XBee "B"

Connect the XBee "B" to the LilyPad XBee, and then use the FTDI board and USB cable to connect the LilyPad XBee to computer.

Open CoolTerm and double-check your settings.

Connect

Turn the switch on the Arduino Fio to "ON". The LEDs labeled "ON" and "RSSI" LED should be lit as shown in Figure 9-30. The "ON" light indicates that the board has power. "RSSI" stands for received signal strength indicator. This light indicates that the radio on the Arduino Fio is making contact with the other radio.

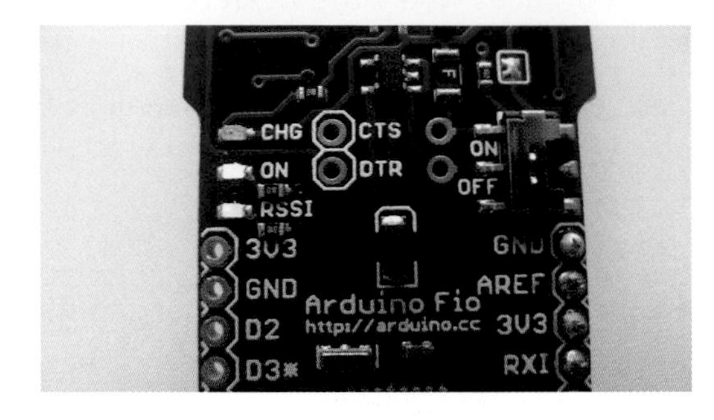

Figure 9-30. *Arduino Fio with "ON" and "RSSI" LEDs lit*

In CoolTerm, click Connect.

The light sensor data should be visible in CoolTerm! You can monitor the serial port through Arduino's Serial Monitor as well. Just be sure to close the CoolTerm connection first.

Receiving data in the Serial Monitor or in CoolTerm is just the beginning. Once you're able to access this data on your computer, you can make use of it in any program or programming environment that is able to manage serial communication. You can import it into Processing, Max/MSP, PureData, and more. Just be sure you always close one serial connection before you open another! A serial port is only available for one connection at a time.

If you'd like to check out an example project that uses a similar approach, look at the Audience Jacket tutorial (*http://bit.ly/UcEnxh*) by the Social Body lab —a jacket that turns an audience of one into a crowd that fills a room.

There are many other ways to use XBees with Arduino. You can use XBees to enable Arduinos to communicate wirelessly with each other. Imagine two sets of shoes where the wearers can feel their partner's footsteps!

It's also easy to scale up with XBees. Rather than having one device that is sending or receiving data, there could be many. Picture a sports stadium full of people wearing networked hoodies with tilt sensors located in the upper sleeve. If all of that sensor data was being sent to a central computer, the act of people raising their arms to do the wave could be used to control lighting, sound, or more!

Figure 9-31. *"Dream Squawk" by Amy Khoshbi; knobs and buttons on the beak of this bird mask allow a performer to control music and sound wirelessly (photographed by Aubrey Edwards)*

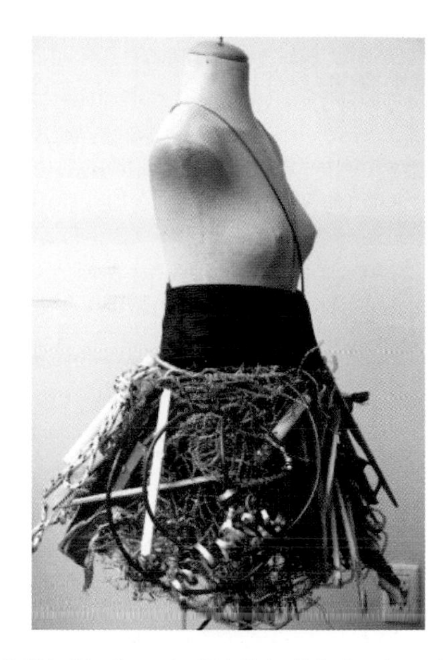

Figure 9-32. *"Earthquake Skirt" by Erin Lewis; seismologic data is retrieved from the Internet and sent wirelessly to this skirt, causing it to shake or shimmy whenever there is an earthquake somewhere in the world*

Figure 9-33. *In "Spin on the Waltz," Ambreen Hussain transmits data from the dancers' bodies to a nearby computer so that their movements influence the music that is being played*

Experiment: XBee Direct Mode

In this example, you'll look at communication between two wearables instead of between a wearable and a computer. XBees don't always need to use a microcontroller—they can also be configured to operate on their own. Using the XBee Direct Mode, simple functionality for sending and receiving is accomplished with sparse circuitry.

In this example, you will create two simple circuits that allow for bidirectional communication through the use of buttons and LEDs.

Configure the XBees

To use XBees in Direct Mode, you'll use some configurations that have already been introduced (in Table 9-1) but you'll also need to add a few more. Table 9-2 includes all of the configurations that you will need.

Using CoolTerm, configure XBees "A" and "B" with the settings provided in Table 9-2. If you need to refresh your memory on how to configure an XBee, see "Configuring XBees" on page 200.

Table 9-2. XBee configurations for XBee Direct Mode

AT Command	XBee "A"	XBee "B"	Function
ATRE	—	—	Factory reset (erases previous settings)
ATID	B0D1	B0D1	Sets PAN ID (same for both XBees)
ATMY	A	B	Sets individual XBee IDs
ATDL	B	A	Sets Destination Low address
ATD4	5	3	Sets digital I/O pins
ATD6	3	5	Sets digital I/O pins
ATIR	64	64	Sets the sample rate to 100 milliseconds (Hex64)
ATIT	1	1	Sets the number of samples before transmission to 1
ATIA	B	A	Sets I/O input to destination address
ATWR	—	—	Saves settings to firmware so they are retained when the radio loses power

Circuit

Now that your XBees are configured, you can assemble your circuit.

Parts and materials:

- (2) LilyPad XBee (SF DEV-08937) with right-angle male headers soldered on (AF 1540, SF PRT-00553)
- (2) XBee Series 1 (802.15.4) with trace or wire antenna (AF 128 or SF WRL-11215)
- (2) LilyPad LEDs (SF DEV-10081)
- (2) LilyPad Simple Power boards (SF DEV-10085)
- (2) 3.7V lithium-ion polymer rechargeable battery (AF 258, SF PRT-00339)
- (2) LilyPad Buttons (SF DEV-08776)
- Alligator clip test leads (AF 1008, RS 278-1156, SF PRT-11037)

Using alligator clips and the parts listed, make the circuits shown in Figures 9-34 and 9-35.

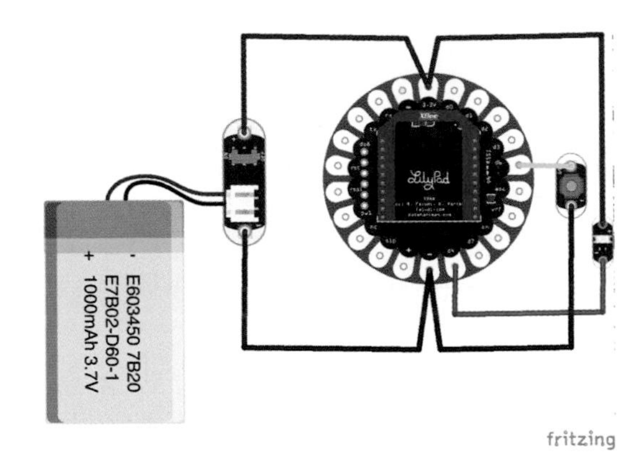

Figure 9-34. *XBee "A" circuit connections*

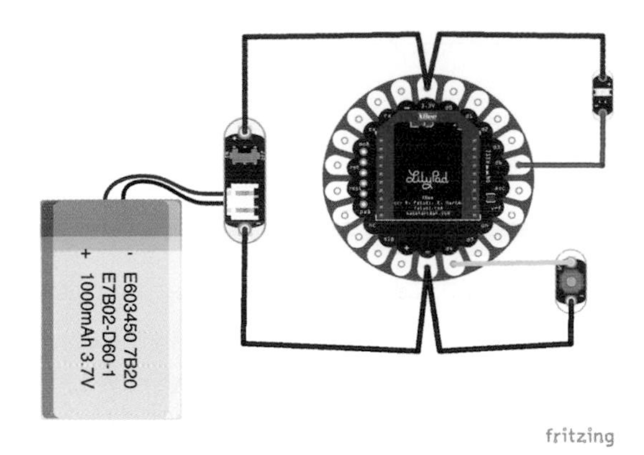

Figure 9-35. *XBee "B" circuit connections*

Battery Options

This example uses a rechargeable lithium polymer battery. You could also use a 9V battery instead. Just connect the red wire of the 9V clip to the "+" pin instead of the "3.3V" pin and the ground wire to the ground pin. Power supplies connected to the "+" pin pass through the onboard 3.3V regulator.

Connect

Now that your radios are configured and your circuits assembled, you should be ready to communicate!

Turn the switch to "ON" on both LilyPad Simple Power boards. You should see both the "ON" and the "RSSI" LEDs light on the LilyPad XBees (Figure 9-36), indicating that they are both receiving power and able to communicate with each other.

Figure 9-36. *"ON" and "RSSI" LEDs lit on a LilyPad XBee*

Now you're ready to test the circuit!

Press the button in circuit "A". You should see the LED in circuit "B" turn on. Then press the button in circuit "B". You should see the LED in circuit "A" turn on. You are now communicating wirelessly without the use of a microcontroller!

Troubleshooting

If the LED in either circuit does not turn on, check the following:

- Is the LED polarity correct? (+ goes to 3.3V, and – goes to your output pin)
- Is the battery connected? And is it charged?
- Is the switch on the LilyPad Simple Power boards turned on?
- Did you solder the pads on the LilyPad Simple Power boards?
- Are all the connections with the alligator clips properly made?

If none of the above appear to be the problem, go back and check all of your XBee settings using CoolTerm.

For a project example using the XBee Direct method, check out the Super Hero Communicator Cuffs (*http://research.ocadu.ca/socialbody/project/diy-superhero-communicator-cuffs*) example made by the Social Body Lab.

Figure 9-37. *Super Hero Communicator Cuffs use the XBee Direct method and conductive fabric switches; when one person puts his hands together, the LED will light on the other person's cuffs*

Other Wireless Options

This is only a taste of what can be done in wireless communication in wearables. The Bluetooth as well as the XBee and Arduino examples demonstrate a simple "cable replacement" technique where the data that normally would be transmitted to you computer over a USB cable is instead transmitted wirelessly. The XBee Direct example reveals more of the radio's onboard abilities but still remains a very simple example.

The possibilities that exist beyond this are immense. The adoption of Bluetooth Low Energy is making it much easier to connect from a wearable to a mobile phone. And there are many other tools that will become more readily accessible and easy to use in the years to come. For further resources on working with wireless technologies, see Appendix C.

Thinking Beyond

Wearable electronics is on the move. In the coming years we will likely see the line blur between wearable and mobile devices. We will hopefully witness significant material developments and the emergence of new connectors that will enable textiles and electronics to more effectively intermingle. Our thinking will mature as to when, where, and how technology can and should be worn. And we will start to learn more about how wearables perform in the social context and what the longer term effects are on our everyday lives and our sense of human connectedness.

This is just the beginning. In all of this I encourage you to use the act of making as a way to imagine our possible futures. The fields that this book touches on are vast and rich. There are many possible journeys that extend beyond these last pages. Let the material that you've learned act as a springboard to catapult you into exciting and unknown areas of exploration.

Tools

The process of making wearable electronics projects requires diverse set of skills and tools. Whether working at a school, art studio, hackerspace, or research lab, it's important to make sure you have the tools and materials to get the job done.

Electronics

Here are some basic items that are useful for the electronics aspects of your wearable electronics practice.

Soldering Iron

Unless you are building circuits that are entirely soft, a soldering iron is an important tool. Soldering irons range from super cheap to quite costly. Spring for something a little nicer than the $15 iron you might find online. At the very least, you want an iron that has temperature control.

Personally, I'm a fan of Weller soldering irons. I use a Weller WES51 (Figure A-1, DK WES51-120V-ND, JC 217461) in the lab and take a Weller WLC100 (JC 146595) with me on the road.

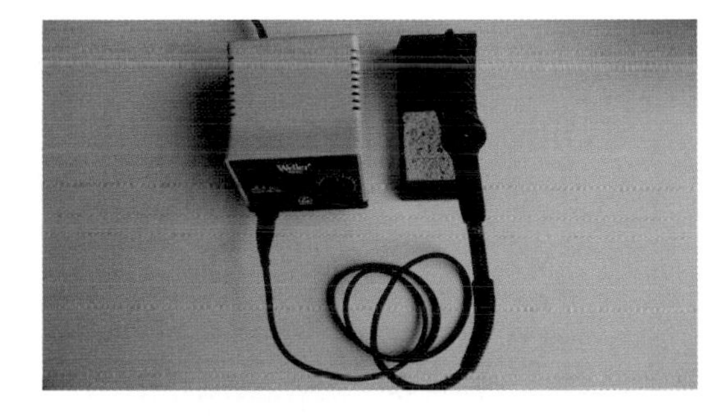

Figure A-1. *A Weller WES51 soldering iron*

Safety Glasses

Keep safety glasses (Figure A-2, JC 2133691, SF SWG-11046) on hand so you can protect your eyes when soldering.

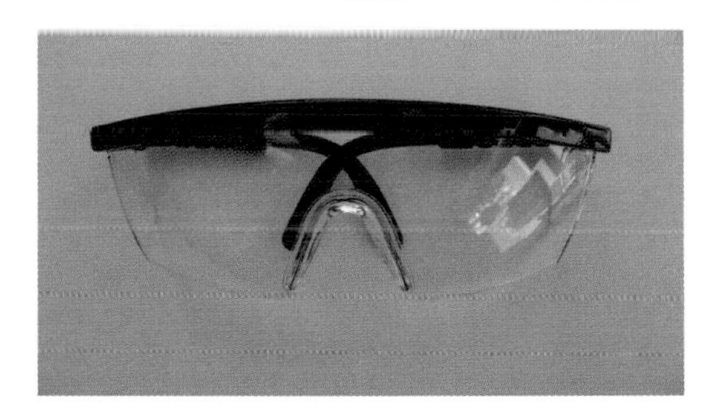

Figure A-2. *Safety glasses*

Desoldering Tools

Sometimes you make mistakes. Use a solder sucker (AF 148, SF TOL-00082) or solder wick (Figure A-3) to undo solder joints when necessary.

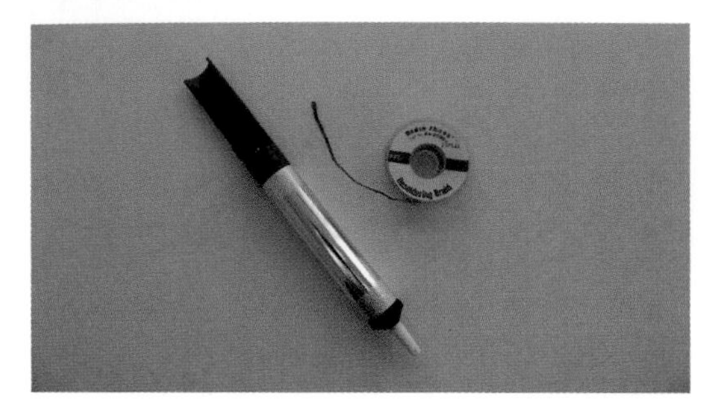

Figure A-3. *A solder sucker and solder wick*

Helping Hands

Circuit assembly and soldering often take more than two hands. Helping hands (AF 291, SF TOL-09317) offer a way to hold components in place as you handle the soldering iron and solder. Traditional helping hands often include a magnifying glass (Figure A-4) for precise work.

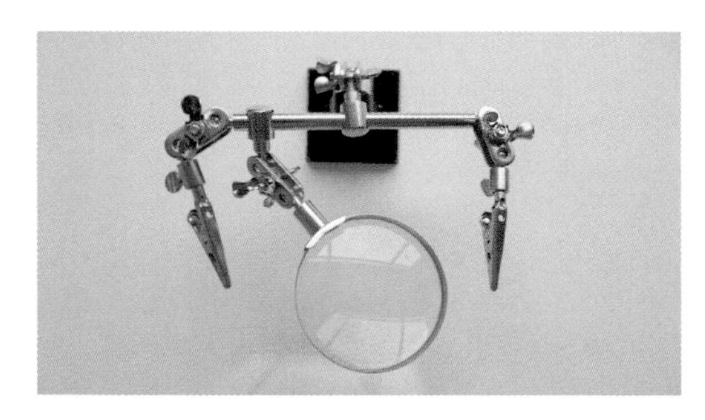

Figure A-4. *Helping hands*

The Panavise Jr. (Figure A-5, AF 151, SF TOL-10410) is a vise specifically designed for use with printed circuit boards (PCBs). These have adjustable arms with a shallow tray that can accommodate circuit boards of a variety of sizes.

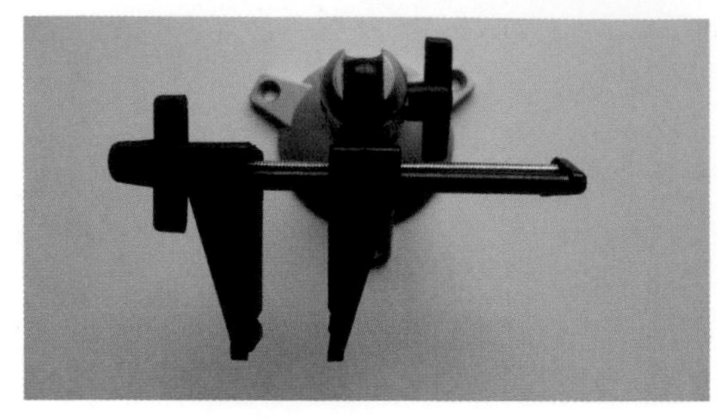

Figure A-5. *Panavise Jr.*

Wire Strippers

Wire strippers (Figure A-6, AF 147, RS 6400224, SF TOL-08696) are essential if you are using wire in your projects. Strippers with a 20–30 AWG range work well for wearable electronics projects.

Figure A-6. *Wire strippers*

Flat-Nosed Pliers

Flat-nosed pliers (Figure A-7, AF 146, SF TOL-08793) can help with bending and manipulating the legs of through-hole components.

Figure A-7. *Flat-nosed pliers*

Small Snips

Small snips (Figure A-8, AF 152, SF TOL-08794) are useful for cutting and trimming wire.

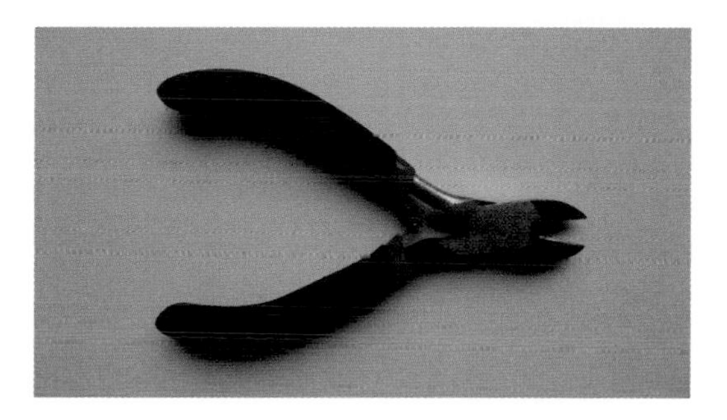

Figure A-8. *Small snips*

Multimeter

Multimeters (Figure A-9, AF 71, SF TOL-09141) are an essential electronics tool for measuring continuity, voltage, current, and resistance.

Figure A-9. *Multimeter*

Heat Gun

A heat gun (Figure A-10, JC GDT-1001A-R, SF TOL-10326) is used to activate heat shrink tubing. This high-temperature gun runs much hotter than a hairdryer, and when pointed at shrink tubing it will cause it to shrink.

Figure A-10. *Heat gun*

Screwdrivers

Small screwdrivers (Figure A-11, JC 127271) are useful for adjusting screws on terminal blocks or opening up some battery cases. The type with a flathead on one side and Phillips head on the other are small and easy to toss in your toolbox.

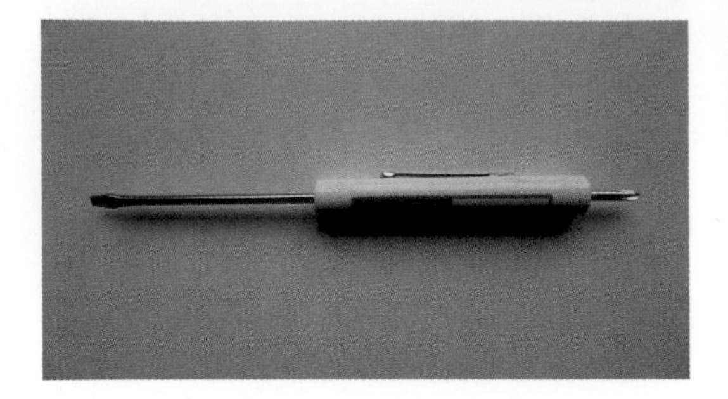

Figure A-11. *Mini screwdriver*

Compartment Boxes

Compartment boxes (Figure A-12) are essential for keeping parts organized. Get them from your favorite electronics supplier or from your local hardware or craft store.

Figure A-12. *Compartment box*

Sewing

These are the tools that will get you up and running with sewing. Most of these can be found at your local sewing or craft store.

Needles

For sewing conductive thread, get sharps with larger eyes, size 7 or similar (Figure A-13). Needles meant for embroidery usually work well.

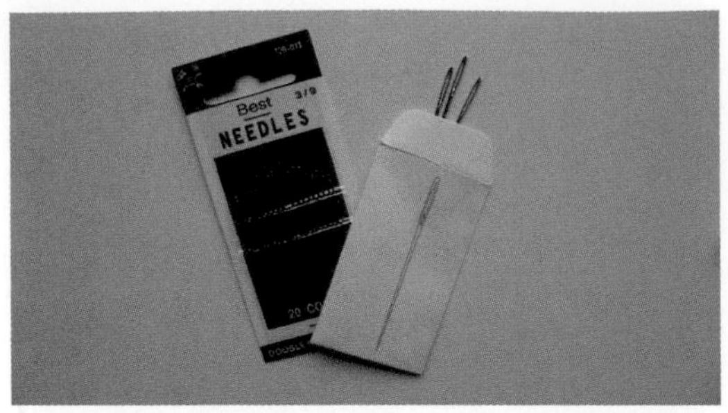

Figure A-13. *Needles*

Needle Threader

A needle threader (Figure A-14) can save you some time and frustration when working with gnarly conductive thread.

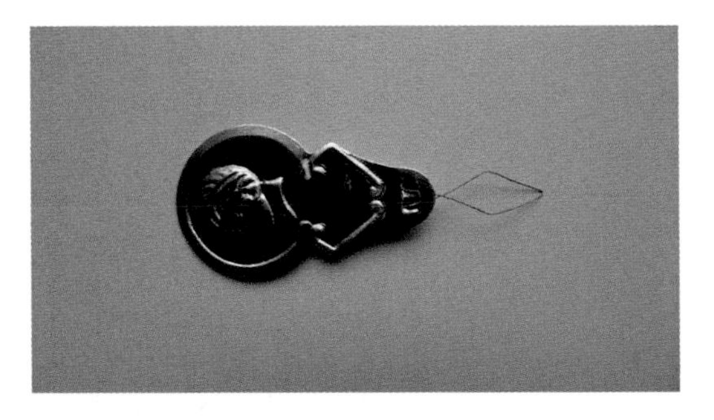

Figure A-14. *Needle threader*

Seam Ripper

A seam ripper (Figure A-15) is the sewing equivalent of desoldering tools. Use this nifty device to remove stitches with ease!

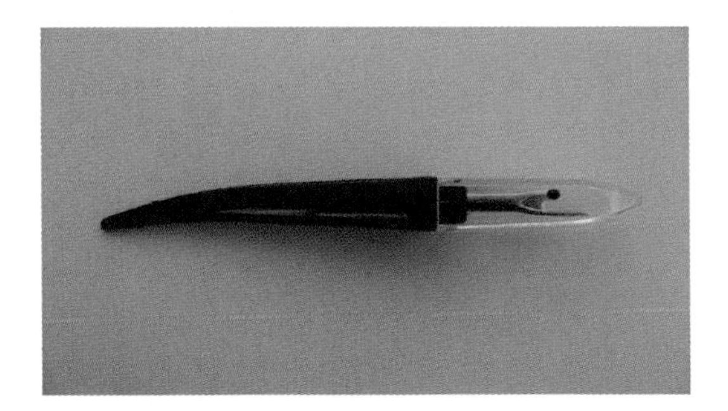

Figure A-15. *Seam ripper*

Pins

Pins (Figure A-16) can be used to temporarily hold fabric or even components in place while you are sewing your circuit together.

Figure A-16. *Pins*

Scissors

Sharp scissors (Figure A-17) are a must, particularly when working with conductive fabric. Consider dedicating a pair exclusively to fabric—they will stay sharper and last longer. Pinking shears (Figure A-18) and a precision knife (Figure A-19) will also come in handy.

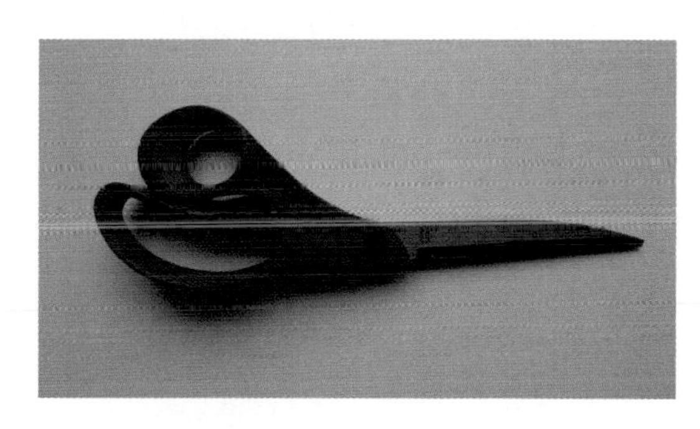

Figure A-17. *Scissors*

Figure A-18. *Pinking shears cut fabric with a zigzig edge to prevent fraying*

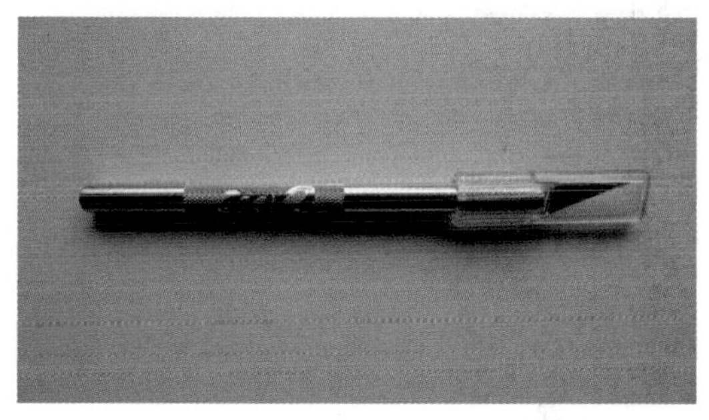

Figure A-19. *A precision knife is useful when cutting small and intricate shapes*

Iron

Ironed fabric is much easier to manipulate, cut, and sew. Irons (Figure A-20) are also useful to melt a specific variety of adhesives meant to be used bond fabric. Varieties include Heat & Bond, Wonder Under, and others. You'll probably want a craft iron (Figure A-21) and small ironing board (Figure A-22).

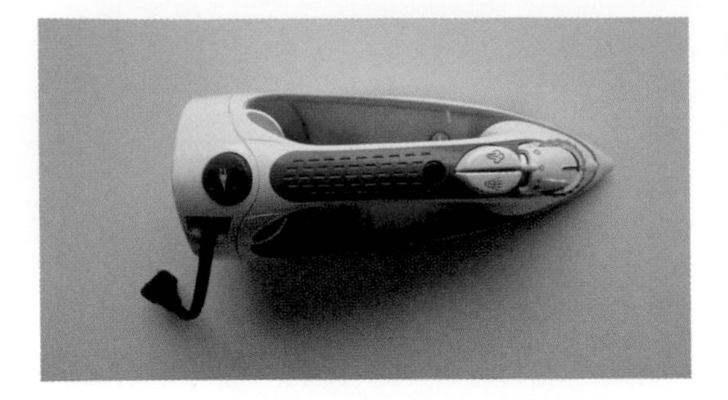

Figure A-20. *A basic household iron*

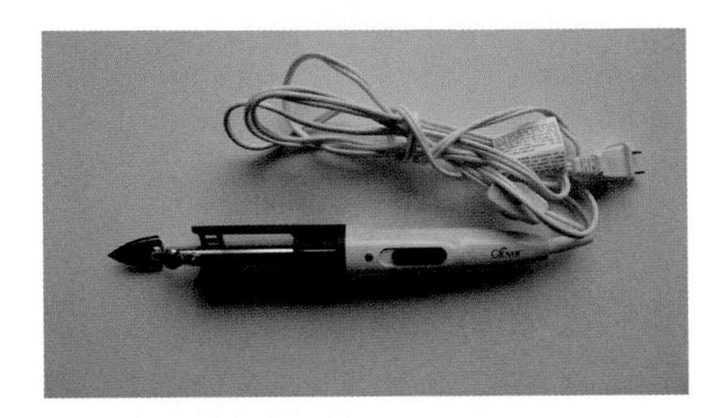

Figure A-21. *A craft iron*

Figure A-22. *A small ironing board*

Measuring Tools

Measuring tools are helpful when cutting fabric to size. Clear rulers (Figure A-23) allow you to see the fabric beneath the ruler.

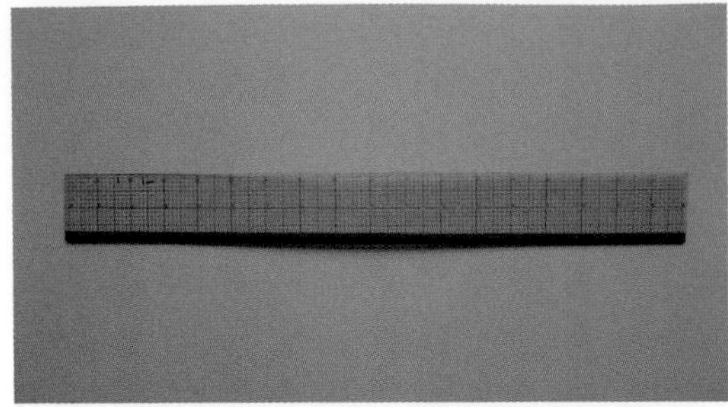

Figure A-23. *Clear ruler*

Thread and Fabric

It is helpful to always have some basic thread (Figure A-24) and fabric (Figure A-24) on hand for prototyping. Simple cotton thread and muslin work well. Having some neat colors and textures available can help jog the imagination and inspire new designs.

Figure A-24. *Thread*

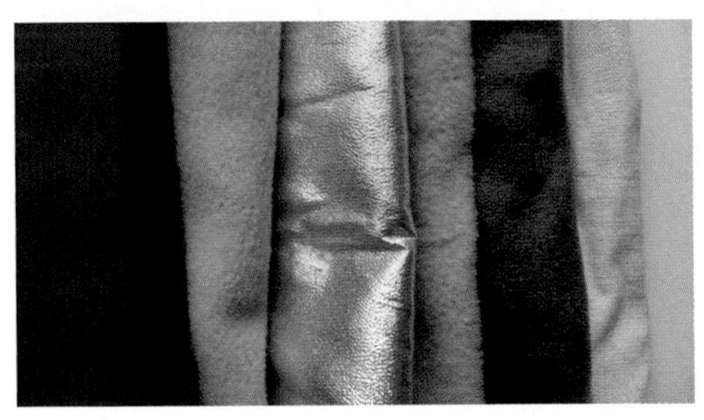

Figure A-25. *Fabric*

Embroidery Hoops

Embroidery hoops (Figure A-26) hold fabric in place when hand sewing. This can make sewing conductive thread circuits a lot easier.

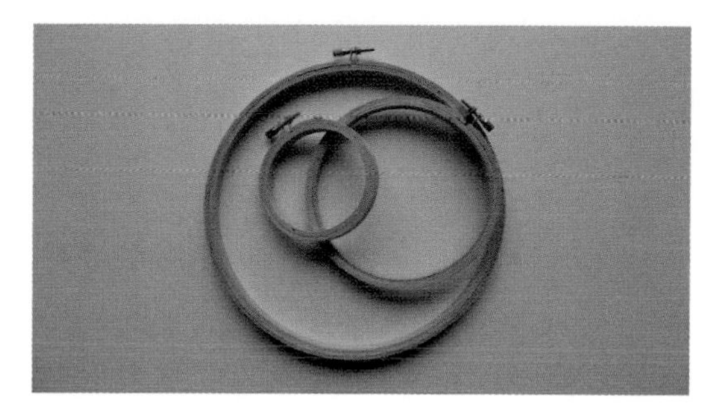

Figure A-26. *Embroidery Hoops*

Documentation

Documenting your work is important, whether it be for class, research, or your own portfolio. Always keep documentation tools in your workspace so you can capture your process and share your work with others.

Camera

Depending on the quality of images you seek, your camera (Figure A-27) can be as simple as a smartphone or as complex as a fancy DSLR. Try to work with a camera that can also capture video so you can capture your LEDs blinking and motors spinning!

Figure A-27. *A camera*

Tripod

A tripod (Figure A-28) provides a stable base that can improve the quality of your images tremendously.

Figure A-28. *A tripod*

Copystands (Figure A-29) can be especially useful when documenting electronics. They provide a stable table view that is difficult to accomplish with a standard tripod.

Figure A-29. *Copystand*

Batteries

B

Bodies are dynamic, mobile vessels in which we travel through the world. It is because of our transient nature that wearables require a portable power source. This power source most commonly takes the form of batteries. This section introduces factors to consider when incorporating a batteries into wearables.

Types of Batteries

Here are a few things you need to know about batteries:

- Batteries convert chemical energy to electrical energy.
- There are two types of batteries: primary (single use) and secondary (rechargeable).
- Even within the same battery type, voltage and capacity for batteries differ slightly based on manufacturer, chemistry, type, and other factors.

Round cell batteries are the type that you are probably most familiar with. They have a cylindrical shape and usually provide 1.5V, depending on their chemistry. This category includes AAA, AA, C, and D batteries. Each type is a different size. Usually the larger the battery is, the greater its capacity. AAA and AA can be used for wearables but tend to be a bit bulky. C and D batteries are too heavy for most wearable applications.

Nonround batteries come in a variety of shapes. 9V batteries are the type from this category that are most likely to be used for wearables.

Coin cell batteries are disc-shaped. 2032s (20mm) and 2450s (24.5mm) are commonly available sizes. These batteries are small and thin—excellent for low-current wearable applications.

Finally, lithium-ion and lithium-ion polymer batteries (Figure B-1) have recently become popular for use in small electronic devices. The ones used in this book are flat, rechargeable, and relatively lightweight, offering 3.7V and a capacity ranging from 150–2000 mAh. They are a bit more expensive, but ultimately a great investment because they can be used again and again in a variety of projects.

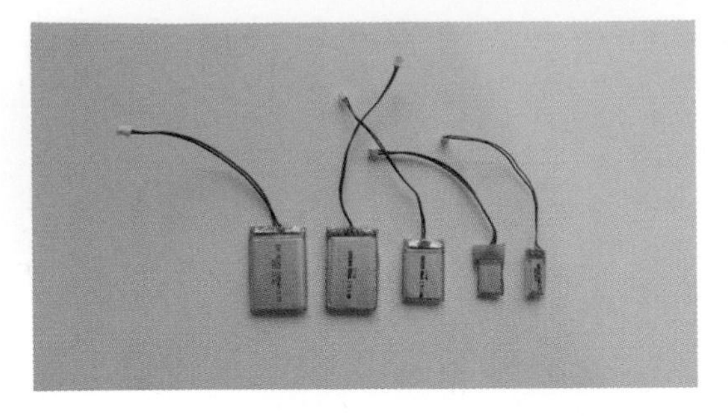

Figure B-1. *3.7V lithium-ion polymer rechargeable batteries with JST connectors; these batteries come in a variety of sizes and capacities*

When possible, try to use rechargeable batteries in your projects. They have less of an environmental impact and are cheaper in the long run, as you won't have to keep buying new ones.

Table B-1. Battery comparison chart

Type	Size (in mm)	Weight (in grams)	Voltage	Capacity (in mAh)
Primary (nonrechargeable)				
CR2032	20 x 3.2	~3	3	~250
AAA	45 x 10.5	~11	1.5	~860-1200
AA	50.5 x 13.5	~23	1.5	~1800-2600
9V	48.5 x 26.5 x 17.5	~35	9	~400-565
Secondary (rechargeable)				
LIR2450	24 x 5	~6.4	3.6	~110-160
Single-cell lithium-ion polymer batteries	12 x 6 x 5 - 5.8 x 54 x 60	~2-36	3.7	~40-2000

Battery Holders and Connectors

A stable connection (Figure B-2) to the power source is essential when creating reliable circuits. Working with the appropriate battery holder or connector greatly improves your chances of making a stable, solid connection.

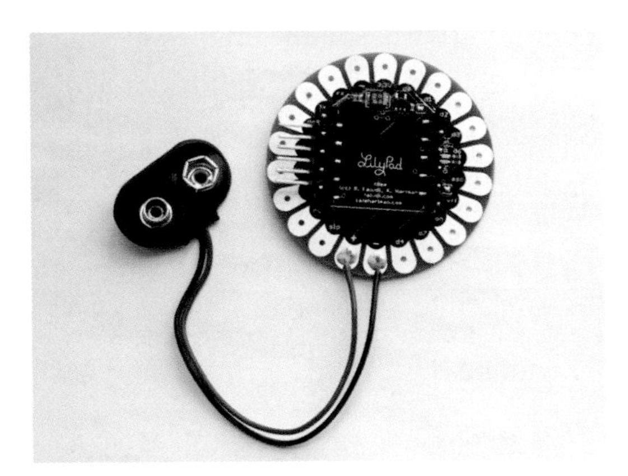

Figure B-2. *To make a more reliable connection, you can always solder the leads of your battery holder directly to the board it is powering; see Figure 5-5 for an example of how to provide strain relief for a soldered connection*

AA and AAA battery holders can accommodate anywhere from one to eight or more batteries. Multiple battery holders usually connect batteries in series, meaning the voltage of the batteries are added together. For example, if you put three alkaline AAA batteries (1.5V) in a 3xAAA battery holder, that battery pack will provide 4.5V.

Keep in mind that battery voltages differ depending on whether the battery is primary (disposable) or secondary (rechargeable). For instance, a primary AAA battery provides 1.5V, whereas a secondary AAA might provide 1.2V. This doesn't make a huge difference when it's one or two batteries, but at four or more, it can become an issue.

These battery packs (Figure B-3) will sometimes feature a door or a full enclosure, which can help to protect the batteries, or a switch (Figure B-4), which can act as an off/on switch for your project.

Figure B-3. *2xAAA and 3xAAA*

Figure B-4. *2xAAA with cover and switch*

Coin cell battery holders (Figure B-5) can either be standalone (like the ones we used in Chapter 1) or mounted on a circuit board (like the LilyPad Coin Cell battery holder or the LilyPad Coin Cell Battery Holder Switched). The latter are more expensive but easier to connect to using conductive thread.

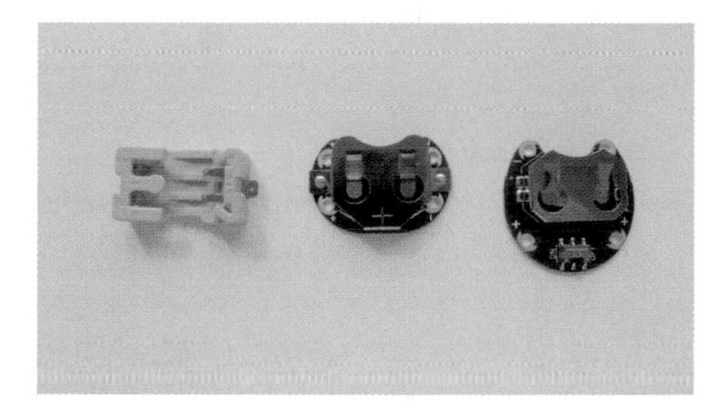

Figure B-5. *CR2032 holders: SMD, LilyPad, LilyPad with switch*

9V batteries feature a snap or clip connector (Figure B-6) that is specific to that battery type (Figure B-7). You can also find full battery holders for 9Vs, but it is the clip that is the most important part.

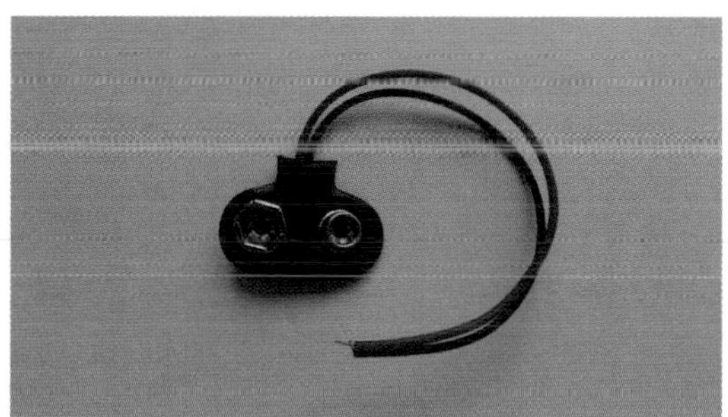

Figure B-6. *9V battery clip*

Figure B-7. *9V battery with clip*

Lithium-ion or lithium-ion polymer batteries often feature wires with a JST connector (Figure B-8). JST connectors are featured on many microcontroller boards, including the LilyPad Arduino Simple and the Flora. If you need a standalone JST connect, the LilyPad Simple Power (Figure B-9) is a good option.

Figure B-8. *JST connector on LilyPad Arduino Simple*

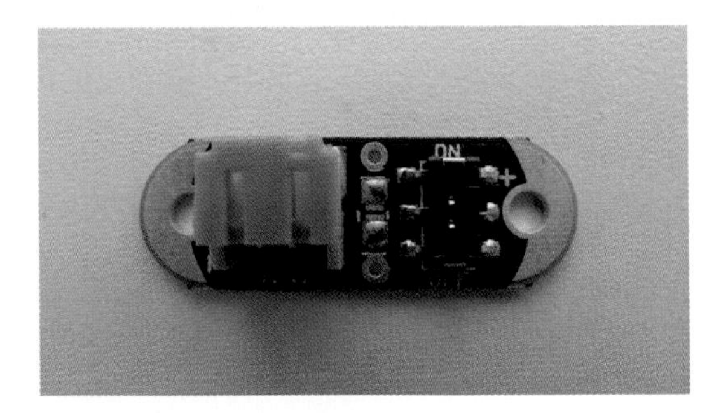

Figure B-9. *LilyPad Simple Power*

Charging LiPo Batteries

Lithium-ion polymer batteries are rechargeable. Some microcontroller boards such as the LilyPad Arduino Simple and the Arduino Fio have onboard charging circuits and can recharge the batteries via USB. Otherwise a standalone charging board can be used:

Some battery connector boards feature more complex circuitry that will actually change the voltage of the source battery so that it better fits an application. The LilyPad Power Supply Board (Figure B-10) uses a step-up circuit to convert the 1.5V provided by a AAA battery to 5V. The LilyPad LiPower Board (Figure B-11) converts the 3.7V provided by a lithium-ion polymer battery to 5V as well.

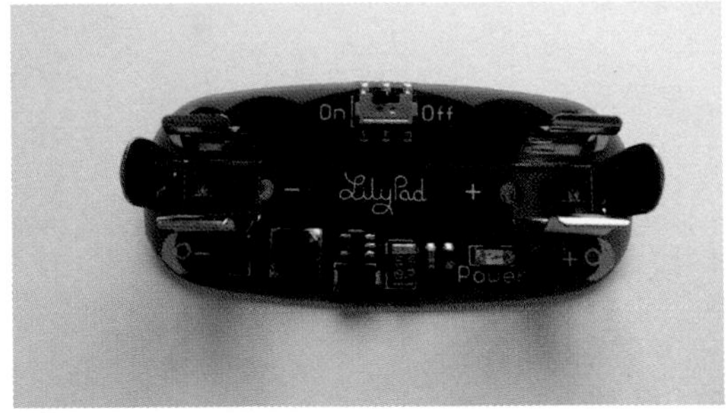

Figure B-10. *LilyPad Power Supply Board*

Figure B-11. *LilyPad LiPower Board*

Finally, you can also create your own battery holders or connectors to meet the needs of your projects. Batteries of the same type and size can be connected. Connecting batteries in series increases the voltage. Connecting batteries in parallel increases the amperage. The following examples show how to use a LilyPad ProtoBoard Small and two JST connectors to connect lithium polymer batteries in series and parallel.

To connect two LiPo batteries in series, first make the cuts shown in Figure B-12 using a ruler and knife.

Next, solder JST connectors in place as shown in Figure B-13. With two 3.7V LiPo batteries connected, sewtab "F" will provide a connection to 7.4V, and sewtab "B" will provide a connection to ground.

To connect two LiPo batteries in parallel, make the cuts shown in Figure B-14 using a ruler and knife.

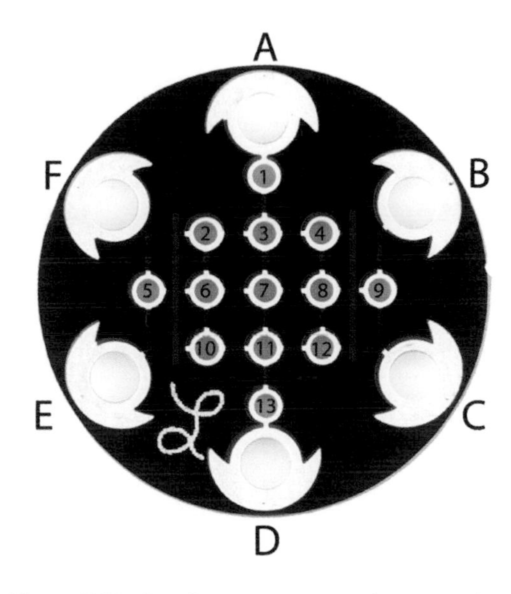

Figure B-12. *Cuts for connecting two lipos in series*

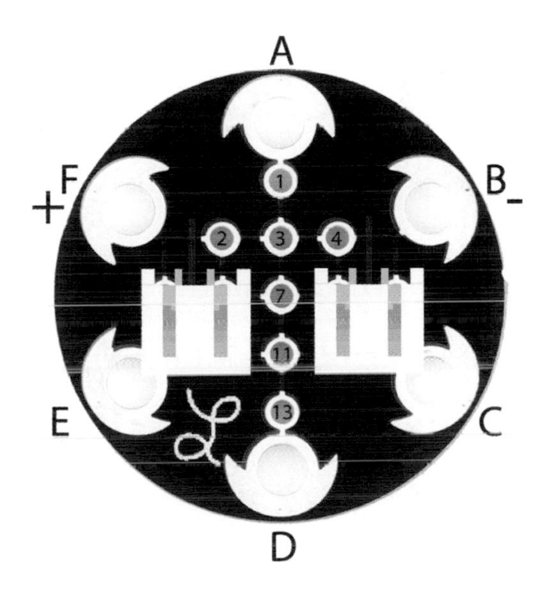

Figure B-13. *Connector placement for connecting two lipos in series*

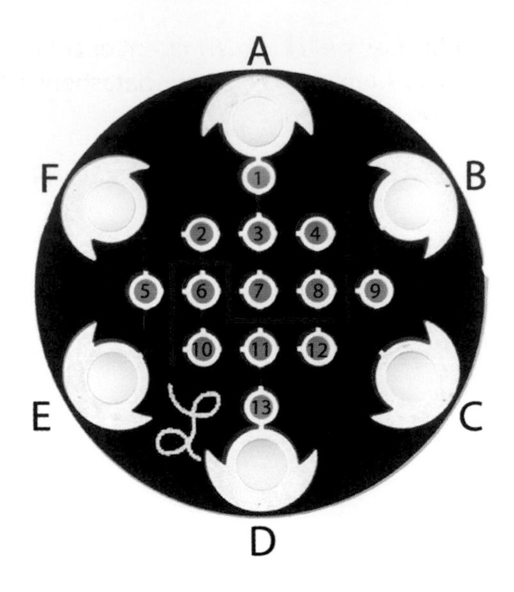

Figure B-14. *Cuts for connecting two lipos in parallel*

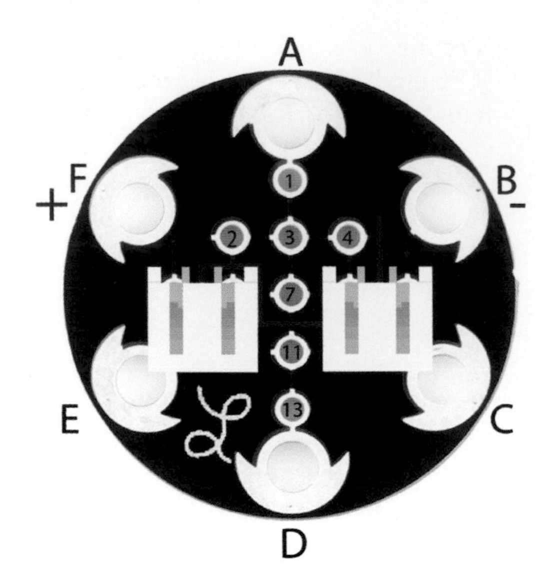

Figure B-15. *Connector placement for connecting two lipos in parallel*

Then solder JST connectors in place as shown in Figure B-15. With two 3.7V lipo batteries connected, sewtab "F" will provide a connection to 3.7V, and sewtab "B" will provide a connection to ground. The capacity will be that of the two batteries added together.

Factors to Consider

There are many factors to consider when choosing a battery or battery pack for your project:

Voltage
: Does the voltage supplied by the battery or battery pack fall within the acceptable range for *all* components you are working with? This includes not only the microcontroller but also sensors and actuators. If not, have you planned to use a voltage regulator or step-up circuit to adjust the voltage accordingly?

Capacity
: Manufacturers rate batteries according to ampere hours (Ah), 1 amp-hour = 1 amp (1000 mA) for 1 hour, 100mA for 10hrs, 10mA for 100hrs. In reality, this may differ depending on how much current is drawn.

Maximum current draw
: Does the maximum current draw of the battery accommodate the highest expected current draw of your project?

Size, shape, and weight
: In additional to their electrical characteristics, it is important to consider batteries from a design perspective. Where will these batteries live in your wearable and on your body? How will they feel? What will they weigh?

Availability
: Where will this project be used, and who will be using it? Is it important to be able to replace the batteries easily?

Intended use

Different batteries are designed for different purposes. Factors to consider include shelf life and whether the use applications will be intermittent or continuous. Consult the battery's datasheet for more information.

Resources

Here are some additional resources that will help you learn where to shop, what to read, and where to learn.

Where to Shop

General electronics supplies:

- All Electronics
- Adafruit Industries
- Creatron Inc.
- Digi-Key
- Jameco
- Maker Shed
- Mouser Electronics
- Newark element14
- RadioShack
- Robotshop
- SparkFun Electronics
- Seeed Studio

Conductive materials:

- Bare Conductive Paint
- Fine Silver Products
- Lamé Lifesaver
- Less EMF
- Inventables
- Plug & Wear

Sewing gear:

- Jo-Ann Fabric and Craft Stores
- Seattle Fabrics
- The Felt Store

For Your Bookshelf

Whether you're looking for instruction or inspiration, these titles will enable you to expand upon and deepen your knowledge of the topics covered in this book:

- *Arduino Cookbook* by Michael Margolis (O'Reilly)
- *Arduino Wearables* by Tony Olsson (Apress)
- *Building Wireless Sensor Networks* by Robert Faludi (O'Reilly)

- *Fashionable Technology: The Intersection of Design, Fashion, Science and Technology* by Sabine Seymour (Walter de Gruyter & Co)
- *Fashion Geek: Clothing, Accessories, Tech* by Diana Eng (North Light Books)
- *Fashioning Technology: A DIY Intro to Smart Crafting* by Syuzi Pakhchyan (Make: books)
- *Functional Aesthetics: Visions in Fashionable Technology* by Sabine Seymour (Springer Vienna Architecture)
- *Getting Started in Electronics* by Forrest Mims III (Master Publishing, Inc.)
- *Getting Started with Arduino* by Massimo Banzi (Make: books)
- *Getting Started with Adafruit FLORA: Making Wearables with an Arduino-Compatible Electronics Platform* by Becky Stern and Tyler Cooper (Make: books)
- *Getting Started with Processing* by Casey Reas and Ben Fry (Make: books)
- *Learning Processing: A Beginner's Guide to Programming Images, Animation, and Interaction* by Daniel Shiffman (Morgan Kaufmann)
- *Make: Electronics* by Charles Platt (Make: books)
- *Making Things Talk* by Tom Igoe (Make: books)
- *Open Softwear* by T. Olsson, D. Gaetano, J. Odhner, and S. Wiklund (Blushing Boy Publishing)
- *Practical Electronics for Inventors* by Paul Scherz and Simon Monk (McGraw-Hill/TAB Electronics)
- *Physical Computing* by Dan O'Sullivan and Tom Igoe (Thomson)
- *Sew Electric* by Leah Buechley and Kanjun Qiu (HLT Press)
- *Switch Craft: Battery-Powered Crafts to Make and Sew* by Alison Lewis and Fang-Yu Lin (Potter Craft)
- *Textile Messages: Dispatches From the World of E-Textiles and Education* by Leah Buechley , Kylie Peppler, Michael Eisenberg, Yasmin Kafai (Peter Lang International Academic Publishers)

For Your Bookmarks

The links listed below range from wearable technology blogs to DIY tutorials. Add them to your bookmarks so you can keep up on the latest in wearable electronics!

Soldering:

- Adafruit Guide to Excellent Soldering (*http://bit.ly/UcKUbk*)
- Make: Weekend Projects Thumbnail Guide to Soldering (*http://bit.ly/UcKVMg*)
- *Soldering Is Easy* (*http://bit.ly/1pEZhUE*), a comic book by Mitch Altman, Andie Nordgren, and Jeff Keyzer

Sewing:

- BurdaStyle Techniques (*http://bit.ly/UcL16G*)
- How To Sew Instructable (*http://bit.ly/UcL4PP*)

Multimeters:

- Make: Video Podcast multimeter tutorial (*http://bit.ly/UcL6qW*)
- Adafruit multimeter tutorial (*http://bit.ly/UcL960*)

General electronics:

- Adafruit Learning System (*http://learn.adafruit.com/*)
- SparkFun Electronics Tutorials (*http://www.sparkfun.com/tutorials*)

E-Textile:

- Adafruit Flora Tutorials (*https://learn.adafruit.com/category/flora*)
- Adafruit Gemma Tutorials (*https://learn.adafruit.com/category/gemma*)
- High-Low Tech Lab Tutorials (*http://hlt.media.mit.edu/?cat=20*)
- How To Get What You Want (*http://www.kobakant.at/DIY/*)
- LilyPad Arduino Tutorials (*http://lilypadarduino.org/*)
- Instructables Soft Circuits Channel (*http://bit.ly/1mYalbX*)
- SparkFun E-Textile Tutorials (*http://bit.ly/1mYan3o*)

Materials research:

- Materia (*http://materia.nl/*)
- Materiability (*http://materiability.com/*)
- Material ConneXion (*http://bit.ly/1mYbRL4*)
- Open Materials (*http://bit.ly/1mYbSyl*)
- Transmaterial (*http://transmaterial.net/*)

Wearables blogs:

- Adafruit Blog: Wearables (*http://bit.ly/1mYaCeK*)
- Electric Foxy (*http://bit.ly/1mYbKz6*)
- Fashioning Tech (*http://bit.ly/1mYbX5q*)
- Make Blog: Wearables (*http://makezine.com/tag/wearables/*)
- Soft Circuit Saturdays (*http://softcircuitsaturdays.com/projects/*)
- Talk to My Shirt (*http://bit.ly/1mYaFqX*)

Conferences and events:

- International Conference on Tangible, Embedded, and Embodied Interaction (*http://www.tei-conf.org/*)
- International Symposium on Wearable Computers (*http://www.iswc.net/*)
- Maker Faire (*http://makerfaire.com/*)
- Smart Fabrics and Wearable Technology (*http://bit.ly/1mYaGva*)
- Wearable Technologies Conference (*http://bit.ly/1mYal6d*)
- Wearable Technology Expo (*http://bit.ly/1mYalD7*)

Hacker and Maker spaces:

- Hackerspaces.org
- Makerspace.com

Where to Learn

Books and online tutorials are helpful, but sometimes nothing can beat a real live human showing you how it's done. Opportunities for learning about wearable electronics, fashionable technology, and soft circuitry have exploded in the last few years, with opportunities ranging from workshops to classes to full-on degree-granting programs. Check out the GitHub repository (*https://github.com/katehartman/Make-Wearable-Electronics*) for this book to find an up-to-date list.

D Other Neat Things

This book focuses on how to develop electronic circuits that live in the wearable context. Here are some materials and prototyping techniques that fall outside of the scope of this book that could be fantastic additions to your wearable electronics projects.

Materials

There is a wide range of interesting nonconductive materials that are smart, responsive, or just plain fun. Here are a few to consider.

Reflective Materials

Chapter 8 presented LEDs as a way to provide visibility. There are also passive materials that can provide visibility without electronics.

Reflective materials (Figures D-1 and D-2) bounce back light. Retroreflective materials bounce back light that is pointed at them with a minimum amount of dispersal, thereby increasing visibility in the dark. These materials come in the form of paints, films, silkscreen inks, and textiles. They can be used in combination with or independently of powered safety lighting.

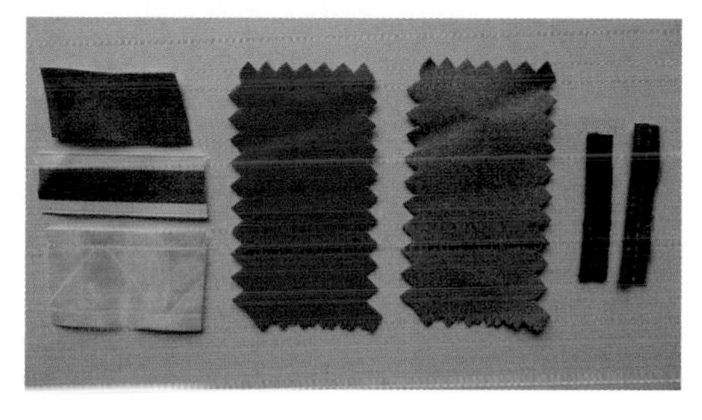

Figure D-1. *Outdoor fabric suppliers such as Seattle Fabrics offer sample packs of reflective textiles and trimmings*

Figure D-2. *Reflective fabric is also available by the yard*

Figure D-3. *Reflective fabric can be cut and embroidered to create different patterns*

Figure D-4. *"We Flashy" by Alex Vessels and Mindy Tchieu uses silk-screened reflective ink (like the pattern on this shirt) to create fashionable designs for safety-friendly garments (photographed by Jonathan Grassi)*

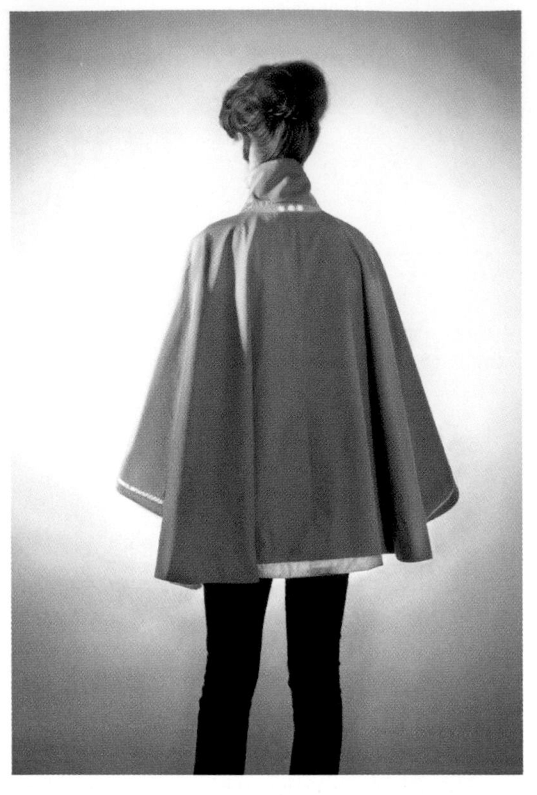

Figure D-5. *"The Vega Cape" by Angella Mackey uses both LEDs and accents of reflective fabric for fool-proof visibility (photographed by David McCallum)*

Glow-in-the-Dark Materials

Glow-in-the-dark materials (Figure D-6) also provide visibility in dark settings but function a bit differently. They absorb light when it's available and slowly release it when it is dark. These materials often have a greenish glow, but sometimes other colors as well. They are available as panels, ribbons, coatings, pigments, sewing threads, and more (see Figures D-7 and D-8).

Figure D-6. *Photoluminescent panel (SF COM-11552)*

Figure D-7. *Powerless Illuminating Polymer Ribbon is a super thin, tear resistant ribbon that can be stitched onto fabric and sewn into garments (IV 24033-08[1])*

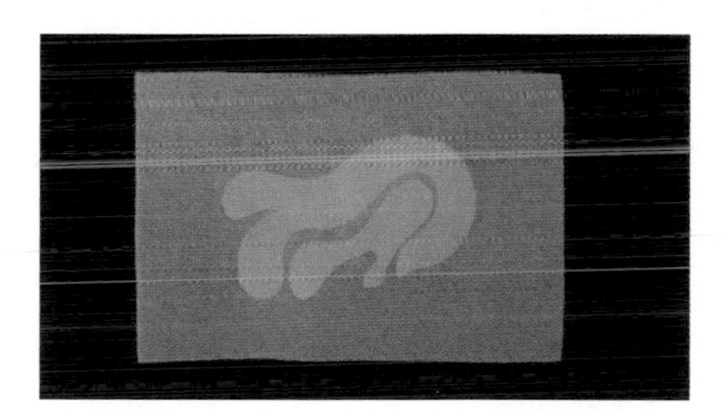

Figure D-8. *Ribbon in low-light environment*

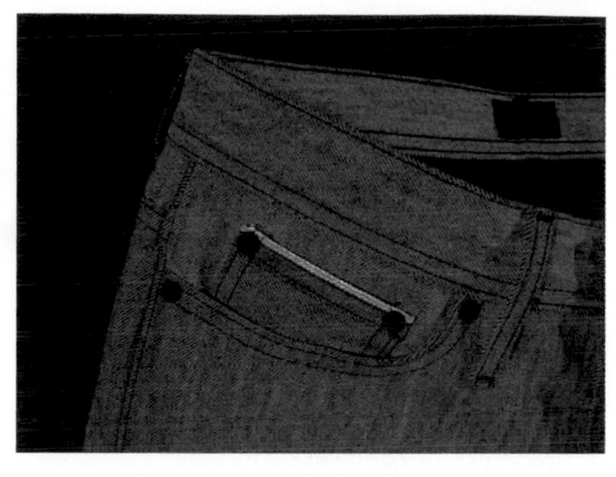

Figure D-9. *Glow-in-the-dark jeans produced by Naked and Famous Denim feature a phosphorescent coating*

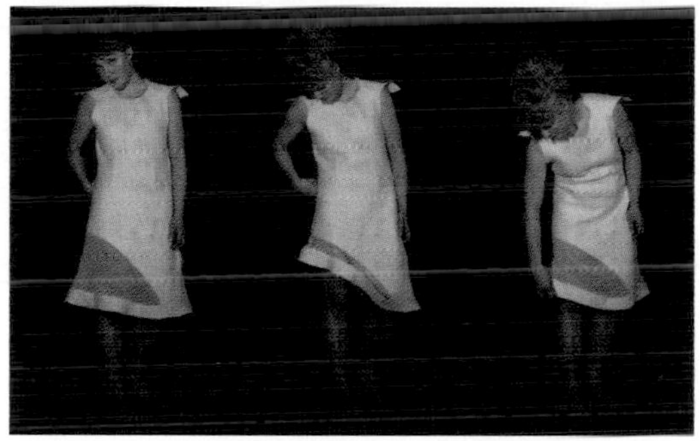

Figure D-10. *"Vilkas" is a dress with a kinetic hemline that rises over a 30-second interval to reveal the knee and lower thigh; the project was created by Joanna Berzowska, Marcelo Coelho, Hanna Søder,*

1. *Inventables (for a complete list of supplier abbreviations, see* *"About Part Numbers" on page xiv**).*

XS Labs (photographed by Shermine Sawalha and Hugues Bruyère; modeled by Hannah Søder)

Shape Memory Materials

Shape memory alloy (RS RB-Dyn-31, SF COM-11899) is a material that shrinks or retracts when heat or current is applied to it. When it cools, it relaxes and returns to its original length or shape. It is most frequently found in the form of wire, such as Nitinol or Muscle Wire. There are also shape memory polymers (IV 21300-02, IV 21123-02).

Figure D-11. *"Enleon" is part of the Skorpion series created by Joanna Berzowska and Di Mainstone at XS Labs; the movement of this kinetic garment is activated by beaded shape memory alloy (SMA) coils and controlled through custom electronics (photo by Nico Stinghe)*

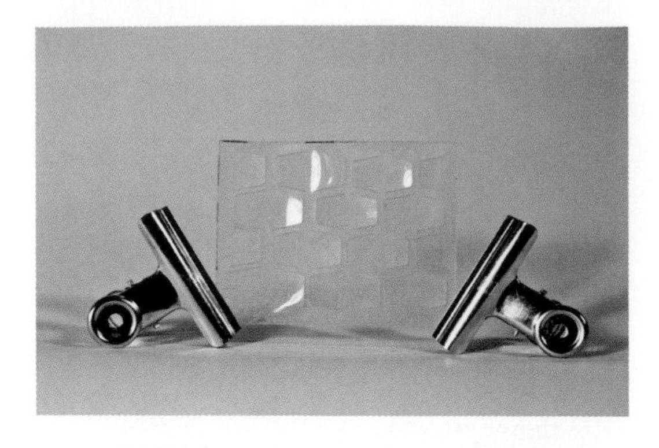

Figure D-12. *Shape Memory Polymer prototype created using the BlackBox DIY fabrication system; BlackBox was developed as a collaboration between Nick Puckett and the BioActive Devices Lab at the University of Kentucky*

Thermochromic Pigments

Thermochromic pigment (Figure D-14) changes color or goes clear in response to a shift in temperature. This temperature change can be environmental, the absence or presence of body heat, or it can even be triggered by running current through resistive heating wire. Thermochromic pigments can be mixed with paint, polymers, and other materials.

Figure D-13. *Temperature responsive vest and collar, created from Shape Memory Polymer for the Beyond Mechanics Workshop at Smart Geometry 2012 (collar by Daria Kovaleva, vest by Daniel Davis and Marc Hopperman)*

Figure D-14. *Thermochromic pigment (SF COM-11558)*

Figure D-15. *"The Hyperdermis Pillow" by Colleen McCarten uses thermochromic ink to create a body-responsive design; the textile starts out black but reveals a fingerprint pattern when it comes into contact with skin*

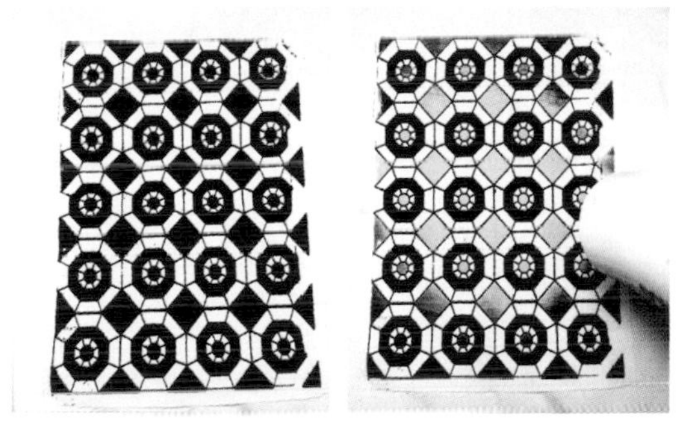

Figure D-16. *This test by Sofia Escobar explores possibilites for using silk-screened thermochromic ink for pattern-changing textiles*

Moldable Materials

When creating casing for wearable electronics, there is sometimes a need for customized plastic or rubber bits to protect your circuit, your body, or both. Moldable rubbers and plastics (Figures D-17 and D-18) often provide a quick-and-easy solution.

Figure D-17. *Sugru moldable rubber (AF 436, MS MKSUMC)*

Figure D-18. *Polymorph moldable plastic (SF TOL-10950)*

Rapid Prototyping Techniques

Rapid prototyping and digital fabrication have also become far more accessible in recent years due to the availability of lower-cost machines, the increase in the number of hacker and maker spaces that enable access to this equipment, and web-based services that will process digital design files and mail back the results.

These tools and processes open up all kinds of new possibilities for wearables. They make it easy to reproduce your own designs or those that have been created by other people.

Digital Fabric Printing

Digital fabric printing makes it easy to customize the materials you are working with. Simply create a digital design file, and you can get it printed on a textile! This can be used to create custom patterns or for circuit layout as well. Spoonflower (*http://www.spoonflower.com/*) is an example of a service that offers digital fabric printing.

Figure D-19. *"The Tornado Dress" by Studio SubTela features an image on the fabric that was created using a digital fabric printing process (photographed by Hesam Khoshneviss)*

Lasercutting

A laser cutting machine can both cut and etch, allowing you to efficiently cut textiles, leather, paper, wood, and other nontoxic materials. This can be useful when creating complex patterns (Figures D-20 and D-21) or when working with multiples. Check your local hackerspace or university to see if they have a lasercutter, or take a look at an online service like Ponoko (*http://www.ponoko.com/*).

Figure D-20. *"Medical Alert [RF]ID Bracelet" by Doria Fan*

Figure D-21. *"Laced-Up Leather" by Hillary Predko*

3D Printing

3D printing (Figure D-22) enables you to print custom objects, connectors, and cases on demand. There are now a wide range of consumer-grade 3D printers available, and they are also popping up in libraries, classrooms, and hackerspaces. In addition, there are online printing services like Shapeways (*http://www.shapeways.com/*).

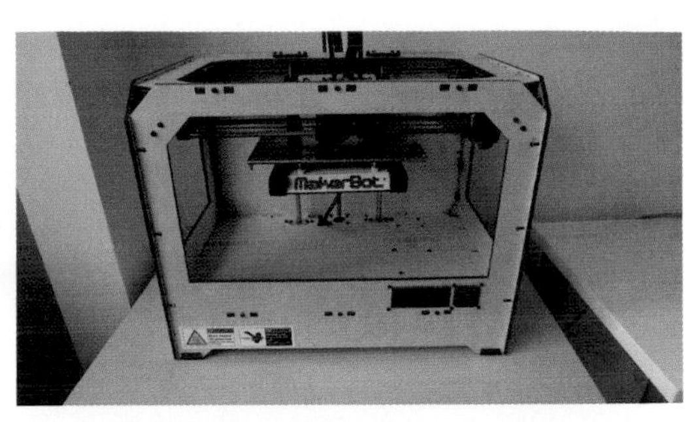

Figure D-22. *A 3D printer*

Figure D-23. *strvct shoes are 3D printed footwear created by Mary Huang*

Printed Circuit Board

A PCB (or printed circuit board, shown in Figure D-24) is a custom-designed circuit board. Unlike a breadboard or protoboard, the connections in a PCB are specific to a particular circuit. Once a custom PCB has been designed, unlimited copies can be manufactured to create rapid and reliable assemblies (Figure D-25).

Figure D-24. *A printed circuit board*

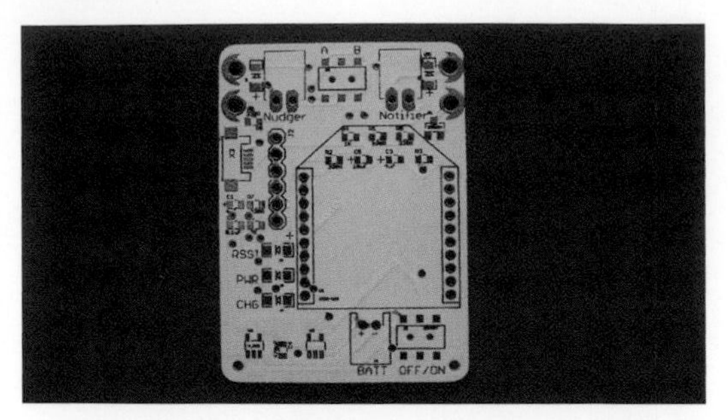

Figure D-25. *A custom PCB for the Nudgeables project by the Social Body Lab*

While PCB design is not covered in this book, keep in mind that it can be very useful for final prototypes of wearable electronics projects. Once you've designed a custom PCB, you can manufacture unlimited copies to create rapid and reliable assemblies.

Circuit design softwares include Fritzing, Eagle CAD, and KiCAD. There is a wealth of board printing services available. OSH Park (*http://oshpark.com/*) is a great one to get started with for small orders.

Microcontroller Options

E

The examples in this book introduce you to a few microcontroller options, but there are many more available. Here are some boards that are excellent brains for a variety of wearable electronics projects.

LilyPad Sewable Microcontrollers

These modules take the traditional LilyPad format and are meant for use with conductive thread. They are all Arduino boards and can be programmed using the Arduino IDE.

LilyPad Arduino 328

The LilyPad Arduino (Figure E-1) is the original board around which the LilyPad toolset is based. The current version of this is the LilyPad Arduino 328 Main Board. The number "328" refers to the version of Atmel chip that sits on the board.

Figure E-1. *LilyPad Arduino 328*

The LilyPad Arduino is a basic Arduino board in a LilyPad package. The LilyPad Arduino contains the same digital and analog pins that you would expect to find on a board such as the Arduino Uno. However, in order to reduce the size of the board, the components necessary to program the LilyPad Arduino have been broken omitted, so an FTDI board or FTDI cable is required for programming.

People who are new to both electronics and wearables sometimes use the terms "LilyPad" and "Arduino" interchangeably. It's important to remember that these are two very different things. "LilyPad" a system of sewable components. "Arduino" is an open source microcontroller platform. More about Arduino is explained in Chapter 6.

LilyPad Arduino Simple Board

The LilyPad Arduino Simple (Figure E-2) is a simplified version of the LilyPad Arduino. It has a reduced number of pins (three analog, six digital, power, and ground) that are more widely spaced to provide more wiggle room for making connections with conductive thread. It also introduces a handy on/off switch as well as a JST connector for a battery.

Figure E-2. *LilyPad Arduino Simple*

Some of the full functionality of a traditional Arduino Uno board is not accessible in the LilyPad Arduino Simple. For example, the RX/TX pins are not broken out, so this board would be a poor choice to pair with the LilyPad XBee or other boards that require serial communication. Be sure to think about which functions you need access to for your project before selecting this board.

LilyPad Arduino SimpleSnap

The LilyPad SimpleSnap (Figure E-3) is a derivative of the LilyPad Arduino Simple. The major differences are the addition of a lithium-polymer battery, the integration of a snap breakout board system, and a different assortment of pins.

Figure E-3. *LilyPad Arduino SimpleSnap (front) with lithium polymer battery*

A rechargeable lithium-polymer battery is included, so you don't need to worry about integrating an external power source. However, you do need to assess whether the onboard battery will meet your project's power needs.

Prototyping with conductive thread can often be a laborious task, and once it is done, it's impossible to temporarily remove components from the system. The pins on the LilyPad SimpleSnap are outfitted with female snaps. A breakout board with matching male snaps (Figure E-4) can mate with it, as shown in Figure E-5, and is used for all conductive thread connections to external components. The result is that the Arduino and battery portion of the package can be removed when a project is washed, or if the board is needed for use in another project.

Figure E-4. *LilyPad Arduino SimpleSnap (back) and LilyPad SimpleSnap Protoboard (front)*

Figure E-5. *Snapping LilyPad Arduino SimpleSnap to LilyPad SimpleSnap Protoboard*

The LilyPad Arduino Simple and LilyPad SimpleSnap have a slightly different assortment of input and output pins. The Simple offers three analog (A0, A1, A2) and six digital (3, 5, 6, 9, 10, 11), and the SimpleSnap contains four analog (A2, A3, A4, A5) and five digital (5, 6, 9, 10, 11).

LilyPad Arduino USB

The LilyPad Arduino USB (Figure E-6) is an updated version of the LilyPad Arduino Simple that features the ATmega32U4 chip. Because this chip has built-in USB support, this board does *not* require the use of an FTDI board for programming, which means you have one less accessory to worry about.

Figure E-6. *LilyPad Arduino USB—ATmega32U4 Board*

The LilyPad Arduino USB uses a USB micro cable, not USB miniB.

Adafruit Sewable Microcontrollers

Adafruit offers a few microcontroller options specifically meant for e-textile applications. These boards feature Atmel chips and are programmed using a modified version of the Arduino IDE. See the Adafruit website for details.

Note that these boards use I2C for their input and output devices. This means that components such as sensors or LEDs are chainable, meaning that you can connect far more than the number of input and output pins on the boards.

Flora Main Board

The Flora Main Board (Figure E-7) is the primary microcontroller option for the Flora toolkit. It can be programmed directly via USB and features an onboard JST connector that can be used for battery packs ranging from 3.5 to 16V. It has an extremely robust power system and is designed specifically for use with fabric.

Figure E-7. *Flora Main Board*

Gemma

The Gemma (Figure E-8) is a tiny cousin of the Flora. It is programmed via USB and features three digital pins (two with PWM) and one analog input pin. Because of its small size, it is great for lightweight or tightly spaced applications.

Figure E-8. *Gemma*

Other Microcontrollers

Not all wearable circuits are made using conductive thread. Beyond the world of e-textile toolkits there are many other microcontroller options. Here are a few that are useful for prototyping and producing wearable electronics.

Arduino Uno

When it comes to prototyping, the Arduino Uno (Figure E-9) is the most common Arduino board that people will first encounter. It is helpful to be familiar with this board because it appears often in the wealth of Arduino tutorials that exist on the Internet. While not great for wearing, the Arduino Uno can be a roomy and robust option for working out your circuit details in the pre-wearable stage of developing your project.

Figure E-9. *Arduino Uno*

Arduino Micro

The Arduino Micro (Figure E-10) is an excellent choice when building out a soldered circuit on protoboard. It has a small footprint and low profile. It can be programmed directly via USB.

Figure E-10. *Arduino Micro*

Arduino FIO

As you saw in Chapter 9, the Arduino Fio (Figures E-11 and E-12) is a great option for wireless projects. It features a footprint for an XBee radio or other wireless device, as well as a JST plug for a battery and battery charging via USB. It does require an FTDI cable or board for programming. There is also a different version of this board (Fio v3) available through SparkFun Electronics.

Figure E-11. *Arduino Fio, front*

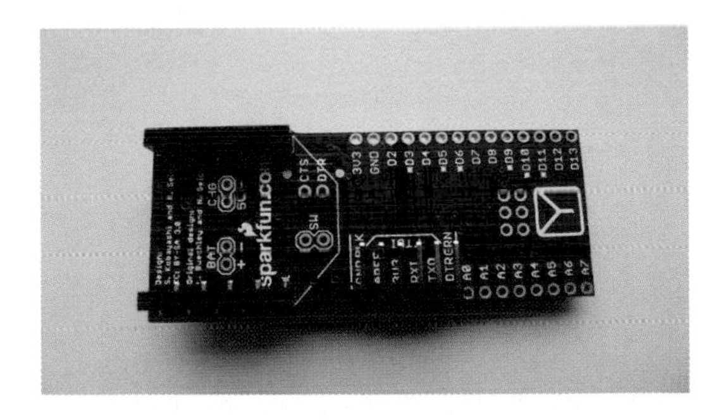

Figure E-12. *Arduino Fio, back*

Table E-1. Microcontroller comparison chart

Board type	Programming interface	Output voltage/input voltage	Analog I/O	Digital IO/PWM	Battery connector	Sewtabs	Extras
LilyPad Arduino 328	FTDI	2.7-5.5V/ 2.7-5.5V	6/0	14/6	No connector	Yes	n/a
LilyPad Arduino Simple	FTDI	2.7-5.5V/ 2.7-5.5V	4/0	9/4	JST	Yes	n/a
LilyPad SimpleSnap	FTDI	2.7-5.5V/ 2.7-5.5V	4/0	9/4	Comes with attached battery	Yes	Pins break out via sewable snap board
LilyPad USB	USB	3.3V/3.8-5V	4/0	9/4	JST	Yes	n/a
Flora	USB	3.3V/3.5-16V	0	4	JST	Yes	n/a
Gemma	USB	3.3V/3.5-16V	1	3	JST	Yes	Has 3 I/O pins, one of which can be set to analog
Arduino UNO	USB	5V/7-12V	6/0	14/6	Power jack	No	n/a
Arduino Micro	USB	5V/7-12V	12/0	20/7	No connector	No	n/a
Arduino Fio	USB	3.3V/3.7-7V	8/0	14/6	JST	No	XBee Socket

Index

Symbols

A

B

C

D

E

F

G

H

I

L

M

N

O

P

R

S

T

U

About the Author

Kate Hartman is an artist, technologist, and educator whose work spans the fields of physical computing, wearable electronics, and conceptual art. Her work has been exhibited internationally and featured by the New York Times, BBC, CBC, NPR, in books such as *Fashionable Technology* and *Art Science Now*. She was a speaker at TED 2011 and her work is included in the permanent collection of the Museum of Modern Art in New York. Hartman is based in Toronto at OCAD University where she is the Associate Professor of Wearable and Mobile Technology in the Digital Futures program. There she founded and directs the Social Body Lab, a research and development team dedicated to exploring body-centric technologies in the social context. She is also the Un-Director of ITP Camp, a summer program for grown ups at ITP/NYU in New York City. Hartman enjoys bicycles, rock climbing, and someday hopes to work in Antarctica.

Colophon

The cover and body font is Benton Sans, the heading font is Serifa, and the code font is Bitstream Vera Sans Mono.